BLIND TRUST

Also by John J. Nance

Splash of Colors

BLIND TRUST

DISCARD

JOHN J. NANCE

WILLIAM MORROW AND COMPANY, INC.
NEW YORK

TO HOWARD CADY

Library of Congress Cataloging-in-Publication Data

Nance, John J.
  Blind trust.

  Bibliography: p.
  Includes index.
  1. Aeronautics—United States—Accidents—Human factors.   I. Title.
TL553.5.N26   1986   363.1′251   85-13709
ISBN 0-688-05360-2

Printed in the United States of America

First Edition

1 2 3 4 5 6 7 8 9 10

BOOK DESIGN BY PATRICE FODERO

AUTHOR'S NOTE

This story is for everyone, since airline safety affects everyone. Whether you fly or not, you have, undoubtedly, friends and loved ones who periodically trust their lives to airlines.

Airline flying in the mid-eighties is not *un*safe—it is simply not *as* safe as it should be, can be, and must be. Nevertheless, the airline industry and the U.S. government are avoiding—hiding, in some cases—the truth, for fear the truth may frighten away passengers.

The pages that follow are not meant to scare you away from the airline system of the United States. They are, however, meant to illustrate beyond doubt the extremely serious fact that we have a national crisis upon us: a significant deterioration in the margin of safety—the *potential* for fatal airline accidents.

And above all, they are meant to show that airline accidents kill and maim human beings rather than statistics; thus, *no* accidents or fatalities are ever acceptable, regardless of how few there are in any given period of time.

This book is also meant to spur you to action, for only public pressure on our government can spark the proper studies and debates that can lead to bedrock changes in the way the airline system is monitored, controlled, and operated—and only through rapid, substantive change can the current precipitous slide in airline safety be checked.

The stories presented in this book actually occurred, and every person referred to is—or was—without exception real. There are no villains, only fellow humans and human institutions—flawed in various degrees, as are we all.

—J.J.N.

CONTENTS

7

INTRODUCTION

The ultimate cost of those $99 airline tickets may be measurable in more than services lost and leg room sacrificed. The true cost may be paid in passenger lives, because through haste and ignorance, Congress has inadvertently degraded airline safety.

What has occurred in the airline industry since the Deregulation Act of 1978 may not seem a closely guarded secret, but the true extent of the deterioration in airline safety is exactly that. Like the townsfolk in Hans Christian Andersen's immortal tale "The Emperor's New Clothes," there are too many in the industry and in government with such a vested interest in deregulation's success that they dare not admit to any imperfection, including the obvious and the ludicrous—until and unless the fatality rate forces the issue.

But there *are* imperfections in the new order, and they threaten to reduce drastically your degree of assurance that when you (or a friend or loved one) climb on an airliner in the United States, you will alight at the other end alive and well (though perhaps not as well fed as in years past). Commuter crashes in New England, a spectacular disaster in the Potomac, the federal shutdown (and subsequent restart) of airlines that had operated with safety problems undiscovered and unaddressed for years, and the growing concern of aircrews and operational managers throughout the industry are but the tip of the iceberg—the symptoms of a deeper malady that has been repeatedly ignored.

When Congress drop-kicked the airline industry out of the fold of regulated, federal protection—leaving it to the vicissitudes of free-market competition—it placed the responsibility for maintaining airline safety on the back of the Federal Aviation Administration, without giving the FAA the money, the laws, or the capability to meet the challenge. The system was not adequately safe in 1978; the FAA even then was unable to ensure airline safety (though the public was led to believe otherwise). Only experience and the stability of regulation kept it safe.

The established mainstream airlines had always kept their operational safety levels far above the federal legal minimums (though Congress in 1978 was blind to that fact) by passing the costs of safety-related functions through to the ticket prices. Safety, therefore, was not a strictly cost-accountable item—until 1978.

But by pitting the various airlines against each other, and by helping reduce (or destroy) profits, Congress forced the industry to make *everything* cost accountable—including safety. Congress then blindly dumped the impossible task of preventing a decline in safety onto the FAA. The FAA, however, can enforce only the minimum safety standards, which have always been far *below* the industry norm! As money has become tighter in the industry, the result has been a predictable and guaranteed decline in the margin of airline safety as airlines increasingly drop their standards to the legal minimums in order to save money and stay competitive.

Supposedly, the FAA watches them all and keeps them safe. In reality, the FAA is too undermanned and ill equipped to have any idea what actually goes on at the heart of the average airline. In effect, no one is watching. As of 1985, the airlines are on the honor system with respect to safety—and not all are honorable.

To make matters worse, the incorporation of smaller commuter carriers and inexperienced new carriers as a backbone of the airline system serving smaller communities (even though they are often unable as a group to rise to major airline standards) has further eroded the passenger's peace of mind.

The airline business is a human business. Humans think and act subjectively, try as technology may to mold us into objective perfection. Humans fail in the cockpit, on the ramp, in the executive office—and in Congress. Such failures are predictable, and if we do not build in enough of a safety buffer to absorb them, accidents will result. It is that very safety buffer that is being thinned drastically and dangerously by the free-market forces of cost accountability unleashed by deregulation.

What follows, then, is a true story of human nature lagging behind the demands of the technocracy it has created—a tragic anthology of some who have been caught in the lethal riptides of this hastily-conceived national experiment with unfettered competition.

Chapter 1

BETWEEN A ROCK
AND A HARD PLACE

Jim Merryman was deeply upset, pacing back and forth like a caged animal and gesturing in short, staccato movements of his large hands—when he wasn't rubbing his chin, that is.

And Phil Simpson was alarmed. In the many years he had known Jim, up to the present moment in May 1979, he had never seen him so agitated.

"I really don't know how much longer I can stand this, Phil—I really don't!"

Simpson watched the pained, exasperated expression on Merryman's face as he stared somewhat absently at the ramp area through the glass wall that formed the north side of the waiting room in Downeast Airlines' terminal.

"What do you mean, Jim?"

"Stenger"—Merryman almost hissed the name with visible distaste—"is . . . impossible. Absolutely impossible. He keeps the pilots upset all the time; he yells at me over everything and undermines any decision I make. I don't really have any authority, you see, just all the damn work. I work my ass off around here, and for what? For nothing!"

The thirty-five-year-old chief pilot for Downeast Airlines kept pacing as he talked in a low voice through clenched teeth. Simpson listened closely, noticing with increasing alarm that his young friend was looking everywhere but at him. That was unusual too. Jim Merryman always looked you straight in the eye.

"All our experienced guys are leaving and the weather has been so lousy I can't get any training accomplished. I've got to interview and hire and train and check out a bunch of new, green pilots to replace our experienced ones who're leaving. Stenger is all over me about that—blames *me* for their leaving—and all the time he's been pressuring these guys about everything. The atmosphere around here encourages them to bust minimums—both takeoff and landing minimums, Phil! Stenger seems to think it's great when one of our pilots takes off [in zero-zero fog]—with passengers! They're tired of it and I'm tired of it, and I'm exhausted. The whole thing's impossible!"

Phil Simpson watched Merryman fume and pace the linoleum tiles of the small terminal area. Two other Downeast employees—both of them very young—hovered near the ticket counter that lined the south wall. They could hear Merryman's voice, but they weren't really listening to him. Bob Stenger, the owner of the airline, was out of sight—but definitely not out of mind. His mercurial presence hung like a specter over the six-foot-two, auburn-haired pilot.

Stenger was well known to Simpson. Since 1962 Phil Simpson had been chief inspector for the Maine Aeronautics Commission, and for more years than he cared to recall he had been familiar with tales of Stenger's explosive temper and his tendency to alienate people—especially his employees. More important, though, he was all too familiar with Stenger's attitude toward the concept of strict compliance with the rules and regulations that governed commuter-airline operations.[1]

There are federal regulations that dictate the point at which the clouds over an airport have dropped too low and the visibility has deteriorated too much to allow commuter-airline pilots to take off or land. As Simpson knew well, they were established for a very good reason: to keep potentially foolish pilots (with a burning de-

---

1. Air-taxi and commuter-airline operations are governed by Part 135 of the FARs (Federal Air Regulations), whereas large airline operations are governed by the more restrictive and demanding rules of Part 121. Thus, large airlines are said to be "121 carriers," and air-taxi/air-commuter operations are known as "135 carriers."

sire to do their boss's bidding and complete the flight so as to keep their job and peer esteem) from trying to depart or land when the visibility is too poor to be safe. Phil Simpson knew the rules, but he had no authority to enforce them.

Bob Stenger, as head of Downeast, had to promise the government and the public to live by those rules in order to get and keep his operating certificate in order to carry the public in his airplanes, collect their money, and make his fortune. He had to promise as well to require his pilots to obey those rules. And Chief Pilot Jim Merryman had to live by the same regulations in order to keep his pilot's license. Of course, to the vast majority of unsuspecting passengers with places to go and schedules to keep, that's the end of the matter: The rules are there for legitimate reasons; the airline wants to stay in business; therefore (the flawed assumption goes) the airline will surely obey them.

But too often those rules are interpreted by people under intense financial or personal pressure, and such rules for minimum takeoff and landing weather often mean canceled flights and lost revenue for the owner of a small air-taxi outfit. It is no state secret that too many such owners operate on a very thin margin—on the ragged edge of solvency. Flights canceled and revenue lost due to bad weather can spell disaster on the bottom line.

Downeast Airlines' home airport was located on the coast of Maine and plagued by sea fog much of the year—hardly a situation conducive to uninterrupted daily air service, and hardly a business without its financial risks. Even in 1968 (as he watched Bob Stenger get the Boston to Rockland route established) Phil Simpson had been well aware that a commuter airline, if operated out of Rockland strictly by the regulations, might have a tough time making a profit. The weather conditions were just too quirky.

Robert Stenger, however, had beaten the odds. He had made a substantial income running his commuter-airline and charter service for the previous eleven years, regardless of wild weather and fog. Somehow, Downeast flights always seemed to get through. Bob Stenger was proud of the dependable air service he was providing to the area, and even prouder of the profits. He considered his success the success of a hard-nosed businessman determined to watch every penny in order to make substantial money. And, as Simpson knew, Stenger viewed the natural enemy of his scheduled operation—the foggy Rockland weather—a little differently from the way he viewed the Federal Aviation Administration regulations.

Robert Stenger knew all about sea fog and approach minimums, and he had scant respect for any pilot of his who would cancel or divert a passenger-carrying, money-making flight because the actual cloud ceiling and visibility were slightly below legal minimums. A real man—a good pilot—would find a way to get the job done and get in. Those who could not or would not were considered "pantywaists" or cowards, and that attitude permeated Downeast operations.[2]

Merryman turned his commentary to his personal life—what there was of it. He and his wife had divorced, and Merryman could be with his young son only on weekends. Jim Merryman was extremely devoted to his boy—they took trips, had a roomful of "toys" father and son played with together, and were inseparable during their weekends at the Merryman farm in Brunswick, Maine. With the wreckage of his marriage still littering the landscape of Jim Merryman's emotions, little Jimmy was a living (but last) vestige of what father, mother, and son had once shared as a happy family. That part of the past was still vital, and Jim clung to his boy

2. Downeast Airlines evolved from a fixed-base operation called Mid-Coast Airways. That area of Maine from around Rockland south to Portland, which literally forms the southeastern border of the state and has been known for at least two centuries as "Downeast Maine" spawned the name of Downeast Airlines). Robert Stenger had created, owned, and managed the fixed-base operation since 1960. He had become involved with aviation after leaving the U.S. Army as a sergeant in the mid-fifties and using his Veterans Administration educational benefits to pay for flight lessons leading to his commercial pilot's license. In 1968, after Northeast Airlines began its merger with Delta Airlines and simultaneously applied to the Civil Aeronautics Board to discontinue service on the Boston to Rockland route (Rockland's airport is actually named the Knox County Regional Airport), Northeast-Delta's executives began searching for a company to sponsor as a replacement carrier. Stenger's operation, which already had an air-taxi certificate from the FAA under Part 135 of the Federal Air Regulations, applied for the route as a scheduled air-taxi carrier and subsequently qualified as a Part 135 commuter airline. Bob Stenger reasoned that the route could be a potential gold mine, since the area would need continued air service for the well-heeled residents who swarmed in each summer season to their cabins and mansions around Rockland (swelling the local population to forty thousand with their seasonal migration). With the exception of the problems inherent in operating from an area often blanketed with that legendary coastal fog, he was right—Downeast Airlines would indeed prove to be a financial bonanza. In fact, he too was destined to join the ranks of the well-heeled that his airline was formed to serve, largely on the profits from Downeast operations (though his tendency over the years to display his wealth would help keep him out of the social ranks of the surrounding area—dominated by members of substantial "old-money" families).

with a love and dedication that most fathers attempt and few achieve.

Now, however, Jimmy's mother wanted to impose new restrictions on the cherished weekend visits. That battle alone was tearing him to pieces.

Just as suddenly as it had started, the soliloquy ended with a muttered "See you later, Phil," as Merryman, wearing the handsome uniform Downeast required of its pilots, turned and disappeared through the door to the office to get ready for his afternoon flight to Boston.

Phil Simpson stood there a moment a bit stunned. Merryman hadn't been talking to him, he'd been talking at him. He'd never seen Jim that distracted.

Phil Simpson's position with the state of Maine and his extensive contacts as a native "Mainer" gave him a wealth of knowledge— the inside story on the good, the bad, and the marginal. But the fact that he did not possess the legal enforcement power to deal with some of the problems he encountered saddled him with immense frustration. It was a matter of federal preemption—the Federal Aviation Administration people had the authority to enforce the rules. Simpson was a state law-enforcement officer, but when it came to potential violations of the spirit or intent of the rules governing flight operations—rules written to keep pilots and passengers safe—it was a federal matter. If he uncovered something unsavory about Robert Stenger's airline or any other aerial practice in Maine, all he could do was turn information over to his acquaintances at the FAA and hope for some action.

Now again the old dilemma had raised its ugly head. Jim Merryman had just confirmed Simpson's long-held suspicions about Downeast Airlines. But what could he do with information like that? Even if he had Downeast dead to rights on a Federal Air Regulation (FAR) violation, pressing the FAA to act on it could seriously damage Jim Merryman, the chief pilot, even though Merryman seemed as much victim as instigator. Though he was an excellent, careful pilot, Merryman was pressing many of the limits himself. Simpson knew the danger of involving the FAA. Too often it had shown its willingness to get rid of a thorny problem with a marginal air-taxi operator or commuter by taking the easy way out and prosecuting the pilot rather than pursuing the people who might have driven him to break the rules in the first place. His friend might be vulnerable to an FAA scapegoat attack, and Phil Simpson did not want to ruin Jim Merryman's career only to see the real culprit—the boss responsible for such things—escape.

Of course, it was unlikely that the local FAA would even be interested. The FAA General Aviation District Office (known as a GADO) in Portland, which had the responsibility for supervising Downeast Airlines, was reputed to have a good working relationship with Bob Stenger. All Downeast's paperwork was kept in order, the periodic inspections in Rockland never seemed to uncover any significant violations, and even the previous year's complicated process of purchasing a turbine-powered Twin Otter and going through the proper FAA approval procedures to use it in passenger service had been handled satisfactorily.[3]

As far as FAA inspector Robert Turner was concerned, Downeast and Stenger were in compliance with the rules. Simpson knew of several allegations that Turner and others in the GADO had received over the years (often from former Downeast pilots) that Downeast was flagrantly violating the rules in many ways. To Phil Simpson, the FAA's refusal to do anything about it indicated too close a relationship between Stenger and the FAA. He had no way of knowing why Robert Turner could do little if anything to change Downeast, and he had no way of knowing how frustrated even Turner had become. But he *was* sure that Bob Stenger's Downeast Airlines spawned constant allegations and reports, which the FAA apparently did not want to hear—let alone investigate. And all the while, mothers, fathers, sons, and daughters were climbing into Downeast airplanes to fly between Boston and Rockland, trusting that they were as safe as Delta or United.

Simpson and others in the aviation business had no tangible proof of wrongdoing to offer up to the altar of Turner's FAA, but they did have their instincts and their contacts, and Phil Simpson had always been concerned about the Downeast operation. Driv-

3. The Twin Otter is a twin-engine commuter aircraft built by de Havilland Aircraft of Canada and used extensively by small air-taxi and commuter airlines. The unpressurized and somewhat noisy little airliner can carry two pilots and eighteen passengers and is extremely reliable. It is characterized by its high-wing, fixed-landing-gear design. The Twin Otter's engines are relatively sophisticated turboprops—jet engines harnessed to a propeller through a gear-reduction "box." Since they are essentially jet engines, they are far more technologically complex and demanding to maintain than conventional piston engines. Before purchasing the Twin Otter (68DE) from Air Illinois in 1978, Downeast had operated only Piper Aztec and Piper Navajo light twins (conventional piston engines) and one Cessna 182 single engine.

ing back to his home on the south side of picturesque Wiscasset (past the beauty of the southeast Maine countryside), Phil kept thinking about his distraught young friend and what he had said about Stenger's airline. Apparently it was true that there was too narrow a safety margin in Downeast's operations. When an airline cuts corners and pushes its people to the limit, when some of the people it employs are marginally qualified, poorly trained, and improperly encouraged to press the limits, accidents result. As Simpson knew, Downeast already had some sad and bloody experience in that area.

It had been a foggy evening in Rockland some eight years earlier, August 19, 1971, when one of Bob Stenger's captains called him from the Downeast operations office at Boston's Logan Airport. The captain, Dwight French, Jr., was a forty-year-old military veteran who had spent the previous few years behind a desk and who had taken a job with Downeast after retirement to get back into flying. He had 3,100 total hours as a pilot and 625 hours in the type of aircraft he was flying for Downeast—a twin-engine Piper Navajo. He did not have an airline-transport-pilot rating. To fly as captain for a commuter carrier in that type of aircraft in 1971 required only a commercial license and no copilot. Nothing more.

Dwight French knew the weather in southern Maine, but the only type of instrument approach he felt confident with under severe fog conditions was an ILS (instrument landing system) approach such as the one that could guide him with certainty to the runway at Portland. Rockland was impossible this night—the fog was too low for even Bob Stenger's standards, and Augusta was just barely above minimums. Neither Rockland nor Augusta had ILS approaches.

French had decided that he wanted to divert the flight to Portland. Stenger wanted him to go to Augusta. Augusta was closer to Rockland than Portland, and that would minimize the cost and the bad public relations of busing the passengers to Rockland. As usual, it all came down to a matter of economics for Bob Stenger. After a few minutes of wrangling on the phone, French gave in to the boss and filed a flight plan for Augusta.

An hour after leaving Boston with seven passengers aboard, French failed to see runway 17 at the Augusta airport on his first instrument (VOR RWY 17) approach, and began a missed approach—the precisely mandated procedure a pilot follows in instrument conditions to safely lead him back into position for another safely coordinated approach to the runway. Captain French and his passengers could occasionally see the ground flash by beneath them, but they couldn't see the runway. The instrument procedure he was using (the "approach plate" as it's called) sat in the teeth of a small clipboard strapped to his knee. On its face, among the diagrams and numbers, were the small printed instructions to climb to 2,000 feet.

But the trees and the Maine countryside were almost visible below. French felt more comfortable flying by visual reference than he did flying on instruments—he had told several fellow pilots so just the week before. Too many years behind a military desk had dulled his confidence as an instrument pilot.

French kept the Navajo low, hugging the terrain—trying to stay visual—just barely nodding at the concern of the nonpilot passenger who sat in the copilot's seat with a spare headset, shaking his head and saying, "I can't see a damned thing!"

The guaranteed safe altitude was 2,000 feet, but the Navajo's altimeters now showed less than 600 feet. French didn't really know this terrain, and he didn't really believe that any of the hills to the northwest of the Augusta airport were as high as 520 feet. Unfortunately, one was.

In fact, the altimeter on Dwight French's panel read exactly 520 feet above mean sea level at the precise moment when the Navajo to which it was bolted thundered into a fog-shrouded hillside four miles from the field. French never got the opportunity to complete the approach. He and two of his passengers died on impact. Five others survived—two injured grievously, their lives forever scarred by the mistake.

The National Transportation Safety Board investigated the accident from the New York Regional Office at John F. Kennedy Airport. Such regional offices had scant manpower to handle all the accidents in their respective regions, so most accidents couldn't be allowed to consume more than three days of staff time—with one investigator on the scene. The Downeast accident was no exception, especially since the investigator quickly determined that a perfectly good airplane with operating engines and adequate fuel had been flown into a hillside by a pilot who had, essentially, screwed up. It was pilot error, he determined. No one had time to look into Downeast operations, or Downeast philosophies, or Downeast management practices. Pilot error said it all. Case closed.

But not in the minds of Phil Simpson and others in the area who

knew of Stenger's role that night. A marginally qualified pilot had
been pressured into doing something he was unsure he could do—
and his misgivings had been justified.

Ever since then, Simpson had watched Downeast with distrust.
He knew it was more than pilot error. He knew that the underlying
reason that particular pilot erred with fatal results was a vital element in explaining the accident, even if the NTSB wasn't willing to
grapple with such intangibles. Whether there was room for such a
thing in an official crash-investigation report or not, management
pressure had been a major contributing factor in that crash—and
apparently nothing had changed at Downeast in the intervening
eight years. In fact there had been two other noninjury accidents in
1976 and 1977 that had the earmarks of management pressure and
marginal training practices—not to mention an attitude of noncompliance with the spirit and intent, if not the letter, of the regulations. The lessons, apparently, had not been learned by
Downeast.[4]

Phil Simpson maneuvered his car into his driveway, bordered by
a large tree and his son's basketball hoop. His attractive blue and
white house sat at the end of the driveway, but he wasn't seeing it.
The image of an agitated Jim Merryman was still pacing around
frenetically in his head, dominating his thoughts. He knew his
friend was one of the best pilots in the area and very conscientious,
but the worry kept nagging him—not in words but in feeling: How
long could Merryman (or anyone else for that matter) live under

4. One accident was an overrun mishap in July 1977 that involved a Piper
Aztec. There was a question whether or not Downeast had intended to report
it, but presumably due to the presence of a well-known poet among the passengers and his mention of the mishap on local television the next day, Downeast hurriedly called the FAA and NTSB to file the reports required by law. A
second accident the year before involved a pilot who clipped a tree while flying
too low on approach. Trees do not grow to the height of the legal minimum-
descent altitude at Knox County Airport. The legal minimum is 440 feet. The
tallest trees anywhere near the airport are 90 feet high. Therefore, the aircraft
had been flown below—way below—minimums. The conclusion was inescapable. Perhaps that accounted for the haste with which the Navajo was immediately taxied into the Downeast hangar, the hangar doors locked, and repairs
commenced in secret. Even the other pilots were not allowed in to inspect the
damage the next day. The accident was never legally or formally reported until
it was discovered by the NTSB years later. Downeast claimed that it did not
need to be reported because the aircraft had suffered only "cosmetic" fi-
berglass damage to an engine cover. The law, however, required an immediate
report.

21

that kind of pressure before it affected his ability to be a safe pilot—not to mention the effect on his health? A lengthy vacation was the very least the man needed—and now. "He needs to quit!" Simpson thought to himself. "He needs to go back to commercial fishing or something—anything but work for Bob Stenger." But that was impossible. Simpson knew Merryman's plight. He couldn't get away, and he couldn't afford to quit, and he would never give up flying. The realization saddened him deeply, but he could see no solution for the young chief pilot, and the hour was late.

Simpson got out of his car and glanced at his watch. Merryman would be in the air now in the Twin Otter, headed for Boston—with a cabin full of trusting passengers who would have no warning of the emotional turmoil consuming their captain.

For Jim Merryman, the previous year had been a genuine agony. He and his wife had tried to get along—tried to replace what little love was left between them with their common concern over eight-year-old Jimmy—but it hadn't worked, and the resulting bitter divorce had hit him hard, ripping Jimmy away and leaving Jimmy's father in agonizing loneliness.

It wasn't that Jim Merryman was without companionship. In fact, in the opinion of fellow pilots and other male friends (who watched his effect on women with varying degrees of envy), he might have had a bit too much female companionship.

Merryman was a handsome fellow, a "gentle giant" in the vernacular of one of the many female admirers with whom he sought solace. A powerfully built man, broad shoulders on a six-foot two-inch frame, slightly receding hairline bordering an unruly patch of curly auburn hair, traces of a constant smile set on a slightly weathered face, Jim Merryman was generally quiet and conservative, a man of few words, all carefully chosen, a dynamic self-restraint reminiscent of Clint Eastwood in his less aggressive roles. He was not considered an intellectual giant, but his capabilities were extensive and his mind sharp and clear. His knowledge of the world and the people in it was tempered by experience and growing maturity, his approach to other people marked with caution born of pain and disappointment. With his quiet voice and gentle demeanor, he was the sort of fellow a girl could talk to, confide in, and ultimately fall in love with—which is precisely what several of them did (to the point of plotting marriage, with or without his encouragement).

To the rest of his acquaintances, though, it seemed Jim was constantly juggling at least two impassioned affairs. Coming to work from a different direction each day, as the joke went.

But it wasn't enough. Despite the liaisons, Jim Merryman had never felt so adrift. His aching for family had driven him to move in with his brother and sister-in-law, John and Sharon, on the beautiful acreage of the Merryman family farm near Brunswick (where his mother lived in an adjoining house), and the tenacity of his devotion to them and their home was startling. Despite the opportunities and the invitations for other overnight accommodations, Jim Merryman usually drove the fifty-four miles back to Brunswick every night, no matter how late—and no matter whom he was leaving behind.

The farm was an anchor of sorts—his roots were there; his memories of a thousand happy childhood days could be triggered just by glancing around. The demands of late spring, 1979, left little time for reflecting, but he had been doing more and more thinking about his life lately, and the warm feeling that always enveloped him at home—at the family farm—had become important. Whatever the future held, he wanted that same feeling to be with him always.

Jim had sat in the driveway one slightly frosty morning behind the wheel of his brown custom Dodge van, running the engine and delaying the inevitable departure for the agonies and ecstasies of Rockland. As the van idled, the exhaust fumes wreathing its aft end with undisciplined undulating waves and dancing to the throaty rumble of the motor, James Merryman the man sat mesmerized in the past, thinking about James Merryman the boy cavorting through the driveway and the backyard with his two brothers, John and Dick.

The memories were so fresh. He had wanted more than anything in those early years to be a pilot—to fly—to earn respect and feel the visceral excitement of being in control of an airplane in the free, bracing world of the blue Maine sky. (The presence of the Brunswick Naval Air Station a mile away, with its constant flight activity, had been a contributing factor, of course.) John and Dick—and his parents—had grown a little weary of the subject, but as far as Jim was concerned, no airplane ever managed to fly past the Merryman farm unnoticed or unappreciated by at least one resident.

By age ten, Jim had rigged up a miniature airplane fuselage on wheels—a crude simulator—which he incessantly rumbled and rattled with great enthusiasm up and down the driveway—until his folks and his siblings began openly plotting draconian measures to restore the peace. Somehow, of course, they all survived, as did his determination to hammer those crude metallic (and marginally animated) dreams into reality. And through such interests he shared a kinship with every hopelessly smitten pilot the world over who ever built model airplanes, hung around local airports, or begged rides from other soft-hearted aerophiles. It would be years before he knew it, of course—before his horizons broadened to include the smell of aviation gasoline and the realities of flight instruction.

Of all the things young Jim loved, his family had always taken first place, with aviation and his affinity for his native Maine running close behind. Even in high school, even through the enticing dreams of airline cockpits and the future, he couldn't imagine himself living somewhere else—and that, of course, had been a major roadblock.

Even before he had taken flight lessons and earned his private pilot license, before he had decided to give up his brief career as a lobsterman (as the rugged lobster fishermen of the Maine coast are called), Jim knew that an airline career was a long shot. He had no college degree, no military flight training, and no connections with the mainstream of the airline industry. As a commercial pilot flying for various operators around the state, Jim had come to know many people in Maine aviation such as Phil Simpson of the Aeronautics Commission and Bob Turner of the FAA, but they were locals. Jim was a smart pilot, a good and careful pilot, but he didn't have the glamour qualifications the big airlines wanted. Perhaps, though, if he could get enough experience flying on a smaller scale, he could do it.

When Bob Turner had called then, in 1975, Jim was ready. Turner was principal operations inspector for Downeast and was increasingly upset by Downeast's lack of a stable pilot force. It needed good, responsible pilots. Turner thought of Jim Merryman and suggested he apply. Jim liked the idea. The setup seemed ideal, even though he knew Bob Stenger was a difficult master. He could gain experience as an airline pilot and live at home—in Maine—simultaneously!

It had been an answer—and a curse. A year later Jim had flown a charter to Canada for Stenger and ended up with an FAA violation (for an overdue check ride)—a permanent blemish on his pilot record that Jim believed would make him persona non grata at carriers like Delta or Eastern or United. So he had stayed with Stenger, watching his chances for the big time pass with his thirtieth birthday (most major carriers didn't hire pilots over thirty), trying to convince himself that the little boy from Brunswick who had once dreamed of unlimited horizons and airline cockpits had been destined to stay in Maine. He kept trying to submerge the stomach-gnawing feeling of failure in the happiness he had attempted to build into his marriage—a family closeness like that of his childhood, which had been cemented in 1971 by the birth of his son.

Of course he was proud of the fact that he was an airline captain, despite the smallness of the airline and the constant upsets at the hands of the owner. He was proud of how he looked in the Downeast uniform. It was a respected position, and Jim felt himself to be a solid professional worthy of the trust placed on his shoulders. That crisp, professional attitude was something he wore about him like a cloak, presenting a serious, almost stern face to his passengers (though a far more pliable image to his subordinate pilots). Jim had always let compromise win over confrontation when the point was relatively unimportant (though he was immovable on matters of safety, such as carrying enough fuel). He had always erred on the side of a sensitive nature, approaching his friends and co-workers with understanding rather than the facade of unfeeling hardness he saw in the conduct of his boss, Robert Stenger.

The thoughts and the images, the assessment of his life and its sudden lack of direction abruptly evaporated, giving way to the realization that he had lost track of the time. Suddenly he was aware of being too warm (the van's heater had been blasting away). Quite a few minutes had passed as he had sat there leaning on the wheel, lost in time.

Somewhat reluctantly, Jim Merryman, the harassed chief pilot, put his van in gear and nudged the accelerator. The well-cared-for machine bounced slightly over a few potholes as he headed for the main road, turning to the right toward U.S. 1 and Rockland. He would be back early tonight, he promised himself.

John and Sharon Merryman and their children were a close-knit family—warm and loving—and Jim had taken delight in being included in their family activities. However, as spring had crept by toward the summer months of impending pressure and confusion at Rockland, the family watched in apprehension as Jim drifted away from their conversations each evening, as his appetite dwindled and he became increasingly preoccupied with the irritations at Downeast and the ongoing custody battle for Jimmy. Both John and Sharon Merryman were getting very worried.

Sharon Merryman, in particular, could see the mounting distress. Her husband had a job away from the picturesque farm and routinely worked late, so it was she to whom Jim confided his troubles—often leaning against the divider between the kitchen and dining room and pouring out his frustrations over Stenger and Downeast as well as the frustrations over his ex-wife and Jimmy. Sharon had a habit of studying a person's eyes, and what she saw in Jim's bothered her. They were almost glazed at times. He was exhausted and depressed, and nothing seemed to interest him anymore.

The tranquil setting of the Merryman farm in the late spring was a beautiful retreat, with tall maples, oaks, and pines framing the neat two-story farmhouses bordered by lilac bushes. The old chicken house across the driveway recalled the fifty-year history of their homestead, a farm started by Jim and John Merryman's grandfather, who once had five thousand laying hens housed in the structure.

But the chief pilot of Downeast Airlines was upset beyond the soothing effects of landscape, hearth, and home. In May, at the time his mother entered the hospital for a major hip-replacement operation, Jim Merryman had begun coming upstairs late at night after getting in from the Boston-Rockland run. John would be asleep, but Sharon would get up at his knock and slip on a robe—sitting on the edge of the bed in the slight midnight chill of the room while her harassed brother-in-law paced back and forth across the carpeted floor, endlessly chewing over the problems.

"I can't take this anymore, Sharon. This is getting to be a horror show. The summer season is almost on us and we're short on pilots and I can't find new ones, and I'll tell you, no way . . . no way will I work extra days or take extra flights or charters. Five days a week of this is enough."

Sharon watched as Jim started rubbing his chin again—a nervous habit he had acquired in the past few weeks.

"God, Sharon, I'm so tired I could sleep for a week!" He said it with a thin smile, but there was more truth than jest to the statement.

His sister-in-law watched him pace, unsure what to say. He had mentioned again the idea of quitting and going back to lobster fishing. But he loved flying, and she couldn't believe he'd really quit. She understood his dilemma all too well. If he wanted to fly as well as stay close to his son, he couldn't follow the other Downeast pilots in their exodus to jobs with larger airlines in other parts of the country. Deregulation was siphoning the experienced people from Downeast and he couldn't stop them or follow them. It was the Rockland area or nothing. Downeast and Stenger or no flying at all. Jim seemed defeated and trapped.

Jim Merryman stopped pacing and looked at the figure on the end of the bed. "Did I tell you I've been having trouble breathing lately, and a little pain . . . here . . . in my chest?"

He tapped his clenched fist against his chest and looked at her. His expression was one of alarm.

"I know Stenger's responsible. I know it's all the damn pressure and harassment and stress. I know it . . . but I can't stop it."

As with Phil Simpson two weeks before, the image of Jim Merryman pacing and worrying stalked across Sharon Merryman's thoughts, but as the morning of May 30, 1979, dawned, there was a hopeful sign. For once, her brother-in-law had a smile on his face and appeared calm. He had been out past midnight the night before—a disturbing omen—but this particular Wednesday morning, with sunlight streaming through the leaves of the broadleaf maple by the driveway, a blue sky overhead, and a fresh breeze in his face, Jim Merryman had paused at the door of his van long enough to smile and say good-bye to Sharon as she stood at the top of the steps at the back porch of the house. She waved and thought how different he looked, how exceptionally relaxed, and concluded that he must have reached a milestone in his thinking. She wondered if he'd decided to quit, but remembered what Jim had said over the weekend about Stenger. "He's gone on a drinking spree on his yacht, so things should be peaceful for a few days," he had told her. Jimmy had been with them for his weekly visit, and not even the tragedy of the American Airlines DC-10 crash in Chicago on the previous Friday afternoon could ruin the weekend.

On Monday the depression had hit him once again as he got ready to head back into Rockland and the wrenching problems at Downeast, but somehow he had shaken it. Now, on Wednesday, he seemed almost happy.

Sharon waved again—unseen—as the van bounced out onto the main road in front of the house and turned out of sight toward nearby U.S. 1 and the Brunswick Naval Air Station a half mile distant. Whatever the change, it was for the better. Life must go on. Maybe, she thought, he's finally accepted that.

27

Chapter 2

DOWNEAST MINIMUMS

That damned right engine was vibrating again! Jim Merryman felt a small wave of apprehension wash over him, damaging the good mood he had tried to maintain all day—apprehension over the fire storm Bob Stenger would throw at him and everyone else in sight when he found out. Merryman had flown the Otter back the previous Friday from the repair station at Hyannisport, where it had been parked during four weeks of major overhaul work on both engines. It should have run like new, but on the way back, number two (the right engine) had started making strange noises in flight—unusual, intermittent, ratcheting noises that it shouldn't be making.

The bill for the overhaul had been an incredible shock—more than $58,000. Robert Stenger had personally stormed down to Hyannis when the Air New England maintenance shop called to deliver the bad news: The engines were in terrible shape—the "hot sections" nearly burned out. Stenger had expected only a $12,000 bill and was enraged that the engines had been described as being in "the worst condition I've ever seen!" by one Air New England mechanic.

"What's with you people up there?" the mechanic had queried Merryman when the Otter arrived. "Don't you do any maintenance?"

Stenger came down hard on his maintenance chief, Leo Gallant. But Gallant claimed his people had done everything right. Gallant was overheard blaming the pilots for bad operating techniques. Stenger, in turn, was blaming everyone. He had left on vacation in a dark mood, the thought of a $58,000 loss only partially balanced by the prospect of good-as-new engines.

They were not, however, good as new. At least the right one was sick again, and everyone at Downeast was bracing for another explosion if Stenger got back and found that number-two engine might need yet another teardown.

The Otter was cruising at 8,000 feet on its way to Boston with Merryman in the left seat (the captain's seat) and George Hines in the right. It was Merryman's leg. He always liked to fly the outbound leg to Boston Logan and let his copilots get the experience of trying to find Rockland on the way back. Besides, he rather enjoyed flying away from Rockland—and away from the maelstrom that he had successfully ignored through most of the morning and afternoon. Ignored, that is, until one of the Downeast secretaries had reluctantly told him that Stenger and his wife would get aboard in Boston tonight for the return flight. She saw his smile instantly falter and droop. The vacation was over.

The normal vibration and noise from the twin turboprop engines of the Otter pulsed through the cockpit as Merryman's thoughts raced through a gamut of subjects. They wouldn't be in the vicinity of Boston and busy with the approach for a few more minutes. He could enjoy the solitude and the built-in vibrator that the aircraft's engines provided as the jet-driven blades pulled them through the moist air at 140 knots. Stenger refused to consider buying an intercom system for the pilots in the Otter, so communication between the pilot and copilot by any means other than sign language and yelling was difficult. Sometimes, of course, that was a blessing.

Merryman's copilot had been acting rather strangely all day. Usually he was eager to fly anywhere, anytime, but this Wednesday evening he was spooked. Hines had worried long and loud back at Rockland about the bad weather and fog blanketing most of southern Maine, and especially about number-two engine's strange noisemaking, which had caused one other Downeast captain to refuse to fly the Otter earlier in the day.[1] Hines had not really wanted to go on this flight. Before walking out to the Otter at Rockland, he had turned to a fellow pilot and repeated the tired old line about there being old pilots and bold pilots, but no old, bold pilots. It was uncharacteristic, and Jim Merryman was a bit concerned.

George Hines was five years older than Merryman and a jolly sort of fellow—eager to please, and as a new employee of three months, trying hard to be an obedient company man. But Hines was exceptionally weak as a pilot, especially in the Otter in which he'd logged only forty-six flying hours. No one, however, had come forward to tell Merryman just how weak Hines really was.

Jim had no way of knowing that a mere five days before, while he and his son were trying to enjoy the Saturday together on the farm, one of his Downeast captains had been forced to seize control from Hines on an approach into Boston in the very Otter they were now flying because Hines was all over the sky—totally unstabilized and "chasing needles" but unwilling to admit it or ask for help. Hines, it seemed, had a macho desire to press on regardless of conditions or his own limitations.[2] There had been other incidents as well—episodes that showed that George Hines did not understand how to operate in a two-man-crew environment, got "behind" the aircraft on instrument approaches, and had a bad habit of "touching things"—operating flaps and controls—without telling the other pilot. But none of the other pilots had told Jim Merryman any of this. After all, Merryman was the chief pilot, and a demanding instructor who watched his copilots carefully. And hadn't Merryman already certified George Hines as a captain on the Piper Navajos? It would never occur to anyone that Jim Merryman would need protecting from a marginal copilot.

The image of Leo Gallant's haunted face could be triggered by thoughts of the right engine and its continuing saga of noisemak-

1. Captain Jim Fenske, a young but experienced Downeast pilot who was perceived as "a bit arrogant" by the mechanics, started number-two engine after Leo Gallant and his crew had replaced a bleed air valve they suspected was making the complained-of noise. When he heard the ratcheting once again, Fenske shut it down and informed Gallant and Merryman that he wouldn't fly it again until it was fixed. The decision was actually moot, since the fog was impossibly dense and the morning flight would have to be canceled anyway.
2. Maine Aeronautics Commission inspector Phil Simpson also knew George Hines and had flown with him some months before his joining Downeast when Hines descended through minimums in heavy fog to sneak into a nearby airfield without an IFR (Instrument Flight Rule) clearance.

ing. Gallant, as chief mechanic for Downeast, had begged Stenger to send him and his men to Pratt & Whitney's turbine-engine-maintenance school when Downeast bought the Otter they were supposed to maintain, but Stenger wouldn't even consider it.[3] In Bob Stenger's mind, those highly complicated engines—the whirling collections of exotic metal built to high-precision tolerances and enclosing a fuel-fed inferno—could be maintained by any decent mechanic. Five thousand dollars for a maintenance school was too expensive and unnecessary. As he had told Gallant, "You're an aircraft mechanic, you've got the maintenance manuals [on the Otter], and you can call Air New England or [other Otter operators] if you need advice—you don't need a school!"

Gallant ran a good maintenance department, at least as far as piston-engine aircraft were concerned. That was one of the few positive elements of Downeast's operations, and even though Stenger's constant pressure to save money resulted in some maintenance shortcuts of debatable propriety, the overall quality of the work that was done was excellent. No matter how pressured or browbeaten by Bob Stenger, Leo Gallant wouldn't stand for unsafe or clearly illegal maintenance. He valued his license and his reputation more than he valued his job with Downeast, though he tried hard to keep Stenger "happy."

Merryman and most of the other pilots believed it was lack of knowledge on the part of the maintenance staff that probably contributed to the need for an overhaul. The Otter's engines had deteriorated by early spring, and Leo Gallant simply couldn't fix them. It was Gallant who had persuaded Stenger to have the overhaul (and Gallant and his men who did nothing to stem the rumor that the pilots had caused the damage through poor operating techniques). If that noise in number-two engine turned out to be another major bill, Leo was going to catch hell again.

Jim Merryman had been watching the instruments on the panel in front of him almost subconsciously, as professional pilots are

3. Gallant was allowed to attend (at company expense) the initial training that Jim Merryman and one other pilot received in Carbondale, Illinois, from the Otter's previous owner, Air Illinois. Other than observing the pilots (Gallant is a rated pilot himself) on the check rides and attending the ground school, prowling around the hangars, and asking some questions of the Air Illinois mechanics, Leo Gallant received no formal instruction in the care and feeding of the Otter, even though he and his men would be required to keep it in perfect flying condition.

used to doing. His practiced flying of the aircraft and his small corrections were constant, and second nature.

The Otter's instrument panel was adequate in daylight, but it was a challenge at night, or in inclement weather. The Otter, it seemed, had been designed for daylight flying in which constant reference to the instruments was unnecessary. Merryman knew that wasn't really true—the manufacturer had also meant it to be a good IFR (Instrument Flight Rule) aircraft, but it had not been designed with a pilot's human frailties and limitations in mind. Human-factors engineering, it was called, and de Havilland had apparently given scant attention to the subject when it designed this otherwise reliable little bird. The engine controls and prop controls were on the ceiling, the cockpit lighting was dim, and several important controls were so poorly adapted for human use that it was hard to operate them accurately. The flap handle, for instance, was used to extend the flaps during an approach and was a rather critical control. Yet the flap handle didn't have any notches for the various flap settings of 10 degrees, 20 degrees, and so on (partly because the FAA has yet to issue human-engineering design specifications for such items). A pilot could position the flaps anywhere from full up to full down, and the only way to check on their real position was by looking at a tiny flap gauge above the instrument panel. In bright sunlight, Merryman was used to cupping his hand around the little gauge in order to make out what position his flaps actually had reached—at night it was so poorly lighted that he had to keep a flashlight handy to read it. Flap position on the Otter did make a difference. The airspeeds and techniques for landing with 20 degrees of flaps were markedly different from those required to land with 10 degrees.

And then there was the ongoing battle over the color of the panel lights—the little light bulbs that illuminated the instruments from behind and inside their glass faces. Some were red, some were white, depending entirely on what color bulbs happened to be available whenever one needed changing. It was distracting and unsafe. At night, with the light-intensity control turned up high enough to read the dials with the red bulbs, the light from the white bulbs would destroy a pilot's night vision for at least a few minutes. By the same token, if the white was turned down to a comfortable level, the instruments illuminated by the red bulbs couldn't be read at all. Merryman had complained about it bitterly, as had some of the other pilots. Bob Stenger, however, had been very slow to authorize Leo to purchase enough of the expensive little light bulbs to provide his pilots with a single-color light scheme. They had to be ordered at significant expense from de Havilland in Canada.

Stenger apparently didn't believe it made any difference. Of course Bob Stenger never flew the Otter down to 440 feet (or lower) above the trees at Rockland in pea-soup fog while trying to focus on a psychedelic light display of gauges and maintain aircraft control at the same time. It was distracting, and it was dangerous, but they apparently would have to live with it awhile longer.

Someone would, at least. Jim Merryman had spent the past few days in thought and had indeed reached some decisions, though he had only hinted at them. As he had told Sharon, "This won't last much longer."

The sun was getting low on the cloud-filled horizon to the west as the Boston Center (FAA Air Route Traffic Control Center) controller directed them to switch radio frequencies to Boston Approach Control (the FAA radar control facility that guides air traffic in and out of the metropolitan Boston and Boston Logan Airport area), and George Hines acknowledged as the voice of the approach controller cleared the flight to descend to 4,000 feet. Jim Merryman had also been concerned about the copilot's altimeter. His pilots had been reporting that its needle was sticking—jumping several hundred feet at once on descent. One second, according to the gauge, they would be at 1,000 feet above the ground, the next at 800 feet. Such were the examples he was hearing about, but Leo and his crew couldn't find the problem. Replacing an expensive altimeter on the strength of a suspicion and nothing more was out of the question at Downeast, yet a little glitch like that could kill someone. Whatever the problem, it was apparently intermittent. As they taxied off the runway at Boston Logan Airport and headed for the terminal, Merryman still had reason to distrust the altimeter, but not enough evidence to force Downeast to replace it.

Robert Stenger, Jr., the boss's son, was also known as Rocky. He had gone to work in his father's business several years before, and although well educated, was not regarded by the pilots with the seriousness he would have liked to command. Because he was not a pilot, quieter than his forceful father and to some extent caught in the middle, Rocky served as a weather observer and all-around agent for Downeast at Rockland, as he was doing around

7:00 P.M. when a phone call came in from Boston operations with an upset George Hines on the other end.

Rocky gave him the current Rockland weather. They had departed in VFR (Visual Flight Rules) conditions less than two hours before, but the fog had already started moving in and the ceiling was down to 700 to 800 feet with the visibility at three to four miles. Within an hour it would be down to three quarters of a mile. Hines had heard those stories about Rockland's three-quarter-mile visibility all over southern Maine. Since the best instrument approach to Rockland required no less than three quarters of a mile visibility, Rockland seldom ever seemed to report less than that when a Downeast flight was due in. Tonight the observer was Bob Stenger's son, and his dad was due to come home on Flight 46. It would be no surprise (and no comfort) to hear later on that the "official" visibility was holding at three quarters of a mile.[4]

"George, Leo Gallant came in here awhile ago and wanted me to ask you if number two [engine] was operating okay. He just changed that bleed valve, and he's also concerned about the low oil-pressure reports."

Hines launched into his worries over the engine. He had felt the vibrations—the ratcheting—throughout the trip down, and he was convinced something was very wrong. He felt it was the engine, but it could be in the right wing structure as well. Hines was emphatic. "This aircraft should be grounded. It's the best move for all concerned—we don't know what's wrong with it, Rocky, and with all the fog and lousy weather, I do *not* want to fly this thing back to Rockland!"

Rocky Stenger knew only too well what his father would have to say about a flight cancellation for such flimsy reasons.

"I think you guys had better bring the plane back here, and if you miss here [can't see the runway when minimum descent altitude has been reached on the approach], go on to Augusta and I'll send the bus."

Jim Merryman returned to the counter after Hines had hung up. Hines was going to blow everything. He motioned Hines over to a corner of the room and began trying to explain why grounding the Otter at Boston—especially with Bob Stenger due in at any minute—would probably result in someone's losing his job.

4. The fog over the approach end was often markedly different from that at the terminal, so that in some cases, three-quarter-mile visibility could exist at the terminal while zero-zero conditions shrouded the approach area.

"There's no way he'll put up with it, George."

But George Hines did not want to go back. He really wanted out of this flight. He had a very bad feeling about the Otter, and the weather, and on top of everything else, he knew Merryman would want him to fly the return leg. Hines did not feel confident in such conditions with an airplane as big as the Otter. Damn it, he did not want to go!

Merryman insisted, cajoled, explained, and finally got Hines to agree—by promising to put on extra fuel and to divert to Augusta without hesitation if Rockland couldn't be seen at minimum descent altitude. George Hines still didn't like it, but he'd go along.

Merryman called Rocky and told him they'd bring her back when the local thunderstorms over Boston cleared. Bob Stenger had not shown up yet. Once Stenger arrived and the thunderheads departed, so would they—with extra fuel, the suspect engine, the badly lighted instrument panel, the worried and weak copilot, and the progenitor of Merryman's discontent in person: Robert Stenger senior. It would probably be an "interesting" flight.

Young John McCafferty looked down at his seat belt just to make certain it was secure. It was hard to see, although the fading twilight was still casting a pronounced purple glow into the darkened cabin of Downeast 46. The outline of distant thunderheads filled the western sky, warning of worse weather to come. McCafferty had phoned his parents in Searsmont just before departure from Boston—he expected they would be at the terminal when he arrived. His parents were probably as concerned about his flying as he was—he didn't particularly care for small commuter planes (this was his first experience in one), but there was no other way by air into his home area. He was in the back part of the cabin in seat 5C, with three men seated behind him and the majority of the passengers between him and the cockpit. Some of the reading lights were on already, he had noticed. McCafferty looked out the window at the darkened Atlantic to his right and cushioned his head between the seat and the window frame, trying to let himself relax. He was apprehensive and he didn't know why. The loud buzz of the engines made conversation impossible even if he had wanted to talk to the senior gentleman seated next to him. They had exchanged a few pleasantries, then silence. Perhaps he had better get some sleep. Those two pilots up front must know what they were doing.

Next to McCafferty, sixty-eight-year-old Bill Detterline also adjusted his 190-pound frame in the tiny seat and tried to relax. It would be a fairly short ride to his home in Lincolnville, Maine, once he got in to Rockland, but already it had been a long day. The buzzing of the engines was soothing, and he too dropped off to sleep.

George Hines, the copilot, was flying the Otter on the return leg (in accordance with Merryman's custom), and the chief pilot was handling the radios. Off to their right loomed one of the thunderstorms that had delayed their departure from Boston, now lumbering off to sea, spouting a staccato display of angry white flashes as if muttering to itself in the language of lightning bolts and free electrons—illuminating the inside of the monstrous cumulonimbus cloud formation and making the Otter seem small and insignificant by comparison.

Number-two engine was behaving itself for the most part, but the weather was not. Rockland was expected to be under very heavy fog on their arrival. Merryman, in particular, knew only too well who had made the three-quarter-mile observation, and what degree of credence he could assign to it. Otter 68DE would most likely spend the night on the ramp in Augusta, and they would have another hour-long bus ride with a group of unhappy passengers to face—not to mention the prospect of facing an unhappy boss.

Jim Merryman's view of the darkening countryside unfolding beneath the Otter was a vision of subdued beauty, a vista of deep reds and oranges and dark green that he had seen so often on late flights in summer. As he had told friends, it always evoked pleasant memories—most recently the mental image of crashing waves and the Atlantic Ocean under a full moon, the sweet memory of a night early in May when he and Stephanie Eaton had sat in her van on a beach at Owls Head and talked long and longingly of the months to come. They had been locked in a somewhat torrid affair since Merryman had appeared at the door of her Rockland home several months before, the result of some matchmaking by Downeast operations manager Bob Bradley. Stephanie, a tan and attractive divorcee with blond hair and statuesque profile, several years his senior, had taken one look and fallen head over heels in love. Merryman seemed to feel the same, and in early May they had begun to talk of plans—of a future together. The waves and the wind on the beach that evening had been fresh and wild, like their lovemaking in previous weeks, like the things they had talked

36

about doing in the coming summer. Those memories and promises could replay with majestic clarity in Jim's head, tantalizing glimpses of a more relaxed world without Downeast and . . . without the stomach-churning ratcheting noises in number-two engine—the sounds of reality that would come and go again, breaking the spell.

Merryman called the approach controller at the Naval Air Station at Brunswick, aware that the fellow was sitting less than a mile from the Merryman farm. Brunswick controlled the approaches to Rockland, and Merryman had asked for the latest Rockland weather. The Brunswick navy controller had picked up his telephone tie line to Rockland and talked to Rocky Stenger. Now he was back on the air, relaying the information to Flight 46.

"Downeast 46, I've got some bad news for you."

"What's new?" Merryman asked, resignation rather than humor tingeing his voice.

"Forty-six, Rockland weather is three hundred overcast, visibility [three fourths] of a mile and fog, winds are light and variable, but Bob said it isn't helping any. . . . [He] said the conditions are deteriorating."

"Hey, get the Portland [and] Augusta weather for us please. . . ." Merryman let up on the transmit button and waited for the response. The clock in front of him confirmed that it was 8:30 P.M. exactly. They should be either on the ramp at Rockland or on their way to Augusta in about half an hour.

Three minutes passed before the Brunswick controller came back up. "Downeast 46, Augusta reports, ah, eight hundred scattered, estimated twelve hundred overcast and ten miles [visibility] and, ah, Portland weather is two hundred overcast, one-quarter mile [visibility and] fog, conditions deteriorating rapidly."

"Forty-six, ah, okay, thanks."

"Forty-six, your discretion now to three thousand [feet]."

"Forty-six [acknowledges]. Brunswick [Approach], 46 out of seven thousand."

George Hines repeated the clearance and began the descent from 7,000 to 3,000 feet. His approach plate—the familiar white piece of paper put out by the Jeppesen Company in Denver and covered with diagrams and altitudes and the basic procedure for safely approaching runway 3 at Rockland—sat in his clipboard. It was known as a Localizer Only Approach, with the minimum descent altitude reading 440 feet above sea level. Hines and Merryman both were aware that Bob Stenger disliked that minimum figure intensely. Stenger had maintained for years that his pilots could go down to 300 feet with complete safety, and that the FAA-approved minimums were excessively conservative. In at least one pilot interview that Merryman knew about, Stenger had asked the pilot if going to company minimums would be a problem. The applicant replied "No, no problem." Only later did he discover what "company minimums" meant, and he was shocked.

As Jim Merryman knew so well, there was a considerable folklore surrounding the phrase and the concept loosely known as "Downeast minimums." Nowhere could it be found in the Downeast manuals, nowhere could it be found posted on a bulletin board, and by no one could it be clearly defined, but Downeast minimums were anywhere from 200 feet to 350 feet above ground level, depending on whom you talked to—and when you talked to them. (Anything under 440 feet, of course, was illegal.) If the FAA was around, for instance, there was no such thing as Downeast minimums.

Over the years a system of psychological reward and punishment had instilled the dangerous idea that a Downeast pilot could safely grope through the fog, descending lower and lower over the trees leading to the runway, and be able to see the treetops before running into them. If you could see the treetops, you could, so the belief went, stay just above them until the runway appeared. A good pilot could do that on a regular basis because he would know the Rockland airport and its instrument approach with great intimacy.[5] If you made it in with that method and the flight was com-

5. More than a few Downeast airplanes brushed treetops on foggy approaches over the years, and the first-person stories that tumble out when former Downeast passengers are interviewed (and the ease of finding persons with such memories) beg for frightening conclusions about the number of near-disasters there may have been. One very influential resident of the Camden, Maine, area who also happens to be a pilot—a man whose name would be instantly recognizable—recalled with a shiver the loud "thump" on the wing of a Downeast Navajo on the third, and unsuccessful, approach to a fog-shrouded Rockland airport in 1977. The single Downeast captain firewalled the throttles to full power and pulled up instantly. On landing at Augusta, a large dent was found in the right wing's leading edge. The pilot explained to the passengers that they must have hit a seagull—ignoring the fact that seagulls seldom fly on instruments in pea-soup fog. (In fact, no bird can fly "blind.") In another incident, a passenger recalled the sound of treetops slapping the bottom of a Downeast Navajo's fuselage just as she actually saw the roof of the forest whizzing by inches beneath her feet. That too was on the third try, but there was more to come. After unsuccessful attempts to land at Bangor and Augusta

pleted, Robert Stenger would not have to put up with the complaints and the expense of busing angry passengers to Rockland from Augusta or Portland. Therefore, Downeast would make more money and Robert Stenger would be happy. It was highly desirable, of course, to keep Robert Stenger happy, because when Robert Stenger was unhappy with a pilot, the hapless individual would not be left in doubt. From a mild chewing out to a major, withering tongue-lashing peppered with insults and threats to fire (all of it often delivered in front of one's peers), the Downeast pilot who angered the boss had substantial incentive not to do it again. Getting the plane on the ground in Rockland—not Augusta or Portland or Wiscasset, but Rockland—would be the most reliable way to make the boss happy. Stenger liked an on-time operation. The incentive to give him what he wanted was powerful and persuasive, especially to younger or less experienced pilots who had neither the training nor the instincts to resist, or who simply couldn't stand up to the caustic onslaughts.

But there was a more insidious force that perpetuated this attitude: peer pressure. When pilots who had no previous airline, commuter, or military flight experience found they could meet most of the schedules despite the fog by sneaking around the minimums and the rules, a sort of perverse pride in their own abilities developed. A pilot who could fly an overweight airplane, take off in less than legal weather, ignore mechanical problems in order to bring the airplane back to Rockland with revenue passengers, and otherwise help keep Downeast on time, was more often than not proud of himself. Any pilot who came on board and couldn't do as well was less of a pilot—less of a man. If two flights made it through the fog and landed, but the third diverted to Augusta, the third pilot would typically be greeted with sneers: "What's the matter with you, pantywaist? Can't you cut it? Everybody else could find the runway!"

In the final analysis, Robert Stenger didn't have to order or direct his pilots to perpetuate a system and an attitude of cowboyish

---

(the Downeast agent closed up and went home in Rockland meanwhile), and with fuel running low, the Downeast captain finally found the runway in Portland an hour later. As the shaken passengers began making arrangements for their own ground transportation, the young captain walked in and announced that he was refueling the plane and would be ready to "try it again" in an hour or so. No one took him up on the offer. The year was 1978, and the captain, it turned out, was Jim Merryman.

noncompliance with the rules: Peer pressure did it for him, and that had been the case for eleven years.

Jim Merryman had been weary to start with, and the day was wearing long. Listening to the steady drone of the engines and waiting for the ratcheting noise to once again fill the cockpit—as well as the prospect of having Stenger back in the office—were enough troubles to bring anyone down. Merryman knew he would have to watch George Hines carefully on this approach, but overall he trusted George to fly safely. Besides, though Jim was the instructor as well as the nonflying captain, he knew the Otter was effectively a one-man airplane. Without an intercom it was hard to function as a crew. Better to let the copilot fly it, and just watch him.[6] Hines had obviously been nervous about the upcoming approach, even back at Boston, and there would be the usual myriad of tasks to accomplish in the crowded, dark, and noisy little cockpit.

They were down to 3,000 now, about fifteen miles south of the Sprucehead radio beacon, which marked what was known as the Final Approach Fix for the Localizer Only Approach to runway 3.[7]

6. Neither Merryman nor Hines really had a thorough grounding in "crew concept," or cockpit resource-management techniques, the methods of operating a large airplane with teamwork—the flight duties divided between the pilots. In most airliners, one pilot essentially does the flying, while the other takes care of radio communications and operates the various controls, such as landing gear or flaps, on command from his partner (usually, the copilot flies every other leg). The teamwork means that both pilots are simultaneously monitoring every action, like two identical computers running parallel and checking each other for errors. Downeast's piston airplanes, the Navajos, Aztecs, and Cessna 182, were essentially one-man airplanes. Copilots were utilized, but mainly for window dressing. With no interphones and with captains used to flying solo, there was no crew concept used. The Otter had provided an opportunity to learn to operate as a crew, but the lack of interphone and the lack of training hampered the transition.

7. Sprucehead is an NDB, or nondirectional beacon, essentially a low-frequency radio station broadcasting a signal in a band just below the AM broadcast band. It can be picked up by an instrument in the cockpit called an ADF (automatic direction finder). An ADF has a needle that will point to the magnetic bearing of the station in relation to the aircraft. When an aircraft has passed over the beacon, the needle will reverse position.

The Localizer is a vastly more sophisticated station that broadcasts a special radio beam picked up and interpreted by a special gauge in the cockpit that shows whether the aircraft is right or left of the center line of the runway. When a second, horizontal beam is added to show a pilot whether or not he is following the proper glide path leading to the end of the runway, the two

Sprucehead sat on a tiny island just to the south-southwest of the shoreline. The airport was a mile beyond.

The approach was fairly simple for an experienced instrument pilot. Hines would have to fly the Otter inbound to Sprucehead at no lower than 1,700 feet while aligning the aircraft by using the invisible radio Localizer signal coming from the airport (as displayed by a needle on the instrument panel). When the Otter was within seven miles of the Sprucehead beacon, Hines could descend to 1,400 feet. Once over the Sprucehead he would pull the power back and descend rather smartly all the way down to 440 feet at which point he would level off and hope to see the runway ahead clearly enough to land. Jim Merryman had done it countless times before. Hines was far less experienced.

"Downeast, ah, 46, ah, cleared to cruise three thousand [feet] for approach into [Rockland] Airport . . . Report cancellation [of Instrument Flight Plan] via [frequency] one two three point eight . . . Radar service is terminated fourteen miles to the south-west of Sprucehead beacon, and frequency change over to the Rockland Unicom frequency [Downeast company radio] is approved."

Merryman responded. "Forty-six, okay, thank ya."

He paused a second. The fog ahead was a prime consideration. If they did need to go on to Augusta, and he had guessed they would, he ought to have an instrument clearance waiting.

"Brunswick . . . looks like we're probably going to have to miss the approach here at Rockland. . . . We're going down [to take a look] . . . but maybe you can pull us out a clearance for Augusta."

"Downeast 46, [it's] on request."

Hines switched to the Rockland Unicom and said simply, "Sprucehead inbound," into the microphone, a message received by Rocky Stenger. All the while he was descending to the 1,400-foot minimum they had to maintain until over Sprucehead. The fogbank was visible ahead, and the glow of lights underneath the blanket of whitish gray could be seen in the distance. One of them clicked the microphone button five times, rapidly, the remote, ra-

---

together are called an ILS, an instrument landing system. Rockland did not have an ILS (which would have permitted a lower minimum descent altitude than 440 feet) because Downeast's pilots seemed to get in often enough as it was.

dio-controlled method for turning on the Rockland approach lights. The lights, however, didn't turn on, but neither Merryman nor Hines could know it—they could not see the airport environment in the fog ahead.

The navigational needle pointing to Sprucehead swung 180 degrees as they passed overhead and Hines retarded the power and started down. His hand reached for the flap lever and moved it partway through its arc. The flaps obediently ran to the 20-degree position. Merryman's habit was to look in the direction of the flap-position-indicator gauge, but usually he couldn't make it out on first glance. A tired Jim Merryman expected it to read 10 degrees. Unseen in the gloom of the ill-lighted cockpit, the needle had moved to 20.

As the Otter descended, Hines was supposed to turn on the right landing light and turn around to check the engine inlet for ice, a procedure Jim Merryman had taught him. Straining to look over his right shoulder at the engine, bathed in the landing light's illumination reflected from the propeller, it was inevitable that Hines would have a struggle holding a steady descent rate with the control yoke while trying to keep an eye on the engine. But on this approach, another gauge—the rate of climb (or descent)—had begun to tell of an unusually fast approach to the minimum descent altitude, now coming up fast, and apparently neither pilot had noticed. The throttles were set for a descent made with only 10 degrees of flaps, but the flaps were at 20 degrees.

The Otter was now less than two miles from the end of the runway, traveling at an airspeed of approximately 110 miles per hour, and descending faster than either of the pilots realized. Only seconds had elapsed since they had descended below 500 feet above sea level. The normal approach path from Sprucehead to the end of the runway passed over a shoreline just ahead, unseen in the dark and the fog—and studded with sturdy deciduous and coniferous trees reaching nearly eighty feet in height.

In the cabin in seat 5C, John McCafferty was awake. He had been fighting to keep his ears clear as the unpressurized cabin of the Otter descended and was unable to relax because of the same apprehension that had gripped him in Boston. Most of his fellow passengers, including the man next to him, were still asleep. The loud buzz of the engines had now been reduced to little more than a soothing hum. He thought about his folks. They should be waiting up ahead at the terminal, and this flight should be on the ground in . . .

The thought was frozen in his mind in midsentence. Something out the window had caught his eye. The young man pressed his nose to the glass and saw dark shapes ahead—trees. The plane had broken out of the fog and was now below it and heading straight for a wall of dark trees! McCafferty instinctively ducked his head in his lap and folded his arms over his head—all in an instant—snapping into what the little safety cards called the "brace position." Suddenly he understood his apprehension. He knew what was coming.

The sound of engines—somewhat like motorcycles—caused Elmer and Jane Smick to look toward the ceiling of their beachfront home. Rockland Airport was close by, but this was too loud for an airplane—they never flew this low. The sound increased steadily and the profoundly frightening noise of rushing wind—the deep vibrations of many frequencies screaming by like a huge steam locomotive thundering along at top speed—transfixed them, as the sixty-five-year-old Mr. Smick started to raise his hand, gesturing a question mark toward the seismic sound waves shaking their home. The clean snap of something overhead marked the crescendo, and the noise began receding as a new sound—the sound of treetops falling on the roof—took its place.

James Merryman's head snapped bolt upright, every fiber of his body adrenalized by the sound and the jolt and the terrifying sight of trees that had come from nowhere to fill his windscreen. This couldn't be happening! The time compression of adrenaline slowed the sequence down as his right hand shot up to the overhead throttles and jammed them forward, simultaneously pulling on the yoke as the stricken Otter began a sickening left roll. There was nothing but trees ahead.

The nose responded, microsecond by microsecond, in agonizing slow motion, and began pitching up as a second, then a third shattering impact hacked away at the Otter's wings, as sturdy tree trunks batted the thin metallic structure of the intruding aircraft out of the sky—out of their airspace. The Otter's engines were screaming, but the nose was now pointing down—down at the base of those trees. A final round of impacts with those heavy trunks was inevitable, and as they began, like a salvo of heavy artillery at point-blank range, the battered fuselage slammed into the ground, the nose and the cockpit disintegrating in the muddy soil just inches from a ten-foot granite ledge, the incredible impact crushing and scattering the instruments and the seats, the charts and the structural members of the nose section, accelerating the fragile, dislodged bodies of the passengers into the front of the cabin. The shattered fuselage whipped around backward against a tree as the broken remains of the Otter screeched and thudded into the forest floor in a horrible cacophonous noise, spending its kinetic energy in one monstrous frenzy of destruction, and crushing the life from its trusting human occupants—passengers like Bill Detterline, David Scott, Barbara Sheldon, twenty-three-year-old Doreen Goguen and eleven other human beings with individual stories and futures and dreams and loved ones—human beings whose bodies lay broken and mangled and lifeless when the jumbled remains of Downeast Flight 46, Twin Otter 68DE, finally skidded to rest in the darkness, 350 feet from the shoreline of a place ironically named Otter Point.

Just as quickly, the wreckage was smothered in an eerie silence, broken only by a faint hissing of liquids on hot engine parts, and the sound of small pieces of shredded metal falling from nearby branches.

And within that twisted confusion, one—only one—found himself alive.

Chapter 3

THE LEGACY OF OTTER POINT

The Reverend George Stadler looked at his wife and daughter, startled, as he shifted toward the front of the couch to get to his feet. The heavy vibrations—the crunching, high-pitched grating, scraping noise—that had rocked their beachfront home a heartbeat before had lasted only a few seconds, but it had left three somewhat shaken people in its wake. It had been so sudden, so forceful, had he been alone he would have wondered if he had imagined it—it was so quiet now.

But his wife and daughter had heard and felt it too—jolted out of the end of a rather boring episode of *Good Times* on TV, now sitting alert, like him, with their hearts pounding and the echo of that awful sound still reverberating in their memories. Whatever had caused that sudden rolling thunder, he hadn't imagined it.

As Eunice Stadler got up from her chair and her nineteen-year-old daughter, Johanna, began looking for their flashlight, the Lutheran minister flirted with the explanation that a moose had rubbed violently against the house.

"No, it sounded more like a big ship hitting the rocks." Johanna had found the flashlight and was moving toward the door for her coat, her father right behind. The minister looked at his watch and noted the time: 8:59 P.M.

The fog, which had been rolling in on the tide for the last hour, had finally enveloped their home. The security lights of a neighbor across the river had long since disappeared, the visibility less than two-hundred feet as George Stadler and Johanna groped through the chill, damp air toward the beach and stood for a moment listening for voices, or noises—something to explain the horrific sound. It seemed to have come from the east, across the small cove, but there was no noise. Nothing, except the lapping of the waves below them.

Rocky Stenger looked at his watch again. It was 9:01 P.M. The flight should be taxiing up to the terminal by now, but he could hear nothing. He picked up the phone to call the navy at Brunswick.

"Have you heard from Flight 46? Is he on his way to Augusta?"

"No, Bob, we haven't. . . . He hasn't come up on frequency yet. Did he miss there?"

"We haven't heard him." Rocky thanked Brunswick and hung up, a cold feeling of fear beginning to gnaw at him.

Rocky Stenger paused. His mind wasn't quite ready to cross that Rubicon, but what other explanation was there? If 46 had missed, he would have flown by overhead—and they had reported over the beacon three miles out. Rocky had confidently walked into the main lobby area of the terminal just after that call at Sprucehead and told the people that their friends and loved ones would be there in "one minute and fifty-nine seconds." Several had been nervous about the thick fog outside, but he had dismissed it. Those same people were now watching him, trying to determine whether something was wrong.

Rocky pushed open the glass door to the ramp and walked through into the damp, foggy air outside. He had reported three-quarter-mile visibility to the crew and to the world. The fog on the approach end of the runway, though, was quite often much worse than whatever existed at the terminal, and there was certainly fog at the terminal. Rocky didn't want to think about that. He wanted to hear Otter engines. His mind was racing with possibilities, but there was no sound out there that might be coming from a Twin Otter—none of the all-too-familiar noise of a Downeast aircraft flying by overhead, unseen in the thick fog.

Rocky stood there a few minutes, shaking within. He didn't want the people in the terminal to see his panic. He didn't want *them* to panic. BUT THERE WAS NO SOUND! NO ENGINE NOISE!

After what seemed like hours, Rocky pushed open the glass door and walked back inside, trying to avoid the eyes of those in the waiting room. The "one minute and fifty-nine seconds" had been over a full five minutes ago.

Rocky picked up the phone again, dialing the sheriff's dispatcher's number, speaking as softly as he could.

"Uh, this is Downeast Airlines, and, it looks like, ah, we've lost radio contact with one of our planes . . . our Twin Otter . . . Flight 46. He, ah, may be down."

The Brunswick controller began calling Flight 46 on the last assigned frequency. As he listened in vain for a reply, several calls were coming in to the sheriff's office near Rockland miles to the north. There had been this awful noise just before the hour.

The controller made contact with a navy P-3 Orion, a submarine hunter (the military version of a Lockheed Electra), which was flying north into his area. He told the pilot they were missing a passenger bird at Rockland. Would he run up and fly over the area—try to pick up a signal? The navy pilot changed course immediately.

Rocky Stenger was in a daze. Brunswick had alerted Boston Search and Rescue Center; calls had come in to the sheriff and been relayed to Downeast that something, *something,* had happened south of the runway. Rocky was sure he knew what. The conversations with Hines and Merryman were echoing in his head. The people in the waiting room still didn't know, however.

Eunice Stadler was also an ambulance crew leader in the Rockland-area community of South Thomaston. She turned on her police scanner and heard the first call about Downeast 46 and the questions over where it could be. Eunice knew immediately. No ship had hit their rocks. The bone-chilling vibrations that had invaded their living room had been the death knell of a passenger airplane.

As Eunice ran through a mental checklist of her ambulance duties and emergency gear, Johanna, also a trained emergency medical technician, called the sheriff's dispatcher. They would have to coordinate with the rescue crews at Rockland, Owls Head, and South Thomaston, and arrange a fast rendezvous with their ambulance crew. Eunice knew through intuition that they had a mess on their hands.

The sheriff's and police radio calls had been monitored by a local radio newsman, who began phoning counterparts at stations in Portland and Augusta. Within minutes the news was flashed over southern Maine.

In their home in Brunswick, the sound of the bulletin seemed to echo off the walls for a few seconds as John and Sharon Merryman exchanged shocked glances. In a nearby hospital, Jim Merryman's mother also heard the news flash. A Downeast flight was missing. They all knew that there was only one—and that their Jim was the captain.

As the navy P-3 passed back and forth above the murky cloud cover shrouding the Rockland area, receiving the signal from the emergency locater transmitter from 68DE, the sounds of ambulance and fire-equipment sirens whined through the fog, searching for the silent jumble of shredded metal still hidden in the dark of the forest, yards from an obscure dirt path labeled Fire Road 44—marked only by the small sound of labored breathing from a figure some yards from the remains of the fuselage.[1]

He remembered the impacts, and then the small, soft glow through what had been the main cabin door. His legs were broken, as were his arms, but somehow he had rolled off the lip of the broken door frame and fallen to the ground. It didn't hurt—nothing hurt, though something wet kept blinding him, flowing down his face. Mercifully he was in deep shock.

A tiny flame flickered in his peripheral vision, or so his mind warned him, and the fear of a fire or explosion drove him past the shock and the need to comprehend what had happened. He had to get away—that was all that filled his mind. Better action than thought. He crawled, hand over hand, pushing aside—things—in the dark dampness of the forest floor, soaked in jet fuel, as he tried to haul himself as far away as possible. He tried to listen for sounds of the others, but no other voices reached his ears—no

---

1. Most passenger aircraft and now even private aircraft carry ELTs, emergency locater transmitters. In the event that a violent deceleration occurs, or the high-impact radio is somehow torn from its mountings, it activates. The battery-operated transmitter begins broadcasting a rapid, sirenlike sound on frequency 121.5 megahertz (military versions also use 243.0 MHz). This can be pinpointed, depending on the terrain, with direction-finding radio equipment.

other sounds. There might be no one alive back there to help him. No time to think about that, just accept it and—MOVE. Get away—get away. There would be help. Mustn't be too close when it blows. He had managed to crawl twenty yards away by the time the first rescue units stumbled onto the scene.

Surrealistic was the only way to describe it. All the training, all the practice of emergency-preparedness procedures (there had been a countywide practice just the week before), and none of it had adequately prepared Eunice Stadler for what she was experiencing, what she was seeing in the headlights of her ambulance as it wound its way toward what had now been confirmed as "the crash site." There on a one-lane dirt road was a growing traffic jam of sightseers, parking anywhere, walking toward the site in bizarre outfits: a woman dressed to the teeth in a cocktail dress and high-heeled shoes, teetering along in front of the ambulance; a man with a camera; another with a little boy in tow. It was as if Owls Head had turned out for an impromptu midnight carnival. It was a mockery of what Eunice expected to find ahead.

Finally alighting at the end of Fire Road 44, she, Johanna, and her fellow crew members began an arduous trek several hundred feet through a marsh of mud and jet fuel, dragging their equipment and portable lights, summoned by a light in the woods and a voice calling "Over here," a voice from the Owls Head support unit, in the vicinity of a dark, ten-foot-high granite boulder where the shadow of an aircraft tail could be seen leaning against a tree. Lying on the ground was sixteen-year-old John McCafferty, conscious, talking to his rescuers, his scalp almost ripped away. Johanna Stadler knelt down and snapped open her kit, trying to dress the wound while her mother checked the boy's legs.

Eunice left McCafferty and made her way to the ruined fuselage where a young man stood in a daze. He said he had been inside, and that it was useless to look—they were all dead. She looked anyway, along with a doctor and the young man, who reentered the cabin.

In what remained of the interior, the bodies of the passengers were spread out over the seats, all of which had been jammed into one corner, as if "a giant trash compactor had moved through the airplane, breaking everything apart." The pilots' bodies were crushed. It was a scene Eunice Stadler would be able to see for years just by closing her eyes. The aftermath of an airline accident. The mute evidence that people, not statistics, die in such tragedies.

As Eunice Stadler emerged from the wreckage, the flash of a camera strobe illuminated the scene. A man with a small pocket camera and no official reason to be there was taking pictures of the plane and its silent contents—closeup shots of the mangled remains of George Hines and Jim Merryman. The sight of the photographer going about his picture taking was more sickening to Eunice than the subject matter.

There appeared to be only eight bodies in the cabin, so tightly had the impact jammed them toward the front. The rescuers began searching the woods and the treetops for others, unaware that all the victims were really there. The need for information was great, and they pounced on Rocky Stenger when he arrived at the scene around 9:40, asking for information on how many were aboard. Rocky, transfixed by the awful sight of the wreckage and the confusion, seemed uncomprehending for a moment.

"Boston is closed."

"What?"

"Our people at Boston have gone home. I don't know how many were aboard."

It was past 10:00 P.M. when the phone rang in Phil Simpson's home in Wiscasset. Simpson's friend at the Augusta FAA Flight Service Station was on the other end.

"Uh, Phil, Downeast has lost the Otter."

"What?"

"It looks like it went in somewhere short of the runway at Rockland around nine P.M. So far they think there's one survivor—a passenger—and seventeen dead . . . including the crew."

"Oh, my God!" Simpson rubbed his head and closed his eyes, trying to get a grip on the information. "Who were the pilots?"

The answer came back faster than he'd expected, but the first name wasn't a surprise. It was and it wasn't.

"I'm sorry, Phil, but it was Jim Merryman . . . and the copilot was a fellow I think you know too . . . George Hines."

"Both dead?"

"Yes."

Simpson thanked him and replaced the receiver. His wife had been standing there and was looking at him quizzically, concerned.

He met her eyes. "Downeast's Otter has crashed. Seventeen are dead, one passenger survived. Jim Merryman was killed."

Simpson was beside himself. The memory of that last conversation with Jim haunted him—replayed in his mind constantly. The years of concern over Bob Stenger and his operation had come to this. Simpson promised himself that when the NTSB arrived, he would be there. Downeast was not going to get away with this one.

Just before eleven, Stephanie Eaton gave up waiting for Jim Merryman and decided to go to bed. He was supposed to have come by when he got in from Boston, but obviously he had diverted the flight because of the fog. She wasn't surprised. The fog had been rolling in as she left a late shift at nearby Penn Bay Hospital. The sheriff's roadblock she had seen down the road on her way home raised no questions in her mind. Accidents and fog were commonplace events. Undoubtedly, she figured, Jim would call tomorrow from Brunswick.

When news of a passenger airline crash is received at the Federal Aviation Administration's communications command post on the third floor of the FAA building on Independence Avenue in Washington, one of the duty officer's first moves is to activate the National Transportation Safety Board system by calling *its* duty officer at home or wherever. If the crash is of sufficient significance—if it has killed enough people or met several other criteria—the board member on duty has to consider whether or not to scramble the Go Team or leave the investigation to one of the NTSB regional offices. In the midnight hour of May 30, 1979, in Washington, there was no question on the status of Downeast 46. Seventeen people killed in a commuter-airline accident meant the Go Team. The board had been watching the commuter industry with great alarm lately as crash after crash heralded the fact that adequate airline-safety philosophy and Part 135 operators were not necessarily synonymous entities. And here was another one. The beepers and home phones of those on the Go Team went into alarm.

In the early hours of the morning, still shrouded by fog, the ambulance carrying the Stadlers and their crew moved slowly past the incredible number of parked cars that the sheriff's deputies had let in. In the back, in plastic bags, were the mortal remains of Merryman and Hines.

Eunice's crew was exhausted, traumatized, bloody and tired—upset by the sightseers and the lack of control, horrified by the sight of the county sheriff lighting a cigarette while standing up to

51

his ankles in jet fuel. Now, walking down the center of the road was an elderly man, nattily dressed, a dark beret on his head, a 35mm camera slung around his neck. The ambulance braked to halt, the man standing squarely in front, bathed in the headlights. He moved to the window and asked with a large grin if they were carrying survivors?

That, quite simply, was too much. The man seemed disappointed—the crew was astounded. How could a person stop an ambulance for a picture if he truly thought it was carrying injured people? Eunice buried her head in her hands. The whole thing had been a ghoulish nightmare.

NTSB investigator Alan Diehl was talking with his wife about plans for the upcoming weekend and getting ready for bed in their suburban Washington townhouse when his beeper went off. The Go Team had been activated, the accident was in Maine, and he was going to head the Human Factors Team of the investigation. Steven Corrie would be the IIC (investigator-in-charge). Diehl was to be at Washington National Airport before six in the morning to board the FAA's Grumman Gulfstream turboprop aircraft that would fly them to Rockland.

Dr. Alan Diehl had a pretty good idea of the work ahead. The Human Factors section in which he worked for the NTSB had traditionally been concerned with questions of survivability and medical/pathological considerations that could be proved by hard evidence: How much deceleration could the human body take and survive? Were the pilots free of drugs and liquor? Exactly what injuries killed and/or injured the passengers and crew? Were the seat bases and the seat belts adequate? Diehl knew that those areas were still the only legitimate focus for the head of the Human Factors Team on an accident investigation, but as a psychologist he knew only too well that there was always a deeper dimension. There were people involved in these crashes, and their actions couldn't always be explained with the same precision the board could use in describing the failure of a metal part, or the seizure of an engine. Sometimes it was the performance of the humans that was the real cause. Nevertheless, the protocol made room only for a Human Factors Team—and these would be the issues facing him in New England in a few hours.

Thirty-six year-old Diehl was a relatively inexperienced accident investigator who had joined the NTSB almost by happenstance in 1977 (a casual visit to NTSB acquaintances triggered an invitation to bid for a job). Al Diehl was a bit of a maverick in the disciplined ranks of the bureau. By that time he held a commercial pilot's license, with degrees in engineering, management and psychology. He had also worked as a human-factors specialist for two major manufacturers, Ling Temco Vought, and Cessna, where he had been largely responsible for persuading Cessna to install shoulder harnesses as standard equipment, and utilized human factors criteria in cockpit design (a first in private aviation).

Al Diehl had a tendency to look for subjective explanations when no mechanical problems could be found—a habit not well received among many of the more experienced staff members. The NTSB looked for hard answers supportable by hard facts—never mere speculation based on circumstantial evidence. If a human being had failed in the performance of his flight duties, that simplistic but objective fact was usually enough.

Many of the NTSB staffers were still in Chicago trying to piece together the horrible crash five days before (on May 25) of an American Airlines DC-10, Flight 191, which had slammed into a vacant lot just west of O'Hare field. Two hundred seventy-three humans had been obliterated by that impact, and FAA Administrator Langhorne Bond was already rumored to be considering a suspension of the DC-10's certification to fly as an airliner, a momentous move that would ground all DC-10s nationwide and cost untold millions. The national media was bannerlining all aspects of the crash and reviewing the multiyear saga of the DC-10 and its spotty safety record. The media were also focusing attention on National Transportation Safety Board member Elwood T. "Woody" Driver, who had appeared on national TV holding up the broken bolt his team had found next to the runway in Chicago, seeming to indicate that it alone had caused the tragedy.[2] Partially

2. Elwood Driver was one of the politically appointed board members, though he was highly regarded by the NTSB staff. The board members make the decisions, but most of the work, and the overwhelming majority of the field investigations, technical research, and writing, is done by the professional NTSB staff members who are U.S. Civil Service workers, usually of upper-middle-management rank (GS-13 and above). The NTSB was created out of the Bureau of Accident Investigation of the Civil Aeronautics Board (in 1966), and the bureaucratic independence of the board was ensured when Congress in 1974 passed the Independent Safety Board Act, following revelations that the Nixon White House staff had attempted to control the NTSB in 1972–73. (ALPA was instrumental in lobbying Congress for such an act.) The seasoned professional accident investigators (who had been with the forerunner of the NTSB at the CAB) and their protégés had by 1975 achieved world leadership with their amazing technical detective work and accident reconstruction. There

as a result, the activity around the FAA building (in which the NTSB offices are located) in the previous five days had resembled a kicked-over anthill. The tragic fact that seventeen people had been wrenched out of this world by a crash in New England wasn't going to divert much attention from the Chicago disaster, but it had to be probed. In fact, it was one of the worst commuter airline accidents in history.

Diehl rechecked the items in his prepacked "crash kit," set his alarm clock, and went to sleep. Because of the recent death of his infant son, he hated to leave his wife alone. He was hoping for a short investigation and a rapid return.

As the telephone calls crisscrossed the Rockland, Wiscasset, Augusta, and Brunswick area of southern Maine, one after another of the fellow pilots, employees, and friends stood at bedside tables and wall-mounted phones, barefoot on cold floors, rubbing their eyes and trying to comprehend what the voice on the other end was telling them. The Otter was down—destroyed. Jim Merryman and George Hines were dead, and fifteen other people had been killed. Those pilots who had flown with Merryman couldn't believe it. The engine must have failed or torn loose—that damned

---

was a considerable amount of professional pride among such men as Frank Taylor, Bureau of Accident Investigation chief, and C. O. "Chuck" Miller, former head of the same department, along with a host of others. That pride gave birth to several basic rules, one of which was: Never, ever, speculate on the cause of an accident in private or in public during the investigation, lest such speculation prejudice your own (the investigator's) point of view. The corollary was simple, especially in political Washington for staff members who worked for politically appointed bosses (the board members): Never speculate to the press (in any form) or announce premature conclusions on the cause of an accident. Woody Driver had gone to Chicago and stood before national TV news cameras, held up a bolt at the request of the TV crew, and violated both of those cardinal rules. Even though his words made it clear that no decision had been reached on the cause of the crash, his gesture drowned out the words in a flood of visual imagery. He represented them all, and he had mightily embarrassed the staff—whether his conclusion was right or wrong. (In fact, the bolt was not the cause, and not much of a symptom. It had been a typically byzantine chain of failures that had transformed American's Flight 191 with its human occupants into a smoking heap of space-age rubble. Woody Driver's premature announcement, made in the heat of the investigation and under the gun of the press, would look sophomoric later on when American's maintenance techniques and pilot-procedure training, the DC-10's hydraulic-system design, and faulty FAA surveillance and certification throughout the entire affair were unmasked as the principal causes.

number-two engine! No way, NO WAY, would Jim Merryman have flown into the trees without a major mechanical failure. He was far too careful. Those who had flown with Hines and regretted the experience felt a different twinge—the cold edge of recrimination for something left undone, something important left unsaid. But not Merryman, for crying out loud! Jim Merryman was the chief pilot, the chief instructor. He wouldn't have let Hines kill him!

At 5:00 A.M. Robert Turner's phone rang. The FAA principal operations inspector for Downeast Airlines, the man most responsible for assuring members of the public like Bill Detterline and John McCafferty that they could trust their lives to Bob Stenger's airline, felt as though he'd taken a body blow. He had been personally responsible for getting Jim Merryman and Downeast together. He knew George Hines. And, of course, he knew Downeast.

In fact, what Bob Turner had known about Downeast, and how long he had known it, would become matters for speculation in the press and in the FAA headquarters and NTSB hearings. Even in those first few minutes, with the phone still cradled in his hand, Turner knew he might have to start circling his own wagons. Whatever had happened with the Otter, he knew a great deal about what had been going on at Downeast for eleven years. And he also knew a great deal about what had gone on at the Portland GADO to protect Downeast during those years. Turner had been around the FAA long enough to know that his bosses were going to scramble to cover their hides if the crash had anything to do with deficiencies on Downeast's part. That scramble, predictably, would be at his expense.

Robert Turner dressed, put on his coat, and headed for Rockland.

A little group of somber people—principally employees—began gathering at the Downeast terminal around midnight, milling around in small numbers and talking in low tones, none of them exactly sure what they should be doing or saying. The grief-stricken people who had been waiting for the arrival of Downeast 46 had scattered to various locations after the crash site had been located but before the news that only John McCafferty had survived.

The word circulated briefly among Downeast people throughout the area that Bob Stenger himself had been on the Otter—and was

55

dead. They had all heard by that time about his reservation on Flight 46. The thought that he might have been killed in his own Otter brought different reactions, including the phrase "poetic justice" from several of the pilots. While the dazed employees speculated and the rescuers worked at the crash site, a rental car sped through the night on its way from Portland to Rockland. The driver had heard the news, called Rockland, and was headed back as fast as he could go, using the driving time to work out some of the details of what must be done on arrival. Robert Stenger and his wife had phoned the Downeast counter in Boston early in the evening and discovered that Flight 46 was booked full and the weather was deteriorating. If he got on the Otter, Stenger knew that he and his wife would be bumping paying passengers and losing money. Besides, he figured the flight wouldn't make Rockland anyway because of the fog—especially since Jim Merryman was flying. The Stengers had continued on to Portland aboard a Delta jet instead, missing Flight 46.

Early Thursday morning, June 1, a man purporting to be a Downeast representative left a hospital in Augusta, the one to which the bodies, including Jim Merryman and George Hines, had been transported. The man carried with him a sheaf of papers—weather-report sheets, flight-plan documents including weight and balance computations, a pilot logbook, and a copy of the Localizer Approach to Rockland. He had appeared around 8:00 A.M. and told the state medical examiner, Dr. Henry Ryan, that the airline needed the documents. The request rang no alarm bells in Dr. Ryan's mind. It had been a Downeast accident, the body was that of a Downeast captain, and the papers were theirs. He had already packed up the personal effects and papers found on the mangled remains of James Merryman's body. The papers were bloody, but if Downeast needed them for the investigation, they could have them.

The papers, presumably including Jim Merryman's personal logbook, were taken immediately back to Rockland, inspected, and destroyed. Merryman had told Stephanie Eaton a week before that he was meticulously documenting everything he did at Downeast that he felt was marginal—including excessive flight hours. The logbook might be a key to his state of mind, but it was obviously important to someone that it never be found. There was no time to waste—the NTSB was on the way from Washington.

The FAA Gulfstream made several passes over the accident site before landing at Rockland in midmorning with the Go Team members on board. The time en route had been spent in organizational meetings and a few catnaps. Once on the ground the team members split up, the majority heading for the nearby Samoset Resort Hotel where headquarters were to be set up and meeting rooms retained for the investigation. It was a practiced routine, and the ones left behind at the airport were charged with going straight to the accident site and securing the wreckage from the Knox County Sheriff's Department.

Al Diehl had put his bags in his room and come back to the conference area to work out the initial schedule when Phil Simpson introduced himself. Simpson wanted to know if Diehl was aware of the 1971 Navajo crash.

"No, I'm not."

"Then I'll get you a copy of the report—today—this afternoon."

Diehl was puzzled. "Why so urgent?"

Simpson fixed Diehl with a worried look, responding in his quiet, clipped Maine accent, tinged with overtones of Oxford, England. "Your field office investigated from New York, and they missed an awful lot. You see, Downeast is owned by a fellow named Robert Stenger—and we've been watching him for a long, long time in this area. The man pushes his employees too hard, encourages them—pressures them—to break rules, and he's had other accidents you don't know about."

"Where's the FAA been?"

Simpson told Diehl about the FAA's "surveillance" of Downeast. "In fact, Dr. Diehl, on at least two occasions, Robert Turner, the principal operations inspector, or others in the FAA office in Portland, have been told in detail about all this and refused to take any action because, as they claimed, they didn't have any proof."

Diehl wasn't sure where this was heading.

"You believe the background has some connection with last night's crash?"

"I know it does. I knew the captain—very well." Simpson smiled thinly and held up his hand, palm out. "Let me get you that report first, then let's talk. I have to drive up to Bar Harbor today on a state matter."

Simpson took his leave as Diehl turned back to the planning

session, thoroughly curious about the strange turn this was already taking.

To the few law-enforcement personnel still at the site, the little boy looked completely out of place standing before the wreckage of Downeast 46. What he was doing there they weren't quite sure—somebody's little boy brought by his parents to witness a real-life horror story, they supposed. The lad had been wandering all around the shattered aircraft, periodically pointing to various parts, talking to the man and woman who had showed up a half hour before with a pass signed by Bob Stenger.

The little boy was acting as if he were visiting a movie set, walking around in animated excitement, though not exactly smiling. One of the deputies watched him for a moment, wondering if the boy had any idea that people had lost their lives in this place of beautiful trees and terrible destruction—wondering what possessed his parents to bring him.

Almost fourteen hours had elapsed since the same trees had appeared in Jim Merryman's windscreen, and now it was daylight—a crisp, partly cloudy Maine morning. The sky was blue, the fog gone, and the broken bodies long since removed. Overhead a helicopter made repeated passes, snapping pictures of broken treetops and wreckage distribution. The aft fuselage of the Otter lay mockingly on its side, the tail balanced in the tree branches, sunlight bathing it in a spotlight of brightness contrasted against the sky overhead. The Otter's shiny white and blue paint job seemed almost unsullied as it sat in the middle of the forest, stabbed into the marshy earth like some gargantuan dagger.

The man and woman stood silently, not even exchanging glances. They had driven fifty-four miles to come to this place, a journey neither could avoid, nor fully comprehend. The sight of the Otter's remains seemed so unreal, as did the sight of the boy moving around the wreckage. They had checked with a psychologist friend before bringing him, so it was probably for the best. Yet it would surely give him bad memories.

The boy, who looked about eight years old, began talking once again, looking and pointing to the confetti of metallic pieces that littered the ground, the remains of the Otter's cockpit.

"That would be the control yoke, and this is where he was sitting"—his voice diminished to a lower volume, the last of the sentence trailing off—"right here."

The little boy finally stopped moving, standing silently before the wreckage, all alone, the slight breeze rippling through his hair, hands at his sides, the beauty of the assaulted forest surrounding him so incongruously. It seemed to hit him for a moment as he looked up at the stark whiteness of the tail, then back down at the rubble in front of him. It was a vision that he would indeed remember all his life, since, some fourteen hours ago, on that very spot, his father had died.

His father. His dad. His best friend and companion in all the world, who could do no wrong.

Jim Merryman's son, Jimmy Merryman, stood there trying to understand it all, but the enormity of it—what this place meant— just wouldn't sink in.

John and Sharon Merryman watched transfixed. For some reason, almost as a reflex, John Merryman raised his camera and clicked the shutter. He had taken other shots at the site of his brother's death, but that one of Jimmy in front of the Otter's remains would produce a powerful and gut-wrenching image that would haunt them all: a freeze-frame of innocence face to face with brutal reality.

Chapter 4

PIECES TO THE PUZZLE

"Call me at home."

"What did you say?"

Al Diehl was standing in the small terminal at Downeast, a few feet away from the small office that had been set up as a temporary interview room, talking casually to one of the pilots who had been the subject of a formal NTSB after-crash interview—standard procedure in such disasters. It was Friday morning, June 2, 1979, and a lineup of Downeast personnel were waiting for their turns. It was all very formal—made more so by the additional men in the small interview room. Robert Turner, the FAA's principal operations inspector for Downeast (who was responsible for monitoring—surveilling—Downeast and its "compliance" with the rules and regulations), sat with Downeast owner Robert Stenger. For the employees, the atmosphere had hardly been conducive to glib outpourings of inside information.

Now, suddenly alone with Diehl outside Stenger's earshot at the water cooler during a break in the questioning, one of the pilots had lowered his voice and mumbled the request.

"I said call me at home—I'm in the book. I don't want to talk here."

"Okay. All right, I will."

"Don't mention this."

"No, certainly not."

The young pilot turned to go back in, leaving Diehl half stunned. This investigation was getting very strange.

That evening Diehl rang the man's home number and was invited out to his house. Away from the threatening presence of Bob Stenger and his loyalists, an incredible litany of abuse began to unfold. Diehl found it hard to believe at first—it was always possible he had run across a disgruntled employee whose bias against the boss might lead to some wild, fabricated tales. But then, there had been Phil Simpson's statements on Thursday, and the disposal of the flight papers that very morning.

The stories were indeed wild, but they began following a pattern. The things related by the young pilot who didn't want to talk at Downeast (and in subsequent days by others both on and off the Downeast payroll) began to put an entirely different light on the crash. Suddenly, for Al Diehl, the loss of Otter 68DE was no longer just a case of a pair of airmen screwing up and descending into the trees, it was a case of an accident caused by human-performance failures caused in turn by a long history of management pressure and downright disregard (if not contempt) for the rules.

In fact, the image that was developing from what he was being told was incredible! It was difficult to believe that a company could be so callous. Downeast was, after all, a public airline, invested with a public trust, operating with the blessing of the federal government. But the image these people were painting was of an airline operating with only the owner's financial well-being in mind, and the public be damned. You pay your money and you take your chances. A few passenger deaths in a crash every now and then are to be expected—goes with the territory. And besides, the company is insured.

As the scope of his interviews away from the Downeast offices widened, it became apparent that Jim Merryman had been a very troubled, pressured man, driven by Bob Stenger, who held him accountable for everything and gave him authority over very little. One after another, witnesses revealed glimpses of Downeast's activities that, if true, amounted to horror stories, all the more so for the FAA's long-term inability (or refusal) to see the unsafe practices and put a stop to them. Pilot training, for one (the bedrock of aircrew flight-safety activities) was said to be poor to nonexistent. Some pilots told of being hired and given a "ground school" in the Navajo or Aztec that consisted solely of being handed an owner's manual and told to read it overnight. Even the Otter, the most complicated aircraft the company owned, was supposedly entrusted to new pilots who had received less than a day or two of informal ground school.[1]

The growing list of alleged abuses amazed Diehl: pressure on weather observers to falsify weather reports, confirmed stories of Stenger berating his weather observers for using too many ninety-cent weather balloons to check the height of the cloud cover, and the incredible tale of a Piper Aztec that had been used for months in passenger service despite a right engine that would go into feather and shut down without warning.[2]

Pilots who had lived under Stenger's system and had enough integrity at least to feel uncomfortable with the rule breaking told of habitual, chronic overloading of the different Downeast aircraft in Boston and Rockland, along with incidents involving falsification of the weight and balance forms provided to the pilots. Diehl was

1. Although it did not come out until the formal hearing months later, the pilot-training records at Downeast bore little resemblance to reality. On their face, it looked as if all the pilots had received the required ground-school courses and training flights. According to the testimony of those who supposedly received such instruction, the records were false. To the FAA, they looked excellent.

2. The Rockland area is strategically located adjacent to a host of small islands and exclusive residential areas, many owned by some very wealthy and influential people (such as Tom Watson, former U.S. ambassador to the USSR and former chairman of IBM). Downeast pilot Al Brandano told Diehl of a time in 1977 when a crown prince and princess from one of the Scandinavian countries came to the area to visit some influential American acquaintances. Since Bob Stenger's airline and charter company was the closest to the home they were visiting, a Downeast Aztec was chartered for several days of sightseeing. On turning to final approach at a small, four-thousand-foot strip on North Haven Island, the right engine of the Aztec suddenly went into feather, that is, the propeller blades changed their pitch to align with the airflow, thus producing no drag and no thrust—and also stopping the engine. Brandano's heart leaped into his throat as he pulled the feather lever in reaction. The princess looked at him quizzically.

"Please, why did you do that?"

"Oh, we do that all the time to save fuel," replied Brandano. It was a dumb answer, he thought, but what else could he say? We're going to crash and die? The princess accepted his answer as Brandano guided the craft to a safe landing. Though a maintenance team was sent from Rockland, they could find no problem with the engine. The unexpected feather sequence occurred at least three more times in the following months before the aircraft was removed from revenue service and the cause was found. The wrong bearing had been installed during the last overhaul, and Brandano suspected that Leo Gallant had known it all along.

told about the "cement-plant approach," a wild "follow-the-coast-line-to-the-runway" backdoor procedure (totally illegal if flown in instrument conditions) that some of the pilots reportedly used to sneak into the field when the fog was around 200 feet off the ground. There was a long and incredible anthology revolving around the concept as there was around the phrase Downeast minimums, and figures of 200 and 300 feet above sea level (instead of the legal 440-foot minimum) were alternately tagged as the lowest altitude Bob Stenger allegedly expected his pilots to reach before giving up and flying to an alternate—and even the alternate was often dictated by Stenger on the basis of economic convenience. As far as takeoffs were concerned, no visibility restriction seemed too severe for a Downeast flight. Stenger, one pilot reported, had on several occasions stood and watched with smiling approval as a flight departed in visibility that was lower than one quarter of a mile and patently illegal.[3]

There were accusations as well of constant harassment of the pilots and threats of firing for any infraction, consistent over-scheduling and fatigue-inducing duty hours, refusal to furnish the pilots with copies of Federal Air Regulations and other equipment, and even an apparently successful attempt to conceal a landing accident in 1977. It all began tumbling out—sometimes with the pre-

3. To Bob Stenger, FAA-imposed takeoff minimums must have seemed, quite simply, a nuisance. Stenger was not in the habit of directly ordering a captain to depart against his will. He used withering disdain and manipulated peer pressure instead. Those who stood up to him usually prevailed. Those who were intimidated took off—regardless of who was on board. In one instance, U.S. Senator Edward Kennedy was flown out of Rockland on a Downeast flight allegedly with Robert Stenger's blessing. The visibility that morning was nearly zero.

There were several ironies involved in the Kennedy flight, the most significant being that Ted Kennedy had suffered a broken back in the crash of a chartered Beech Queen Air a decade earlier. Though as a senator he constantly needed to travel by air, his sensitivity to air safety might be presumed to be greater than that of the average traveler. (Of course, he was one of the sponsors of airline deregulation from 1976 through 1978, an issue that involved serious and substantial questions of how badly airline safety might be affected by the financial pressures of a free-market airline environment). In the Downeast case, Stenger personally held the departing flight for fifteen minutes waiting for Kennedy to arrive—which was not unusual for an important passenger. But Stenger scoffed at the suggestion that he was giving Kennedy apparent VIP treatment. He had, he said, held the flight only because it meant an extra fare!

amble: "Boy, I'm glad you called! I've been waiting to tell this to someone for a long time!"

Of course, another thread was becoming apparent in the stories Diehl was hearing: If the pilots themselves, through peer pressure, had permitted and perpetuated the "system" at Downeast, and that seemed to be the case, then they were complicit as well.

As the rest of the Go Team members wrapped up their areas of investigation and departed on June 7, Al Diehl stayed on, conducting interviews with friends and family of Jim Merryman and George Hines, talking to girlfriends, ground crewmen (most of them teenage boys), ex-wives, and anyone else who could help shed light on what had shaped Merryman's recent past. The leads to former pilots and employees were furnished from various sources, although Stenger and Downeast were stonewalling most requests. Bob Stenger had been getting constant reports on "that fellow from the NTSB" who hadn't left when the rest of the team had gone home, and who kept asking questions all over the county. Stenger could sense he was in trouble. He knew it the moment Al Diehl sat down across from him and asked about the flight papers taken from Jim Merryman's body. "Oh, those," Stenger had replied. "Well, they were kind of bloody, so we threw them out that morning." Stenger must have known that Diehl understood the significance of that act, and what it might have been meant to conceal. The papers were gone, but the suspicions raised by their disposal would spur the investigation into new areas.

Stenger had "gotten away with the previous crashes," as several pilots put it. His business had been relatively unscathed, and each time the NTSB people had been around only three days. Now, however, ten days had passed and at least one of them was still in town playing detective—a role that several other very angry people at Downeast were beginning to play as well.

One particular Downeast secretary was white-hot mad. She had cared for Jim Merryman, and she felt Bob Stenger's pressures and practices had set up his death. She was determined that Stenger was "not going to get away with it," even if it meant her job. She began watching Stenger, listening to his conversations, standing at closed doors, reading memos, and reporting back to Al Diehl— who had returned to Washington finally on the tenth of June. A massive round of negligence lawsuits was sure to be filed, and everyone at Downeast was worried that the company might fold, and their jobs evaporate along with it; nevertheless, the flow of information continued.[4]

After long, agonizing sessions of late-night telephone calls trying to persuade Downeast personnel (present and former) to commit their experiences to the record, pilots such as Al Brandano and Harold Hamre wrote follow-up letters chronicling the abuses and their personal experiences, as did Kurt Langseth, who quit on June 7, convinced the airline would fold and that it was obviously too dangerous an operation to justify his continued presence.[5]

Kurt Langseth and many of his fellow pilots were just as mad as the secretary who kept calling Diehl in Washington. They had all known and respected Jim Merryman and understood his impossible position at Downeast. But they were also cautious, and much persuasion was needed to get their stories in writing.

Of course, the pilots themselves had put pressure on Jim Merryman from the other end of the spectrum. Their irritation over Stenger's decisions and treatment of them, their inexperience, their reluctance to be led by Stenger's representative (as they considered Jim) all served to alienate them from their chief pilot. Merryman was truly caught in the middle, neither fish nor fowl. Nevertheless, when faced with his death, those who had been cool to his attempts at leadership slowly came to his defense as they perceived it. There would be a public hearing in September, and Diehl—after endless hours on the telephone from his office and home—would end up with more witnesses than he needed, ready, waiting, and willing to place the entire story on the public record.[6]

4. The secretary told Diehl that Stenger had spent days going through all the company records he could find, tearing out any pages that contained entries that might give him trouble. The pages were taken home at night and, according to her, burned. The information was never confirmed.
5. Langseth had not received the direct pressure from Stenger that other pilots reported—pressure to break minimums, fly with overloaded airplanes and mechanical defects, etc. In fact, Langseth felt Downeast was an improvement over the commuter carrier for whom he had flown in the Midwest, Brower Airlines (by then out of business). At least, he felt, Downeast had fairly good maintenance. Brower, Langseth charged, had had no maintenance to speak of, and the airplanes were always in marginally flyable condition. The pressure to make money regardless of the methods was the same at both Brower and Downeast, however. Langseth had left Brower out of fear for his life. (He would later join Air Wisconsin, an airline that, as he put it, "would fire you or give you thirty days off for doing the same sort of rule-breaking which would elicit a hearty pat on the back from Bob Stenger."
6. In the middle of the summer, Al Diehl had remarked to departing veteran

What Al Diehl did not have, however, was massive support in the NTSB. Only his boss, Human Factors Division Chief Gerrit Walhout, had stood behind his desire to stay on the investigation. By early July, Walhout was being pressured to get the Human Factors Team report finished and out. Diehl explained he would need a few months, perhaps more, to do it right, but that met with sheer derision in the ranks (as did his question of whether the NTSB could grant the Downeast pilots immunity from FAA punishment if they testified to FAR violations). Downeast, they complained, might have had some elements of management pressure in the causal chain, but basically it was pilot error. The pilots screwed up and hit the trees. Besides, said some of the detractors among the staff, the owner of the airline had probably learned his lesson. This time, after all, his airline had killed seventeen people. Surely the embarrassed FAA wouldn't let him get away with all these abuses in the future.

Diehl and Walhout were unconvinced, but they needed something more—another element of motivation—to buy the time required to do the thorough investigation that Diehl knew was needed. The Downeast accident must not be marked off as simple aircrew error, even if the final report hinted at some of the underlying problems (such as mechanical and lighting defects and inadequate training). That wasn't enough. This report had to get to the basic problem of management abuse and all the violations of spirit and intent that flowed from that poisoned well.

But Diehl knew, as did his boss, Jerry Walhout, that an indepth, long-term probe into the human causes of the crash involving the performance, attitudes, emotions, and motivations of aircrew and manager alike couldn't be done without getting into the traditionally forbidden area of circumstantial research and subjective, imprecise conclusions—an endeavor that would go against the grain of the ever-precise NTSB engineering ethic. They needed

---

NTSB investigator Gerrald Bruggink that he had three pilots who were ready to put the story of Downeast management pressure on the record. Bruggink, who had pioneered and promoted the concept of expanding NTSB investigations into the areas of human performance and motivation even before Chuck Miller (and had been overruled by the staff and the board members as many times), was packing his office possessions as Diehl unfolded the story. Bruggink gave Diehl a wry look and shook his head. "You got three pilots, Al, you've got nothing. Management will just deny everything. You want a case, get a lot of [witnesses]." Diehl, much to the consternation of his wife, redoubled his efforts, spending most evenings on the phone, chasing down long-lost employees and asking for information. Most of the hours and the phone calls would never be compensated.

time—they needed a justification—and with the crash on the seventeenth of June of another Twin Otter engaged in commuter service (and an unexpected phone call from a former Downeast pilot on the second of July), they would get it.

As Al Diehl and Steve Corrie conducted their investigation in the Rockland area on June 5, 1979, a thirty-two-year-old copilot for Air New England, Richard D. Roberti, sat with a group of other newly hired pilots, listening to a briefing by a representative of the Air Line Pilots Association. ALPA had recently begun to represent the Air New England pilot group, but as the representative had rather innocently warned them, "You guys will not be ALPA members until your one-year probation period is up. In the meantime, if you cross a captain and he wants you fired, we can't protect you—you're gone. So, be careful."

Roberti mentally filed the briefing away. He had logged more than four thousand hours of flight time and was an experienced pilot, but this was his first substantial airline job and he was determined to get through his first year unscathed.

Captain George E. Parmenter had also logged a lot of flight time—over twenty-five thousand hours to be exact—and served as a senior vice-president of Air New England which had its home base in Hyannis, Massachusetts. Captain Parmenter was one of the founders of Air New England. Although he loved flying and the operational side of the airline best, he usually flew a desk. Nevertheless, he stayed legally current and occasionally was called upon by pilot-crew scheduling to go on the line and help out for a day or two at a time. The captain was an interesting and forceful man, sixty years of age (and thus prevented from flying for anything but a commuter airline under Part 135), and possessed of an airline-transport-pilot rating as well as a long history as a Marine Reserve officer and pilot. Captain Parmenter was not the sort of man one trifled with, and that was doubly true for a newly hired copilot unprotected by the union. It was this team—Captain Parmenter and First Officer Roberti—who set out on the morning of June 17, 1979, to fly a full day of revenue passenger flights, substituting for a Twin Otter captain who had (along with many of the other ALPA pilots) called in "sick" in what was apparently an illegal job action.[7]

7. Captain Parmenter took the union activities against his airline quite personally. He had told his operations and crew-scheduling people that he would fly anywhere and anytime to keep the airline on schedule. He was not going to accept a union shutdown of Air New England.

Air New England's Flight 248, the last leg of the day for Parmenter and Roberti, finally got under way from New York's La Guardia Airport at 9:32 P.M., heading directly for Hyannis's Barnstable Airport in Massachusetts. The two men had been on duty since 8:45 A.M., and the flight back to Barnstable was their fifteenth leg of the day. Both pilots were tired—but Captain Parmenter was a bit beyond mere fatigue. He was furious, exhausted, and hungry.

Earlier, when Parmenter and Roberti arrived at Air New England operations at Hyannis around 6:30 P.M., they had thought their day was over. Crew scheduling had different plans—another crew had failed to show up, and they would have to take the flight to La Guardia and back. Parmenter was "visibly upset" by one account, "apoplectic" by another. He had consumed a total of one Danish roll and one cup of coffee since 7:00 A.M., he was exhausted, and he wanted to go home. Nevertheless, he collected Roberti and stormed out to the aircraft.

Now, on the way back to Hyannis in their de Havilland Twin Otter, Parmenter was sullen. Unbeknownst to First Officer Roberti, he was also in substantial physical distress.

Captain Parmenter had been medically disqualified from Marine Corps flying years earlier because of high blood pressure, and had lost his FAA certification temporarily at the same time. Nevertheless, after numerous attempts, he was reissued an FAA first-class medical certificate three years later by a friendly FAA-certified doctor (AME, aviation medical examiner) in Hyannis. The same doctor had continued to recertify him every six months after that, even though his blood pressure and electrocardiogram tracings were marginal. In fact, the captain had suffered a heart attack at some point (a myocardial infarction, which had healed) and had successfully concealed it, had continued to suffer from hypertension, and had become marginally hypoglycemic. In addition, he had been using two prescription drugs constantly for his hypertension (polythiazide and allopurinol), but had never admitted it on the certified reports he filled out with each medical exam. No one, apparently, was going to tell Vice-President and Captain George Parmenter that he was medically unfit to be an airline captain.

Most of the Air New England pilots knew by experience that no one could tell the captain much of anything in the air. He was well known for never responding to checklist items read by the copilot and equally renowned for flying the airplane as if he were solo, with no patience for crew-coordination techniques.

It was such a man that First Officer Dick Roberti was supposed to monitor as they descended to the ILS glide slope on approach to runway 24 at Barnstable Airport. The noisy, dark cockpit of Air New England's Twin Otter 383EX was hardly conducive to conversation, but with this captain—company vice-president, sage aviator, gruff leader, angry over the circumstances that had placed him on these extra legs—Roberti was little more than a passenger. Nevertheless, as Parmenter sat with a death grip on the control yoke and the aircraft began to increase its rate of descent, the copilot dutifully made his required call-outs (whether they were heard or not).

"Five hundred feet above [minimums]." The Otter was passing through 800 feet above sea level and was some 500 feet above the decision height for the ILS approach—the altitude at which the pilot must either see the runway well enough to land safely or immediately pull the nose up while applying power and gain altitude for a missed approach. At 500 feet, Dick Roberti could see nothing but fog outside. Other flights just before them had broken out at about 300 feet above sea level, so this wasn't unusual—except for the high rate of descent, which was very unusual. Roberti watched the altimeter unwind. What on earth was the captain doing?

"Two hundred above," he intoned, then four seconds later, "One hundred above." Roberti glanced up from the instrument panel. Still nothing. His eyes shot back to the panel, past the ILS needles, which were now showing full-scale deflection—they were far below the normal glide slope for the approach!

"Minimums . . . uh, decision height." Still no response, and Parmenter held his rate of descent. "You're one hundred below, Captain!" Roberti jerked his head up to look outside, thinking the captain must—MUST—see the runway by now. His hand was near the control yoke—the rate of descent continued unabated—but Roberti couldn't bring himself to take over. Surely Parmenter must know what he was doing. Surely he was going to level out. Roberti fought to decide what to do but suddenly all he could see were trees as the Otter slammed into a thick forest, decelerating violently with the multiple impact of branches and foliage and an occasional tree trunk into various sections of the disintegrating wings.

The Otter, without its wings and a portion of the left side of the cockpit where Captain Parmenter had been sitting, skidded to a halt in the thick underbrush. Roberti was alive, and amazingly, so were his six passengers (although two were badly injured). Unlike Downeast's crash, there had been no final, head-on impact with the ground—the deceleration had been survivable. The captain, however, was gone—figuratively and literally—thrown through the fragmenting remains of his instrument panel. If he had been alive just before impact, he was no longer.

Richard Roberti had been charged with the responsibility of being the next in command—the safety pilot—the monitor to safeguard and ensure that the passengers reached their destination safely even if the captain went to sleep (or became incapacitated) at the controls. Because of the circumstances, however, he had let George Parmenter fly his perfectly good aircraft into the ground because he couldn't risk being assertive.[8]

The effect of the Air New England accident on the NTSB was to fan the flames of concern among the staff and the board members over the commuter industry. The FAA had been replying to the numerous NTSB recommendations for commuter-airline reform (and thus massive changes in Part 135 of the FARs) with the same bureaucratic promise: They were studying the situation and would be issuing a completely new and rewritten Part 135 shortly. This

8. Tragically, Roberti's reluctance wasn't unique, nor was it to be in the future. The reluctance of other pilots to challenge a captain had killed more than one crew and their passengers, and would do so again—and that was true for the major Part 121 carriers as well as the small Part 135 commuters. (In fact, an Allegheny Airlines accident in New Haven in the early seventies was a classic case of this syndrome.) The NTSB in its subsequent report on the accident (the Blue Cover report as it's called, which is a distillation of the facts for the general public) said: "The poor altitude and pitch control exhibited by the captain and the steep descent rate that the aircraft achieved should have alerted the first officer to the existence of an abnormal situation. However, a flight simulator study of subtle incapacitation conducted by United Air Lines [which currently leads the industry in assertiveness training and other advanced techniques based on human performance awareness] demonstrated that recognition of the phenomenon by the other crewmember is a difficult task ('Study of Simulated Airline Pilot Incapacitation: Phase II, Subtle or Partial Loss of Function,' *Aerospace Medicine,* Vol. 42, Number 9, September 1971. This study culminated in a training film). In the United simulator study, when the captain feigned subtle incapacitation while flying the aircraft during an approach, 25% of the aircraft hit the 'ground.' The study also showed a significant reluctance of the first officer to take control of the aircraft. It required between 30 seconds and 4 minutes for the other crewmember to recognize the captain was incapacitated and to correct the situation. The first officer of flight 248 had 1 minute and 9 seconds from the outer marker to impact. It is quite possible that the first officer also was suffering from fatigue which dulled his senses and reactions." There was a similar accident on April 22, 1966, in Ardmore, Oklahoma, when a management captain for American Flyers, a nonscheduled airline based in Fort Worth, Texas, crashed a Lockheed Electra into an Oklahoma hillside, killing seventy-two people. The captain had concealed a major heart problem and had apparently suffered a heart attack at a critical moment in the approach phase of the flight.

sort of nonsense had been going on since several commuter crashes in 1972! Finally, in 1978, the new Part 135 was issued. Nevertheless, on May 30 and June 17 of 1979, there was still no massive enforcement action to require the commuter air-taxi industry to grow up and act like an airline industry with stringent standards and licenses, maintenance and training. Year after year planes had crashed and innocent, trusting passengers had been maimed or killed, but still the "airlines" of the commuter industry—the Bob Stengers, the Captain Parmenters, and the Brower Airlines—were allowed to fly, effectively, under the old rules.

The year 1979, however, was rapidly becoming too bloody to ignore. Aside from all the smaller ones investigated by NTSB field offices, the main NTSB team in Washington had tackled three commuter crashes before Downeast, involving a total of five deaths and ten injuries.[9]

Then came Downeast and Air New England, to be followed by more. Pressure was building on the NTSB for a major probe of the commuter industry for the purpose of focusing public attention on the FAA's reluctance to do anything. (In political Washington, a special report by a somewhat independent agency such as the General Accounting Office or the NTSB may be the only way to force a bureaucracy to move off dead center and stop studying a problem to death.) That same pressure, plus a telephone call on July 2, gave Walhout the ammunition he needed to keep Diehl on the case.

During the on-scene investigation in Rockland, Al Diehl was unable to get one particular pilot to talk to him at all. The fellow was still on the payroll at Downeast and didn't want anything to do with the NTSB or the FAA. Despite constant entreaties from Diehl and other pilots, Richard C. Mau (who had originally been scheduled to ride as an observer on Flight 46 with Merryman and Hines) kept his mouth tightly shut.

Until July 1, that is. That particular Sunday evening he refused to take off as captain of a Piper Navajo on the flight back from Boston to

9. The crashes were: an Allegheny Commuter flight (Nord 262 aircraft) that crashed in West Virginia on February 12, killing two and injuring eight, as a result of the captain's decision to take off with snow on the wings; a Rocky Mountain Airways Twin Otter crash in Cheyenne, Wyoming, on February 27, which injured two, due to lack of proper crew training and poor maintenance (the second Rocky Mountain Airways crash in four months); and the ditching off Los Angeles on March 10 (another Nord 262) of a Swift Aire Lines flight after takeoff, in which three people drowned because of lack of crew training and flawed emergency-procedure execution.

Bangor with a line of thunderstorms rolling across upper New England. There were three Navajos scheduled to fly, and when all three captains called in to Rockland for advice, Robert Stenger ordered them to take off anyway. Mau refused. The other two captains were willing, but in the ensuing fight over the telephone, all three were canceled. The next morning Richard Mau was fired by Bob Stenger, who claimed that he had lost a thousand dollars as a result of Mau's cancellation. Richard Mau no longer had a job to protect, so he picked up the phone and called Al Diehl.[10]

Diehl and Walhout circulated the story of the firing. One month after his management practices had contributed to the deaths of seventeen people, it was apparent that Stenger was still operating as he had for the previous eleven years. Things were to be done his way, or else, even if his way meant an apparent disregard for the rules. The feeling around the NTSB was one of indignation. Stenger had learned nothing. He must be stopped.

Suddenly Jerry Walhout had the support he needed. Al Diehl was taken off the Go Team and assigned full time to the Downeast case. With that assignment, the first full human-performance investigation in the history of the National Transportation Safety Board began—and the implications for the future would be seismic.

10. Mau was described later by Stenger and other Downeast employees as a very marginal pilot, though others considered him quite competent. Downeast mechanic Frank Soleno had watched in fascination one afternoon as Jim Merryman had tried to teach Mau to land the Navajo. Richard Mau kept forcing the aircraft down at too high a speed, landing nosewheel first. Yet others, including Merryman, seemed to have no subsequent worries about Mau's flying. After the crash of Downeast's Twin Otter, Downeast would try to portray Richard Mau as an incompetent instrument pilot, begging the question of why he was hired and retained in the first place if this was true.

One thing seemed consistent: Richard Mau appeared to his co-workers as a bit of an eccentric. His constant references to the science-fiction show *Star Trek,* his extreme self-assurance, and his way of looking at things in general irritated other Downeast pilots and mechanics and infuriated Bob Stenger, who considered him immature. Nevertheless, Stenger kept him on—until the incident in Boston.

Later interviews with another pilot who always refused to give a formal statement revealed one final piece of the puzzle over what could have distracted Merryman and Hines—the practice of turning on the right landing light and having the copilot check the right engine inlet for icing or other problems after breaking out on an approach. (The light would reflect off the back of the shiny propeller illuminating the engine.) Sure enough, the documentation of the wreckage had shown the right landing light—only the right one—was on at impact. The discovery could have explained why George Hines had permitted the rate of descent to get away from him.

Chapter 5

THE HUMAN FACTOR

It was so obvious—so painfully obvious—but most of those engaged in commercial aviation couldn't see it, or wouldn't. Human beings engaged in a human enterprise are subject to human failures. Pilots and controllers and maintenance people err and cause accidents because they are human, and we imperfect humans are all prone to make such mistakes. Discovering that a human error—pilot error or otherwise—has occurred is merely the starting point. To have any hope of preventing such an error from causing such an accident again and again, the *reason* the error was made in the first place must be discovered, and the underlying cause of that human failure must be revealed and addressed in future operations.

Yet throughout the history of commercial-airline accident investigation the pattern had been the same: The discovery that a human failure was the immediate cause of an accident almost always put an end to the investigation. Pilot error. Maintenance error. Procedural error. Controller error. As if there were no need to look further—no need to see that such accidents are the effects of other causes, causes yet to be addressed.

When people die of a little understood disease, public-health authorities move heaven and earth to find the cause and eliminate it so that others won't fall victim to the same thing. Certifying the mere fact of death-by-little-understood-malady in itself solves nothing. Yet when people die of a human error in aviation, the investigators, the regulators, and the industry in general simply sign the death certificate and walk off, never thinking to look for the cause in order to effect a cure.

As in the Downeast crash, the probable cause of accidents is, distressingly often, the fact that "the pilots erred." What then? Order all pilots to go forth and err no more? Warn them sternly that if they do err—and survive—they will indeed be lined up against the nearest concourse wall and shot, before live Minicams? How does a federal agency or an airline company persuade, cajole, convince, or force its people not to err? Or does it perhaps make more sense to try to understand what surrounding influences cause people to commit terrible mistakes and work to eliminate such influences? If an accident is caused by an overly fatigued individual making a fatal mistake, doesn't it make more sense to attack the cause of the fatigue instead of the effect—if, that is, the goal is the prevention of similar occurrences? Then why has this one point been so hard for the industry to understand for over forty years?

Through the late summer of 1979 as Al Diehl sat in Washington and sifted through the evidence of Jim Merryman and George Hines's all-too-human error at Rockland, Maine, the dark, frustrating irony of the long-held, industrywide attitude faced him everywhere he looked—everywhere he probed. It was the traditional view of aviation engineers and technicians, an axiom as they saw it, that what cannot be measured with precision cannot be controlled. The long-held view within the FAA was even more pointed: If you can't write a rule to govern it, you can't control it! That attitude, of course, leaves little or no room for discussion. It's damn hard to write a rule for aircrew members and maintenance people that says, in effect: "Fatigue, emotion, disorientation, preoccupation with personal problems, lackadaisical attitudes, misreading of instruments, faulty manipulation of badly designed controls, improper selection of maintenance parts, failure to understand radio communications clearly, boredom, or mistakes of any sort resulting from human weakness are specifically prohibited in airline operations!"

Moreover, the airlines themselves historically have dragged their collective feet on extensive consideration, study, or implementation of human-factors engineering in aircraft design, as well as on study of the impact of human-performance problems and limitations on the safety of their operation. The principal reasons are familiar ones: ignorance and cost. Ignorance of the vital (and obvious) importance of this area as one of the principal keys to air safety, and a reluctance to spend the money and dilute profits (or add to losses).[1]

1. The phrases "human factors" and "human performance" need definition. Harvard Professor Ross A. McFarland, generally acknowledged as the "father" of the human-factors field as it relates to aviation, published two landmark books on the subject, *Human Factors in Air Transport Design* (McGraw-Hill, 1946); and *Human Factors in Air Transportation* (McGraw-Hill, 1953). Professor McFarland taught that any control or other item on an aircraft (or any machine, for that matter) that must be read, monitored, manipulated, or otherwise operated by a human being must be designed so that the individual's physical, psychomotor, and psychological capabilities and senses are compatible with it, and so that he or she will be able to interact with the machine as the designers intended. That is to say, if an engineer wants to design an instrument that will tell a pilot the oil pressure in his engine, for example, he must design a gauge that the pilot can see and read with a reasonable degree of certainty. A gauge too small, or unlighted at night, or marked off in some foreign language, will most likely be meaningless to the pilot, and thus the purpose of the gauge is defeated. The information transfer will never take place.

Human factors, then, are those considerations of how a human is going to interact with the controls and displays of a machine, air machine or otherwise. To ignore human factors in the design of an airplane is to predispose a tendency toward incorrect operation and possible disaster. (In fact, one variation of the infamous Murphy's Law is used most often in aviation design: "If it can go wrong, it will," as well as the aeronautical engineer's corollary: "If it can be operated, inserted, turned, or read incorrectly, it will be.")

Human-performance considerations, though properly classified as a subcategory of the field of human factors, are even more ethereal, subjective, and difficult to quantify. The effects of fatigue on an aircrew's performance under normal and abnormal conditions; the effect of stress; the effect of economic pressure on a dedicated or extremely dependent pilot force; or the unpredictability of different pilots' responses to the same problem are all human-performance considerations. They are rooted in the fact that the element that determines in any situation how a person will react is the human mind, and the human mind is an analog computer, which will never operate with the precise and predictable certainty of a digital computer. The human mind is subjective by nature. Scientific and engineering disciplines are called "disciplines" largely because they seek to order—to discipline—the subjective unpredictability of the human mind-computer into a predictable and structured method of thinking and acting. The fact that this is a quest that can never be fully consummated is reflected in the language with which we describe a person who pursues the professional discipline openly acknowledged as inherently resistant

At some points it has almost seemed that the senior managements of the U.S. airline system were trying their best to avoid any understanding of the issue, for to understand the nature of human-factors problems and human-performance failures is to realize that system and hardware have to be changed. Change, in turn, means spending money. Buying updated instruments to make cockpit displays more readable and less subject to misinterpretation, for example, or adding more fire extinguishers or emergency exits, or reducing the number of hours aircrews can fly in a month, means the expenditure of considerable sums of money. If the benefit to be derived from those expenditures isn't clear and certain, the motivation for airline chiefs and boards of directors to authorize such expenses becomes weak. Thus the tendency in the industry has always been to be skeptical, if not openly hostile, to recommendations for changes in the name of correcting human-factor/human-performance problems. That is why airlines in 1984 continue to use (for example) de Havilland Twin Otters with difficult-to-read flap gauges. Replacing such gauges costs money, and besides, the pilots are paid to *find* a way to read them. Back we go to the same treadmill: Here's a perfectly good airplane—you are ordered not to crash it!

Certainly there have been enough voices crying for airline recognition of these factors—begging the managements and the regulators to do something to prevent the same errors from causing accidents over and over again. Chief among these proponents of change is a partisan organization that has been responsible for many of the advances in modern airline safety since 1935, but whose pedigree is that of a tough-as-nails labor union—and therein lies the problem. Regardless of the inherent worth of any recommendation it makes or position it takes on air safety, the Air Line Pilots Association is still a labor union, and thus its statements and positions can be too easily dismissed as "tainted" by its own self-interests.

---

to all attempts to put it in order: the medical profession. A physician is said to be engaged in the "practice" of medicine. In the same way, engineers engage in the "practice" of engineering through their continuous attempts to bring their performance up to a level of precise certainty and predictability—a level that is, unfortunately, never quite attainable. So human performance, as a catch phrase, refers to all those myriad unpredictable considerations that aircraft companies, airline managers, chief pilots, maintenance chiefs, and regulatory authorities must forever consider in deciding whether a given set of circumstances in any airline operation is more or less likely to trigger a human-performance failure (which could otherwise be avoided).

The Air Line Pilots Association, founded in 1931, has been deeply involved in research and a solid proponent of human-factor/human-performance concerns for decades. Its leaders correctly understood early on that to dismiss a human-error accident as attributable to "pilot error" and then slam the book on it was to practically guarantee a repetition of the accident. They worked hard through the early years of commercial aviation to pressure the regulators, the manufacturers, and the airlines to improve the way instruments are laid out in a cockpit. The object was to make it far more difficult for a pilot to misread his instruments or make disastrous errors by manipulating the wrong control. They fought as well for shorter crew-duty days and monthly and yearly maximum flight-hour limitations to counteract the worry over fatigue-induced crew errors. ALPA, in fact, became far more than just a labor union or professional association. Because of its tireless and often stubborn stewardship of such advances in air safety, it became more than the champion of its own members' safety and welfare. It became the champion of the flying public's safety as well.

As a labor union, ALPA had grown from the ferment of labor abuse and exploitation in the crucible of the prewar period. That was an age in which airlines were created, expanded, bankrupted, sold, and reorganized with blinding speed, and no airline entrepreneur wanted his ironfisted control over his pilots diluted by the likes of a labor union. World War II, of course, disrupted this evolution, but when the nation got back to civilian life in 1946, the airline managers resumed their business-as-usual pursuit of expansion—as nonscheduled carriers began cropping up like Johnson grass on a Texas lawn.

Dealing with ALPA's leaders was neither a pleasant (nor always a gentlemanly) enterprise in those days. They wanted a long list of reforms for their members, including job security, more money, better equipment, and safer flight-duty rules. And there is no question that it would take whatever it could get. It was easy, however, for airline owners and managers to lose sight of the fact that some of these demands were vital and valid safety items—items that would save lifes.[2]

As the union matured into the sixties and seventies and the con-

---

2. The fascinating history of the Air Line Pilots Association and its struggle through the formative period of the fifties is nowhere better told than in the book *Flying the Line* by airline historian George Hopkins (printed by the Air Line Pilots Association, Washington, D.C., 1982; IBSN 0–9609708–1–9).

tracts became more lucrative, a curious thing happened. Because of the financial bargaining success of the pilots over three decades in a regulated environment, the labor-union members of ALPA quietly took on a more professional, white-collar status within their communities and, more important, within their own thinking. The number of pilots with college backgrounds and multiple degrees increased substantially, and the acceptance of, and understanding of, the psychological underpinning of the field of human factors and human performance grew. ALPA formed a close and productive relationship with the Aerospace Medical Association, which helped give ALPA pronouncements the legitimacy of thorough and scholarly research. The joint studies and the communication led to greater awareness among ALPA's membership, and along with that increasing awareness came a greater, and more eloquent, advocacy that led to numerous advances in accident investigation, standardized cockpit instrumentation, ground-proximity warning systems, and such small but important changes as the little tag you find yourself staring at (and too often ignoring) on the back of airline seats: PLEASE KEEP YOUR SEAT BELT FASTENED WHILE SEATED.

However, the one organization that had the greatest understanding and the greatest chance to establish an early system for combating human-factor and human-performance threats to air safety was, because of its union nature, in an adversarial relationship with the industry and the regulators. ALPA would talk about desperately needed safety measures, and the airlines and regulators would hear nothing but union rhetoric. Unfortunately, there have been many times between 1931 and 1985 when the revenue-reducing rules and limitations the pilots wanted to impose on the industry were viewed as merely an arrogant attempt to feather their own nests, when in fact they would have been of considerable and direct benefit to the safety of the flying public. Of course, not to overstate the case, there were also many instances in which ALPA proposals would have been of more benefit to the bank accounts of ALPA's members than to the public safety.

On the part of the industry—the airline owners and managers, the regulators, and the manufacturers—the fledgling discipline of human factors/human performance was not rejected; it was simply not understood. Since these people viewed commercial aviation as an objective system in which human foibles could be contained by specific rules and procedures, they could see little justification for spending time and money examining their various airline-related operations for potential human-factor problems, many of which

78

seemed to involve unmanly psychological "mumbo-jumbo." The trouble revolved around the malady that still infected the industry in 1979 when Jim Merryman met his death in Rockland: If you can't quantify it, how can you control it? How can you regulate the attitudes of the aircrews, for instance? How can you dictate management style? All that could be done, it seemed, was to provide the crews with a perfectly good airplane, order the mechanics to keep it running, and the pilots not to crash it. Beyond that, to the mind of the nonpilot or nonmaintenance airline executive with no gut-level operational experience, there was zero capability for substantive control of such concerns as whether an employee was in sufficiently good emotional condition to do his or her job. (Certainly back in the early fifties it seemed there were too many airplanes falling out of the sky for mechanical reasons to fritter away investigative time on ethereal matters such as what psychological stimuli might cause a pilot to foul up and bend a perfectly good airliner.)

To be sure, some excellent research was supported and the results incorporated into airline operations, usually through airline medical departments. Pan American World Airways and British Overseas Airways Company (BOAC, the forerunner of British Airways) were pioneers in supporting research into human fatigue and its effects on aircrews.[3] Ultimately, though, it was the eco-

3. The word *fatigue,* when used in such a context, is broadened to mean any human propensity for performing at less than an optimum capacity. It can mean a state of being tired, but it can also refer to a human state of mental confusion brought about by lack of oxygen, or diminished performance caused by emotional trauma due to personal matters. The early research of Professor Ross McFarland and his colleagues at the Harvard Fatigue Lab was devoted to quantifying the various areas of human fatigue and relating those areas to actual industrial environments, including aviation. Fatigue research was an outgrowth of the intense academic and medical interest in what was known as Taylorism, the theory that varying the worker's environment (through light intensity in the workplace, temperature, etc.) could lead to optimum, sustainable gains in production. Taylorism did not take into account human emotions, psychological factors, or social factors. Thus, the work of the Harvard Fatigue group was greatly expanded to examine all relevant factors affecting human performance—fatigue—so that such factors could be adjusted or controlled to produce better and safer work environments, and thus promote more efficient production. It was Professor McFarland who led the way in applying and expanding such research into aviation-related areas. In fact, he served as adviser to BOAC in the late thirties and through the forties, later holding a formal position with Pan American. These airlines applied the results of research into aircrew duty times and circadian-rhythm disruption, for instance, through their

nomic factor that retarded the airlines' ability as a worldwide group to understand that no matter how mechanically perfect the airliners became over the years, accidents simply could not be prevented if the propensity for human failure was ignored or inadequately addressed.

Certainly there was no callous disregard for safety in the industry's foot-dragging on human factors. The goal of accident-free airline operations was unanimously supported by all. In fact, the industry, if anything, was overzealous in the quest for air safety, having had ample evidence through the prewar years that an airline that crashed airliners and killed people would be shunned—and thus run out of business—by a nervous public still unconvinced that man, after all, was really meant to fly. The focus on safety, especially at the larger, more established carriers, such as United, American, TWA, and Pan Am, manifested itself in expanding maintenance departments and training departments charged with keeping the airplanes in near-perfect condition whatever the cost and training the aircrews in the latest techniques and with the latest equipment. That concern also resulted in airline medical departments in the largest carriers to evaluate and to some extent control, through recommendation to the management, the human problems they observed, such as alcohol use, sleep deprivation, and hazardous working conditions in all departments. Efforts such as these would enable the Pat Pattersons (United) and the Eddie Rickenbackers (Eastern) to turn to the public (in advertisements and speeches) and say, "We spend more money on safety than anyone else in the industry! We have the best-trained pilots, and the latest, best-maintained equipment. You can trust us!"

Nevertheless, these same airline leaders were unimpressed with

---

medical departments. This, of course, took both money and the dedication of executives and boards to the proposition that such human-factor research was a direct benefit to safety and therefore to the overall health of the company. The sad truth, however, was that the smaller domestic carriers could not afford such medical departments or specific research or advisers. In large measure they simply did not understand such subjects, and therefore, if any of the fruits of human-factor research were applied to their operations, it resulted from the application of ALPA pressure or government regulation. The restrictions on the amount of time a pilot can be allowed to fly in a given twenty-four-hour period, or a week, a month, or a year, is a perfect example. While the airline medical directors of the large carriers understood the importance of such rules and the reasons behind them, they became part of the general airline system only because they became federal regulations, or were incorporated into union contracts negotiated for the pilots by ALPA.

the concept that grown men would need evaluation or monitoring of their psychological or emotional health. And that prejudice was the norm throughout the industry.

As 1979 wore on, the depth to which Al Diehl was taking the Downeast investigation (his research into the human-performance causes) became a puzzlement to the traditionalists and an irritant to those who had in the past successfully dodged such questions as "irrelevant" to an air-crash investigation. Diehl ran into resistance at almost every turn.

The FAA in particular seemed to be fighting Diehl every inch of the way, but for reasons of self-protection. The FAA vehemently denied that the surveillance of Downeast was deficient. It trotted out stacks of neatly filed FAA surveillance and inspection reports to show that all of Downeast's innermost secrets had been consistently laid bare to the stern and probing eyes of the Federal Aviation Administration. FAA inspector Robert Turner's superiors could sense an impending scandal when one was dumped on their desks. Like all good bureaucrats, they wanted an immediate restoration of the veneer—the image—that the FAA had things under control and the public could sleep tight tonight. Damage-containment proceedings were the order of the day. Fend off that pesky NTSB man and restore the public image first—then wash the dirty linen behind the closed doors of the agency. And if that was not possible, blame the principal operations inspector, Robert Turner, and treat the Downeast situation as an isolated problem totally unrelated to the FAA system nationwide.

Nevertheless, a major failure did exist with the FAA's surveillance of Downeast. The FAA should have officially "discovered" the problems at Rockland and taken steps to correct them. The system, in other words, was and is flawed. But if the FAA refuses to admit the system is flawed in the first place—if it refuses to acknowledge that it *should* have discovered the Downeast problems—there is no way it can openly overhaul the way it monitors small commuter airlines.

This has nothing to do with laziness or arrogance on the part of the FAA personnel. It results from a very basic problem: The FAA too is a human system, and human systems, by definition, become self-protective for reasons of self-preservation. This does not excuse the failure, it merely explains it.

And the FAA had plenty of other assaults on its methodologies to contend with. Among other problems, its role as the steward of the flying public's safety was under a new attack; this one focused on its medical-examiner system and resulted from Captain Parmenter's death in the Air New England crash. The man should *not* have been flying a commercial passenger aircraft but had somehow slipped through the net. As with the principle involved in Downeast, it wasn't the *fact* of the doctor's repeated medical clearances of Parmenter that should have been the central focus of concern, it was the doctor's *reasons* for doing so, reasons that threatened the system.

The only true mystery, Diehl thought ruefully, was why so much research and work on the subject of human performance in aviation was still being, basically, ignored.

It wasn't as if the subject was the result of a late-breaking discovery, innovative and daring. The bedrock cause of Jim Merryman's death—the fact that chronic fatigue and extreme emotional stress on a pilot could cause a strapping fellow like Merryman to lapse into uncharacteristic inattention—was not a revelation. It was a repetition. Specifically, it was a repetition of work begun over forty years before by a Harvard professor, Ross A. McFarland.

Professor McFarland had been fascinated with aviation long before Charles A. Lindbergh disappeared into the mists of a Long Island morning on his bold attempt to reach France in a tiny air machine.[4]

As a scientist interested in psychology and physiology, McFarland's inquiring nature pushed him into research on the effects of flying on the human body and mind. He was directly involved in some of the pioneering work on the subjects of man at high altitudes and oxygen deficiency (hypoxia), including some spectacular and drastic experiments in the South American Andes during the winter of 1935. In World War II he worked extensively with the U.S. military as a naval officer, researching in Pensacola,

4. In fact, Professor McFarland happened to be present at Le Bourget Field in Paris on the night of May 21, 1927, as Charles Lindbergh landed from his epic flight. McFarland, interestingly enough, had been in England at Cambridge University doing research on physiological factors associated with high-altitude flight. Just before Lindbergh's departure, he had gone to Paris in conjunction with his research. The temptation to be present at such a momentous event, and one that foreshadowed the growing necessity for the very category of research he was pursuing, was too much. He caught a ride to Le Bourget to wait for the arrival.

ton Graybiel, and serving in the Solomon Islands in 1945, pioneering detailed research on the fatigue profiles of returning bomber crews. It was McFarland who helped direct aircraft manufacturers' attention to the importance of designing aircraft controls, levers, and switches with ease of pilot operation in mind, rather than expecting pilots simply to adapt their own physiology to difficult-to-operate controls. And it was McFarland who had become the oracle, the grandfather, if you will, of the fledgling field of human factors, which he husbanded into viability, first from his position in the Fatigue Lab of the Harvard Business School (and as medical coordinator of Pan American Airways), and later as professor of aerospace health and safety in the School of Public Health at Harvard.

It was in the fall of 1952, with the Korean War raging and General Dwight Eisenhower on his way to a first term in the White House, that Ross McFarland sat morning after morning in a large chair in his study in Cambridge, a writing board balanced on his lap, working on the preface to his newly completed work, which had taken the better part of six years to prepare—a book he hoped would change the way the airline industry regarded the role of the human beings it employed. The U.S. military—the new separate service of the air force in particular—had been listening to him quite intently, and work was being done in many areas not only to design the machine for better operation by a man but also to study the effects of such subjective variables as alcohol, fatigue, emotional upheaval, and attitude on the efficiency and safety of air force pilots as well as pilots in the civilian airline industry. The industry was beginning to become interested as Dr. McFarland had known it eventually would, but the level of interest was in its infancy and principally involved the long-distance overseas carriers who knew they had human-performance limitations built into their operations.

It was to the nonparticipating commercial carriers—the managers and owners of commercial airline companies, as well as to the regulators—that his new work was addressed. *Human Factors in Air Transportation,* as the extensive text was to be titled, would surely have an effect—McFarland felt certain of that as he finished the longhand draft of his preface and summed up some of the major points he had made:

Not all the problems of flying . . . can be solved by aeronautical engineering. At present, air transports must be operated by men, and the human organism is infinitely more complicated than an airplane and its engines. . . .

The author is convinced that air transports can be operated safely and efficiently only in so far as the human variables are understood and controlled.

McFarland thought back to the preface of an earlier book published in 1946 and aimed squarely at aeronautical engineers.[5] He had written that text to give the engineering community basic criteria for designing machines to be operated by humans in such a way that they would carry humans in some degree of comfort. He knew engineers, and he knew the futility of talking about the needs of the human occupants of an airplane unless he could put these needs in graphs and numbers. It was useless to tell an engineer to be careful to design a "comfortable" seat without telling him the range of dimensions in width and breadth and height that would constitute "comfort." The word "discomfort" was equally useless in engineering terms—it couldn't be quantified. So McFarland had given definitions of "discomfort" in terms of atmospheric pressures at altitude and what rate of climb or descent was acceptable—and therefore "comfortable."

Ross McFarland knew he had to reach more than the fledgling fraternity of aeronautical engineers. Those who designed airplanes for commercial use were controlled by managers and executives, and those airlines that used their creations were also controlled by managers and executives. Management of the human variables was as important as the basic human engineering of the aircraft. To succeed, he had to get the attention of those aviation managers. This new book would address such people directly—if only they would read and heed his words:

The problems raised by human limitations are not restricted to the aeronautical engineers or their biological or medical consultants but must also be considered by the executive and administrative personnel of the aviation industry. Decisions to incorporate such principles in the design of huge and expensive air transports cannot be made judiciously unless management itself has an understanding and appreciation of the basic reasons. This study therefore attempts to give an impartial interpretation of the basic

5. This was a work titled *Human Factors in Air Transport Design,* published in 1946 by McGraw-Hill in New York.

human problems to be understood and dealt with by the executive as well as the technician.

Surely, McFarland thought, the dynamics of rebuilding a vibrant civilian society after the war would catch the imagination of such people all over the world, and these ideas would be incorporated. Sadly, as he sat and wrote in 1952, he did not realize that the new work would be merely round two of a protracted battle, and he had no way of knowing how long a battle it was to become—or how many passengers and crew members would meet death and injury in the next thirty-three years because the industry refused to understand. The forces Ross McFarland was fighting with his words and his logic were a combination of the profit motive, an ingrained resistance to change, and to no minor extent, the sociological imperative of the age, which dictated that when a man made a mistake, the manly thing to do was acknowledge it and move on.

A man was not supposed to make excuses in the fifties, and explaining the motivational causes behind a fatal airborne mistake was regarded as an unworthy—unmanly—search for excuses. At least two generations of American males had been brought up in an age of military discipline, whether directly involved in the military or not. It was proper to look an officer (or the chief pilot) in the eye and say "No excuse, sir!" It was not acceptable behavior to stand before one's superior and explain why fatigue, or bad attitude, or inattention born of boredom caused one to make a mistake.

Then too, aeronautical engineering by 1953 had by no means approached a state of perfection. Air machines quite often demonstrated a distressing tendency to fall out of the sky for various reasons relating to structural failure, inadequate power, or other mechanical factors. In addition, the skies over the United States in particular and the world in general were becoming more and more crowded, with airliners sharing airspace with private aircraft, and more ominously, with the growing fleet of high-performance military fighters, bombers, and transports.[6]

6. Controlling all this was a primitive set of Air Traffic Control Centers under the jurisdiction of the Civil Aeronautics Administration, which in turn was under the Commerce Department. The centers had few radarscopes, and those they did have were unsophisticated and borrowed from the military. Most traffic across the nation, when it was controlled at all, was controlled by men sitting in windowless rooms with rows of paper "strips" identifying airliners flying on Instrument Flight Plans at specific altitudes on specific airways, which

The airline system was rife with mechanical and procedural problems that could be solved by objective solutions and increased engineering precision, but there wasn't enough time or money in the industry (or the regulatory structure) to adequately address even these very real, very threatening troubles resulting from a growing airline system. If the basic, acknowledged problems of building better airliners and a better air-traffic system were not being solved, certainly there would be no ground swell of enthusiasm among the airlines, the regulators, or the manufacturers to spend money on substantial research programs into ethereal areas of human-factors/human-performance worries—areas that might or might not be cost-effective. (The first human-factors lab was formed at Chance Vought—now LTV—in 1956 by Charles Miller, who would later join the NTSB.)

Then too, there was no tolerance in the industry (or indeed in much of society) in the fifties for psychologists and their "mumbo-jumbo" about human motivation and performance. Pilots and operations people were hired to do their jobs without making mistakes. They were required, and ordered, not to let stress, fatigue, boredom, physical incapacity, or psychological aberration affect their performance. They were paid to read instruments and operate controls correctly, and charged with the responsibility of not having "information transfer" problems. What else could a flight manager do? He hadn't the time or the resources—or the training—to "baby" his people. If one couldn't perform, he should be fired. The fact that sometimes death in an air accident resulting from a "weak" human performance failure ended an airman's career before he could be fired was tragic, but not a sufficient motivation to change the system.

No one seemed to notice in this period that regardless of how dire the threatened consequences to a pilot, controller, maintenance man, or other operational employee who screwed something up, screwups still occurred. Aircraft still hydroplaned off the end of runways, pilots still undershot in heavy rain and crashed, and maintenance men still left wrenches in engines, with later disastrous results. Humans continued to fail for human reasons—but in commercial aviation, the consequences of such failures affected more and more people.

---

in turn were marked by invisible radio beacons, which were usually very primitive as well. As would often happen in the repetitive history of U.S. aviation regulatory affairs, the amount of air traffic had already vastly overtaken both the technology of the air-control system and the budget of the CAA. More important, the public had no idea how bad things were becoming.

The mid-fifties were marked by growth. More people were flying on the airlines than ever before. The public increasingly accepted the idea that commercial flying could be safe. That meant more airlines in the skies and more pressure on the Civil Aeronautics Administration's primitive air-traffic-control system to keep the airliners safely separated when airborne. But the CAA was administered by a bewildered Commerce Department and treated like a bastard child by the Congress (which funded it reluctantly and kept it in penury). For many reasons, the CAA was becoming increasingly ineffective, and by 1956 a yawning credibility gap had developed—a credibility gap that separated the public's trusting belief that the system was safe from the chilling reality that it was not. Despite the advances in technology and electronics, the bedrock of U.S. aerial navigation (the routes and the procedures by which airliners flew from one city to another) was still an antiquated concept known as "see and be seen," and the radio navigation aids for pilots to use in finding their positions and staying on prescribed courses were pitiful in both number and capability.[7]

In the meantime, near-misses (midair collisions that had almost, but not quite, occurred) were being reported daily by airline crews, private pilots, and military pilots all across the country. It was obvious to those who flew in this increasingly dangerous environment that the "system" of air traffic control was out of control, and those who flew were engaging in a game of chance.

The only way to change the status quo, however, was to motivate Congress to spend the money to build a better system, and the only means of motivating Congress was to present it with an alarmed and upset electorate. The electorate, however, was neither alarmed nor upset—nor was it aware of the seriousness of the problem. Near-misses were not collisions. Near-misses did not kill constituents. Besides, deficiencies in the system were being "studied," according to the bureaucracy, and only a handful in Washington, D.C., knew that the problem was in reality being studied to death in the inimitable style of American bureaucrats when they don't want to take a politically risky stand.

Time has a way of destroying complacent attitudes with the impact of the inevitable, however, and in the summer skies above the magnificent vistas and lonely buttes of the Grand Canyon in Ari-

7. An argument had been smoldering for six years between the CAA and the navy over which type of distance-measuring navigation equipment would become standard across the United States, and the internecine skirmish was undermining all hopes of improvement in the system.

zona, fate was preparing a singular occurrence, which would at once shatter the complacency and fan the flames of public indignation at the hands of an alarmed press. The events of June 30, 1956, would, with lightning speed, expose the inadequacy of the air-traffic-control system and the CAA, demonstrate the determination of the industry to solve all problems with mechanistic solutions, and prove Murphy correct once again: If anything can go wrong—it will!

Chapter 6

THE WISP OF SMOKE FROM TEMPLE BUTTE

Ted M. Kubiniec, a thirty-seven-year-old staff attorney for Ford Motor Company, was enjoying the comfort of his well-padded passenger seat aboard the United DC-7 Mainliner—letting the loud but soothing drone of the deep, barrel-chested vibrations from the Douglas airliner's four piston engines wash over him as the vista of a nearby thunderhead drifted past. The towering cloud formation, visible through the generous expanse of the rectangular window by his left arm, was intimidating in all its silent menace and Brobdingnagian majesty. He had tried to catch a glimpse of the scenery below after the captain had come on the cabin public-address speakers to say he would be maneuvering around to show them a bit of the Grand Canyon below—now mostly covered by clouds. There had been a small area visible near Grand Canyon Village, but little else. Kubiniec settled back and watched the eastern edge of the thunderhead slide past the left wing, like a fluffy curtain being drawn aside by an unseen hand, revealing an azure sky beyond. The time was exactly 10:30 in the morning on June 30, 1956.

Airline crews loved to show off the canyon. United 718 had lifted off from Los Angeles Airport at 9:04 A.M. en route to Chicago's

Midway Airport, and though there were indications of thunderhead buildup in the vicinity of the upcoming Painted Desert navigational fix, Captain Bob Shirley and Copilot Bob Harms both figured they could find some of the incredible views of the canyon to present to their passengers from their perch at 21,000 feet. The cloud buildups had been more prevalent than expected, however, and the crew had been forced to weave its way among the towering "feathered canyons," vaporous structures thousands of feet high that one eloquent young poet had described as "footless halls of air."[1]

To Ted Kubiniec, watching the beauty of the scene as the two stewardesses began serving a meal to him and the other fifty-two passengers, the technical reasons for all the banking and turning around the clouds were unimportant. The competency of a major airline flight crew was an accepted and respected fact of life for a flying businessman—airline captains were becoming as trusted as ocean-liner captains—and men like Kubiniec were used to looking out the windows of airliners at fantastic views from dizzying heights, all the while listening to a background of important noises. The major newspapers and magazines of the nation were full of advertisments proclaiming the new age of safe and convenient air travel, which would let a businessman "schedule an appointment today for tomorrow on a different coast, and get there with ease." Many executives were still covered by massive air-accident life-insurance policies (purchased as a fringe benefit by their companies to compensate for the on-the-job exposure to aerial mishap), but accidents were the exception. The new DC-7 (along with the three-tailed Lockheed Super Constellation) had ushered in great reliability in airline travel. As of June 30, 1956, the DC-7 had never had an accident or killed a passenger. You thought about those things in the mid-fifties, sitting back in the cabin and trusting the pilots and the system. You simply accepted it all. There wasn't much you could do back in the cabin *but* accept it!

In fact, Ted Kubiniec's flight crew this Saturday morning was having to maneuver the big DC-7 to stay in the clear in order to conform to the mandate of its IFR (Instrument Flight Rules) clear-

1. This phrase is from the poem "High Flight" by John Gillespie Magee, Jr. Magee was a nineteen-year-old American who joined the Royal Canadian Air Force at the beginning of World War II and was assigned to the expeditionary force helping the British fight the Battle of Britain. In December 1941 he was killed during a combat mission in his Spitfire. Later, his compatriots discovered the poem written in an unfinished letter to his parents. It has since become a well-loved aviation classic.

ance. Since their route-of-flight was "off-airways" so as to fly over the canyon, they were to remain in VFR (Visual Flight Rules) conditions—clear of clouds—while in the uncontrolled airspace that surrounded the canyon.[2]

A few miles away, another group of airline passengers sat in the stratosphere in the cabin of another highly reliable airliner watching a similar vista through the windows of TWA Flight 2, a sleek Lockheed Super Constellation—beautifully distinctive with the three vertical fins incorporated in the tail assembly—powered by four powerful piston engines, each driving a huge three-bladed propeller. TWA had departed Los Angeles at 9:01 A.M. for Kansas City—three minutes before United's Flight 718. TWA 2's Captain Jack Gandy and crew had also encountered the heavier-than-ex-

---

2. There were a limited number of aerial "highways" crossing the United States in 1956 on which the CAA's Air Traffic Control Centers would provide guaranteed separation from other airplanes—positive control. Radar surveillance was almost nonexistent. As late as 1955 only one had become operational in the entire nation, and that was a 100-mile-range World War II surplus military radar operating in Washington, D.C. The controllers, then, were forced to use a handwritten system of "progress strips" to keep track of the location and altitudes of the different airplanes under their control. Since there were no radarscopes with which to "see" the aircraft, virtually all the controller's information came over the radio, from the pilots themselves. To separate the planes, the controllers were supposed to keep those flying at the same altitude apart by at least ten minutes and thirty miles. The system as it existed in 1956 dated from the 1930s.

There was also a limited number of radio navigational stations for aircraft to use in getting across the country (there were only 375 VOR stations in 1955). With so few VORs, the ten-mile-wide air routes over the United States formed an uneven flight path, zigzagging rather than giving pilots a more economical straight-line route. (These were called the "Victor Airways," which compare to the jet routes of today in the same way as the fragmented U.S. highway system of 1956 compares with the interstate highway system of the eighties.) Since flying the Victor Airways was decidedly not a proposition of flying the shortest distance between two points, many airline captains on long transcontinental flights in the mid-fifties would fly "off airways" in uncontrolled airspace, so as to fly a shorter, more direct path. But when a flight operated off the airway system, it could not be given positive control and separation services by a relatively blind Air Traffic Control. Therefore, such flights operated under VFR procedures, stayed clear of clouds, and had the entire responsibility for seeing and staying clear of other aircraft. The concept was called "see and be seen," though the previous decade had provided ample evidence that there were all too many times when fast-moving airplanes simply couldn't "see" and avoid other traffic in time (and might not "be seen" since few of the thousands of private and military aircraft flying at that time through the same areas had anticollision beacons and navigation lights).

pected cloud cover over the canyon. The Connie, as the Lockheed Constellation was known in the industry, had been doing her share of weaving as well through the buildups at 21,000 feet.[3]

The edge of the towering cumulonimbus was past Kubiniec's window now, the sun's rays coming from almost directly overhead and playing through the ragged edges of the cloud structure, just behind the sleek form of a distant passenger airliner, which had just come into view and which seemed to be on much the same course. The TWA Constellation's distinctive form and white body set off by red stripes formed an interesting picture against the clouds to the north of United's position. Kubiniec could see it quite clearly as it drew steadily closer, almost as if the two captains had decided to fly in tight formation for a while—although the approach seemed a little fast. The TWA plane was growing larger in his window—to the point that he could almost make out the . . .

3. Shortly after takeoff from L.A., TWA's Captain Gandy had asked for Air Traffic Control's permission to climb from his approved altitude of 19,000 to 21,000 feet in order to clear some angry-looking clouds to the northeast of Los Angeles. (He had to do this by asking his TWA dispatcher in L.A. over the company radio, who then phoned the request to the Air Traffic Control Center in Salt Lake.) The Salt Lake City Center knew that TWA 2's approved route-of-flight (which would take it through uncontrolled, off-airways airspace) would be crossing the projected off-airways route-of-flight of United 718 near the Grand Canyon in Arizona at an intercept angle of about 13 degrees. Since the two airliners would be in the same area within minutes of each other by approximately 11:30 A.M. Pacific Standard Time, the controller could not approve the same altitude for both of them, even though they would be required to stay in "visual" conditions, stay clear of clouds, and stay clear of other aircraft as well as each other. At that time, both airliners were under IFR clearances. Although they would not be under "positive" control and guaranteed separation while in uncontrolled airspace, they nevertheless could not be assigned conflicting altitudes as long as they remained under IFR. Therefore, the controller turned down TWA's request to climb to 21,000 feet, United 718's altitude. That information was relayed to Captain Gandy on the TWA flight deck. Gandy thought it over. He needed higher altitude to stay out of the weather ahead and keep the ride smooth for the passengers. If he canceled his IFR clearance and received "1,000 on top" clearance, he would be under a visual flight plan and could fly any route at any altitude he desired. Gandy promptly canceled his IFR clearance—which was legal under the rules then in existence and in accordance with TWA procedures—and began a climb to 21,000 feet on the very same route-of-flight he would have followed under IFR. Staying clear of other airplanes was now his sole responsibility, since the act of canceling IFR (and leaving the Victor Airways—TWA 2 was filed to leave the airways in any event at the Daggett VOR navigational fix coming up fast) relieved the air traffic controller of any traffic separation responsibilities. Gandy wanted to give his passengers the usual scenic side trip.

It suddenly became apparent that the TWA Constellation was too close—coming in too fast. The two airliners were flying a converging course as if one were on a freeway and the other were merging in from an angled entrance ramp—both at the same speed. TWA's form began to fill half the width of the window to his left as Kubiniec felt his United DC-7 suddenly roll to the right and pitch down, making the occupants of the cabin, Kubiniec included, light in their seats, knocking the two stewardesses, Nancy Kemnitz and Margaret Shoudt, off their feet, meal trays and drinks flying through the air.

The TWA Connie was dropping in the window's frame of reference. From Kubiniec's point of view it appeared the TWA might pass safely under the United in which he was riding. Kubiniec and the other passengers would only be able to hold on to the arms of their seats and hope to see the white, three-tailed airliner slide by beneath them.

Instead, like a monstrous whale breaching without warning through the calm surface of a glassy sea, the heart-stopping sight of the TWA Constellation's tail and fuselage at very close range suddenly loomed into view at the bottom of Kubiniec's window. The United DC-7 was trying to alter course to the right and lose altitude, its left wing raised against the horizon, so it would appear that the TWA was lower. Now it was obvious that it was not. The white and red-striped TWA Connie was closing fast, still on a converging course with the United DC-7 and at the same altitude.

The left wing of their United DC-7 was still up as the TWA continued in, inexorably, like a nightmare apparition that keeps coming despite the agonizing, failing, attempt to run, to move, to get out of harm's way—as their United crew was obviously straining to do with a right bank and a steepening pitch-down that wasn't stopping the onslaught of the TWA. As they watched in agonizing slow motion, the blue and silver left wing of the United DC-7 crunched into the white body of the TWA Connie. The faces of the TWA passengers behind the windows of the Connie were inevitably visible to the occupants of the United DC-7 as the DC-7's left wing continued its lethal march through the TWA Connie in agonizing microseconds that seemed like minutes, disintegrating metal skin and ribs and stringers, ripping through insulation material and carpets and baggage compartments, filling the air in an instant with confetti as the pressurized air of the TWA Connie escaped through the gash.

In less than a heartbeat the United wing structure had smashed through the TWA fuselage just forward of the three-pronged tail structure, severing it, and disintegrating part of the United DC-7's left wing as well. The main body of the TWA Connie, still on a course a few degrees to the right of United's heading, passed mere feet over the body of the United DC-7, but not before the powerful propeller of United's left outboard engine had chopped a series of elongated gashes in the white underbelly of the TWA as United's engine and shattered left wing structure passed just inches beneath the assaulted TWA fuselage—now trailing debris through the ragged, gaping hole that had been the rear cabin—a scene imprinted in freeze-frame horror on the minds of any of those aboard United who had watched the progress of the propeller.

And just as quickly the TWA was gone from view.

On the TWA flight deck the pilots had been slammed to the left as the tail of their Constellation was being torn off by United's uplifted left wing. Then the rush of air—the decompression—raised dust from the floor and confusion in their minds. Before Jack Gandy or Jim Ritner could even begin to react, begin to analyze what had suddenly assaulted them with such a gut-wrenching force, the realities of aerodynamics raised the left wing and dropped the nose—progressively—to nearly vertical, as the partially dismembered Connie accelerated. With the tail structure gone, the center of gravity of the delicately balanced machine had shot disastrously forward in an instant, and the sleek Connie had become little more than a powered rock. The reaction of Captain Gandy and First Officer Ritner would be that of any seasoned pilot—they would pull on the controls, nearly weightless in the instant it had taken for the Connie to enter a hideous, screaming dive toward the cloud cover below. The altimeter was little more than a blur as it unwound, the sound was horrific, the sixty-four passengers trying to hold on to seat backs, loved ones, anything, as they all accelerated through the cloud cover and below the rim of the canyon. Unseen by the horrified passengers, the two pilots were surely struggling with the control column, but with no tail surfaces to command, it was useless. The passengers couldn't see the unyielding rocks of the northeast slope of Temple Butte rushing up in the windscreen, but Gandy could, and instinctively he would have known. The Connie struck with a thunderous roar, ending their collective lives at a speed of more than seven hundred feet per second.

On the flight deck of stricken United 718 there was still hope. The crew had seen the doomed Connie rake across their field of vision, the left wing was shredded and useless beyond number-one engine (that engine itself ruined by the impact), and now the deck angle of the United DC-7 was steepening to 30 degrees down as Robert Shirley strained at the yoke, trying to pull her up, straining with every ounce of strength to roll her back to the right as they began to corkscrew down in a steepening left spiral. 16,000, 15,000, 14, 13, 12. The cloud cover enveloped them and the rising scream of high-speed flight and overspeeding propellers filled their ears. As Ted Kubiniec clung with the rest of the passengers to seat backs and armrests, in shock, nearly weightless, his mind surely rebelled against the idea that they might not make it, might not live through this, yet his thoughts—as those of any passenger caught in such a nightmare—would focus elsewhere. He would think of his wife, Andree, his young son, Ted, and his daughter, Jan, all of them back in Michigan, unaware of what was about to happen. How would they do without him? Could they handle it? They MUST handle it! WHY CAN'T THE CAPTAIN PULL UP?

In the cockpit First Officer Harms squeezed his radio microphone button, and the speakers in the radio rooms at Salt Lake and San Francisco crackled with his voice.

"Salt Lake, United 718, ahh-h . . ." Harms paused, and through the pause Captain Shirley's voice could be heard in the background, several octaves above normal, screaming in a bone-chilling, unearthly yell of tension and strain indecipherable to the radiomen, but understood by their tape recorders:

"Pull up! Pull up!"

Followed by Harms's voice again, high pitched, betraying the greatest stress a human can face:

". . . we're going in!"

As the radio operators instinctively looked at their speakers in disbelief, chills flittering up their backs, hundreds of miles away over the Grand Canyon and virtually unseen, United 718 punched through the bottom of the cloud cover.

Seven thousand, 6,000, 5,000—the needles of the altimeter seemed to be spinning no matter how hard the pilots pulled and hauled on the yoke.

Shirley and Harms were losing the battle. They were inside the canyon now, below the rim, out of clear airspace, among towering rock formations and unforgiving, rocky cliffs.

The stricken DC-7 whined through the last thousand feet above the canyon floor at many hundreds of miles per hour. And finally it was the south wall of Chuar Butte that crawled into the windscreen, interposing itself between the Douglas and any last hope of deliverance—becoming their instrument of death. The impact, not as severe as the final microseconds of TWA 2, nevertheless was far beyond the survivable. Ted Kubiniec and his fifty-two companions, who had expected to alight well fed and relaxed in Chicago that afternoon, along with a crew of five seasoned airline veterans, were, quite simply, obliterated. Other than as a passing reflection in the eyes of assorted mules and other wild animals who may have looked up briefly at the whining sounds of the plunging machines, their passage from this life—and that of TWA 2—went unseen. The rolling thunder of the separate impacts was masked by the rumbling from the thunderstorm to the west.

Since it was 1956, there were no radarscopes from which they had disappeared. No flight-data recorders to chronicle their flight paths. No voice recorders to capture their last words. And no emergency locater beacons to mark their location. Only the scratchy radio call preserved on a few inches of magnetic tape in Salt Lake Center. And, of course, the mute and grisly testimony of the burning wreckage in the noonday sun.

For six hours, all anyone knew was that TWA 2 and United 718 had simply vanished. Many suspected, but no one knew, what had happened—and where it had happened.

Just before dusk, Palen Hudgin heard the news of the missing airliners while sitting with his wife at dinner. Palen and his brother Henry were the owners of Grand Canyon Airlines, a small air-taxi operation that made its money flying tourists into the canyon from the south rim. Both brothers were pilots, but only Palen had been in the canyon that day. He had told his wife on returning home about a strange, lonely little wisp of smoke—an ethereal and unusual sight—rising from a forlorn corner of the eastern canyon. There had been a thunderstorm with ample lightning in the area the day before, so he figured it was a brush fire. But there was one odd thing. The smoke was awfully black for burning brush.

Now with news of the missing airliners, his wife reminded him of the smoke, and suddenly a picture flashed across his mind in chilling detail, the image of that plume of smoke from earlier in the day.

Palen Hudgin rounded up his brother, returned to the Grand Canyon airport on the south rim, and took off in a Piper Tri-Pacer, flying the tiny craft into the great maw of the canyon near the haunting starkness of the confluence of the Little Colorado River Gorge and the main Colorado River. They flew back and forth a few times as low as 200 feet above the river, searching in the waning light for signs of anything unusual. Finally, on the slope of Temple Butte, the two brothers spotted a stomach-wrenching and unmistakable object: a battered but essentially intact metallic structure with a white and red-striped horizontal stabilizer and three vertical fins.

As they flew by, both men stared at it in utter disbelief. In all the world of aviation there was no mistaking the tail of a Lockheed Constellation.

As they climbed out of the canyon in the growing gloom of darkness, Palen Hudgin knew only too well what that wisp of smoke he had seen earlier in the day had marked. It had been the funeral pyre of TWA Flight 2. They reported it immediately.

The next day the pieces that had once been United 718 were pinpointed by the Hudgin brothers less than a mile to the north of Temple Butte, on an inaccessible ledge and down into a deep crevasse, strewn in fire-blackened confusion over the rock. With that awful confirmation, a long and dangerous procession of helicopter missions ensued for many weeks into the area of the two buttes, until the broken remains of passenger Ted Kubiniec, Captain Shirley, Copilot Bob Harms, and the others had been accounted for.[4]

The thunder of the crashes may have played out its vibrations against the unresponsive cliffs of the canyon within a minute or two on June 30, but the thunder of the outrage and shock that swept the United States in the wake of the highest commercial-airline death toll in history reverberated for months in the press from coast to coast. Within weeks the public learned not only that the two airliners collided but that they had both given radio estimates that they would arrive at the same place at the same time: Painted Desert at 21,000 feet at 10:31 A.M. The sector controller at Salt Lake Center had written down both position estimates and posted them by 10:13 that morning, but had made no attempt to warn either airliner. In addition, it was revealed, this was the same controller who had denied TWA 2 the right to climb to 21,000 feet an hour and a half before the crash because there might be a "traffic conflict" at Painted Desert. Where, raged the press, was the common sense and basic responsibility that should have prompted the controller immediately to warn each airliner of the other's proximity?

4. Because of the incredible destructive force of the high-speed impacts of both airliners, none of the bodies was recovered intact. A mass burial of the victims of TWA Flight 2 was held some two weeks later near the south rim. (*The New York Times,* July 11, 1956, page 1)

Lost in the furor over these reports was the fact that the controller had no duty to—and in fact was prohibited by his regulations from attempting to—relay United's report to TWA and vice versa. They were both off airways and under "see and be seen" rules.[5]

Besides, explained the CAA, "Painted Desert" is simply a line running roughly north-south from the Winslow, Arizona, VOR (the 321-degree radial) and is more than 175 miles long, so there was no way the controller could know where along that line each airliner would pass (though he might have guessed that it would be close to the canyon since it was no secret that both captains would want to give their passengers a good look at it).

With the spotlight of accusation focused on the hapless CAA (and the Salt Lake controller in particular), the nation began to find out just how incredibly primitive its air-traffic-control system really was. The realization that two airliners full of trusting passengers, operated by two of the largest and most experienced carriers on earth, crewed by men of high ability and great experience, had been playing—in effect—a game of Russian roulette among the thunderheads over the canyon was a deep and infuriating shock to the public. It was also a shock to Congress—especially when constituents back home began asking embarrassing questions of their senators and representatives. Suddenly it became common knowledge that in all the nation, only one single, solitary surveillance radar was in operation by the ATC. It also became shockingly clear that the radio navigation facilities and instrument landing systems were incredibly inadequate in number and quality.

5. The Salt Lake Center controller, already deeply upset over the accident, was hounded and attacked by the press and blamed publicly by the airlines. The poor man's children were even taunted in school in the Salt Lake City area as a result of the furor focused by the press. The position taken by United's and TWA's defense lawyers in subsequent liability trials was that the air-traffic-control system, and the controller, had a duty to warn both airliners of the potential conflict and had failed to perform that duty. It was hard for anyone to understand how the man could write down identical positions, altitudes, and times for each airliner, place them together on his strip board, and yet make no attempt to warn the respective captains. However, the man who had written most of the air-traffic-control rules then in use by the CAA, Charles Carmody, assistant director of Air Traffic Control, testified in depositions in 1957 that CAA controllers had been warned *not* to provide such traffic-conflict warnings to aircraft that were off airways, in uncontrolled airspace, and under VFR rules because airliners could be at any altitude and at any position, and ATC had no way of knowing for certain where and how high (in the absence of positive control under radar) they really were. Therefore, the issuance of conflicting traffic warnings in such conditions was, in Charles Carmody's opinion, potentially more dangerous than not issuing such warnings.

The problem seemed relatively simple in concept and solution: In order to avoid flying airliners full of people under the caveat to "see and be seen" (since that caveat had now been unmasked as, essentially, a deadly gamble) airliners should be given positive control and separation from other air traffic. That meant more equipment for the air-traffic-control-system, which meant more appropriations. Therefore the problem could be solved by money. There was no need for any deeper, philosophical examination of the concepts of a system that had trusted human beings to "see and be seen" 100 percent of the time with 100 percent perfection.

Within a month, new and substantial appropriations had been made by the Senate Appropriation Committee to more than double the number of VORs (to more than nine hundred) and to purchase eighty-two advanced, long-range surveillance radars. Suddenly proposals and problems locked in the suffocating arms of the bureaucracy for years were solved, decided, funded, and initiated as the political heat intensified to modernize a hideously archaic system. The $5 million that the CAA administrator had thought adequate for "modernization" of the airways before the deaths of United 718 and TWA 2 jumped to $125 million in early 1957. Now the electorate was upset, and the system was going to change!

Of course, it was too late for Ted Kubiniec and 127 other souls who fell with those doomed airliners into the Grand Canyon on June 30. All the warnings and complaints from the airlines, the Air Line Pilots Association, various government officials, and a scattering of others in previous years had gone relatively unheeded until that date.[6]

Equipment wasn't the only problem uncovered by the Grand Canyon tragedy. The organization of the federal control of the system was a mess as well. The CAA, which was responsible for the air-traffic-control system and administered all the regulations and procedures of private and commercial aviation, was a toothless tiger—a weak agency with little or no authority to do anything but

6. In testimony before the Senate Committee on Interstate and Foreign Commerce, Subcommittee on Aviation, on June 4, 1958, Stuart G. Tipton, then president of the Air Transport Association (which represented virtually all the scheduled airlines) confirmed that his members "since 1954, have favored positive control of all aircraft operating in the same airspace." In truth, the pressure from the airlines and the pilots had been building since 1948 to replace what subcommittee chairman Mike Monroney of Oklahoma (during the same hearings) himself called "the old theory of flight since the Curtis Jenny days of 'see and be seen,'" with up-to-date equipment, radar coverage, and positive control and separation.

negotiate with other government agencies to get things changed. Because it was such a confused and weak organization, surrounded by other bureaucratic agencies that held on zealously to their pieces of the overall aviation pie (such as the Air Coordinating Committee, or the Air Navigation Board, and many others with equally confusing names), there was little or no coordination of the overall national air system. What was worse, the very rules and regulations the CAA was supposed to enforce were written for it by still another agency, the Civil Aeronautics Board.

While agencies wrangled and squabbled and nitpicked over every little detail of every proposal to improve the system, aircraft were crashing and people were dying. But the self-protective nature of bureaucracy dictates that a bureaucrat who desires to keep his job must make haste slowly. In such a byzantine structure, no one seemed capable of making decisions without literally years of studying issues to death. In the meantime, one of the most dangerous wild cards in the game, military aviation (air force, army, navy, marines, and Coast Guard, each with its separate air arm), continued to operate high-speed aircraft daily through the same airspace without civilian control or coordination of any sort. Collisions had occurred in the past, and near-misses were all too common. (A staggering number of the 971 near-misses reported in 1957 alone involved military aircraft.)

The situation was intolerable, and in the spring of 1957 a presidential assistant named Edward Curtis ended a year-long study (which had been initiated by the White House before Grand Canyon) by proposing a new, independent Federal Aviation Agency with considerable powers (even over military aviation in peacetime) to replace the anemic CAA.[7]

Ah, but memories are short on Capitol Hill, and the furor over the Grand Canyon crash had been calmed by the new appropriations and the misguided opinion of the press that the problems threatening commercial air safety were well on their way to solution. Action on the FAA proposal began to bog down as 1957 wore

7. During the interim, Curtis urged the president to create an intermediate Airways Modernization Board to try to take control and move a few steps toward the goal of building a modern air system while Congress wrangled and "harumphed" over the proposed legislation to create the FAA. President Eisenhower agreed, and a tough former air force general, Elwood "Pete" Quesada, was appointed to head the board. Quesada was destined to become the first administrator of the FAA and was a major force in bringing the nation's air system into some semblance of order and modernity.

on, and by the time 1958 rolled around, there were very few upset constituents kindling fires under their respective senators and congressmen.

Once again the price of legislative action was paid in the blood of passengers and crew members. A series of midair disasters followed, involving military aircraft colliding with civilian airplanes: one in February in Van Nuys, California, a second in Las Vegas, Nevada, and the third at Brunswick, Maryland, on May 20. In all, sixty-one people lost their lives.

On May 21, 1958, one day after a Capital Airlines flight crashed after being hit by an air force jet over Maryland, a bill was introduced to create the Federal Aviation Administration, and during the summer, airline executives and industry experts trooped to Capitol Hill in support of the bill, which should have been passed a full year before.[8]

On May 22 and 23, just to make sure the point was well taken and the public reassured that Congress was on the ball, emergency hearings were held before Senator Monroney's subcommittee. In June, a more orderly procession of industry representatives trooped to the Hill in support of Senate bill S.3880. Among them was one of the fabled pioneers of U.S. airline building, W. A. "Pat" Patterson, president of United Air Lines.

Sitting amid the historic surroundings and exquisite, dark-stained woodwork of room G-16 in the U.S. Capitol on Thursday, June 5, 1958, Patterson fixed the senators with a determined gaze.

[Concern over positive control and separation of air traffic] isn't something that developed after Grand Canyon. This is something that has concerned the airlines, the operating people, since 1948. We could see it coming, and we have been working on it constantly, but had great difficulty. We had the opportunity to talk to people in these agencies and they listened, but very, very little was done.

Patterson was well aware of the risk to United's business if he scared the public further by openly admitting the system was gravely flawed. Nevertheless, Patterson was one of a breed of ex-

8. In fact, the sponsor of the bill, Senator Mike Monroney of Oklahoma, had been working on the legislation for the previous year. There is no way, however, that Monroney would have been able to line up thirty-three co-sponsors without the fresh influx of public indignation.

ecutive leaders who not only knew how to take business risks but had a greater sense of public responsibility than many of the aviation managers who would follow him thirty years later in the deregulated airline business of the eighties. To sit before that subcommittee in 1958 was not a calculated act contrived by the public-relations department; in Patterson's view it was a duty.

We recognize, Senator, that when we emphasize such importance and bring to the attention of all concerned this great weakness [in the air transportation system], some people would say that is not good business because we can't help . . . but bring about a certain amount of fear as we talk too frankly about a subject. If we bury our heads in the sands, however, we may continue to enjoy a better volume of business, but I would rather forsake what we are going to lose now to try to help get this legislation through, sacrifice what business we are losing now, because we are going to get it back in the future because of our frankness in talking about it.

It was his airline (along with TWA, of course) that had lost an airplane, its passengers, and its crew on the buttes of the Grand Canyon due to what he saw as incredible government negligence. It was his airline that had lost still another plane with all aboard to an in-flight collision (at Las Vegas, Nevada) with a military jet, again because of government bumbling in approving a flight procedure for the military that was essentially incompatible with civilian safety. To Patterson it wasn't a human-factors problem. He understood very well the potential for continued disasters resulting from the impossibility of assuring that aircrews could see other aircraft and avoid them 100 percent of the time. That was the reason that positive control was needed. There was no sense in wasting time on studies of just what pilots could and could not see and when, or of the effectiveness of cockpit window design. Pat Patterson simply had to assume that they couldn't "see and be seen" with certainty (even in the piston-propeller aircraft, and the jetliners would soon be flying passengers at twice the speeds), and therefore, air traffic control had to take on all responsibility for in-flight separation. It was a matter of hardware and appropriations, good regulatory agencies and effective rules.

Patterson continued:

Now I said to our stockholders, and I say again, the findings in [the Accident Board's Grand Canyon report] were an insult to the intelligence of the American people. Our airplane was at twenty-one thousand feet, IFR, a protected flight. [TWA] was at nineteen thousand . . . [he] went through air-traffic control, asked to go to twenty-one thousand. The answer came back "No." Well, of course, we hit at twenty-one thousand. How did he get there? . . . Because the rule was such there is an escape clause that you can wiggle your way out of traffic control by saying that you will take a visual flight climb on top. There was the cause of the accident, gentlemen, a rule that permitted deviation.

Pat Patterson was becoming a bit heated as he continued.

Now, all that testimony was in there, and we come out with a report with engineering diagrams of the cockpit of an airplane that you can't see out. You don't have to draw pictures for me that you can't see in all directions . . . And they said the probable cause of this accident was the failure of the pilots to see one another. Now, after all that testimony [during the accident investigation] my little grandchild could have written that one!

Patterson's testimony, and that of many other aviation-industry executives and government witnesses, helped Senator Monroney validate the need for speedy passage, and by August 23, the bill had sailed through both houses of Congress and been signed into law, creating the Federal Aviation Administration and giving it power over almost all peacetime aviation in the United States.[9]

9. In an incredible delegation of power, the bill prohibited any other part of the executive branch—including the president—from overruling the administrator in matters affecting safety. The same act freed the Civil Aeronautics Board from its administrative connection to the Department of Commerce and caused the new FAA to absorb the Airways Modernization Board. On November 1, 1958, after the statutory changeover period of three months, the FAA took over, with former general and Modernization Board chief Elwood "Pete" Quesada as the first administrator. The much-needed new agency was in place fewer than seventy days when the first commercial jet flight originated with a Pan Am 707 from New York to Paris.

As vitally necessary as the central authority of the FAA was, the new act had a major flaw. The law imposed on the FAA an inherently contradictory duty: It was to regulate air commerce in such a manner as to best promote its development and safety. The drafters had been in too much of a rush—faced with an archaic and confused system that needed immediate attention—to think ahead to what that charge really meant. There would be, however, countless times in the next quarter-century when the responsibility for promoting safety would fly directly in the face of the responsibility to promote the development (along with profitability and stability) of air commerce.

The establishment of the FAA was a major advance, but in the rush to correct a rickety system that permitted airliners to play blindman's buff (as one witness referred to "see and be seen"), a major opportunity for an even greater advance—an even more significant understanding of the human dynamics of commercial flying—had been lost.

Monitoring the proceedings from his home in Cambridge, Massachusetts, Professor Ross McFarland knew well that the point had been missed. For many years he had held great hope that industry leaders such as Patterson, as well as the regulators and investigators, would realize that the *true* cause of the Grand Canyon disaster stemmed from the inescapable fact that the airline business was a human enterprise, and human failures that might lead to such tragedies would inevitably and predictably occur. Human failures (such as the failure to "see and be seen") had to be *anticipated* in order to achieve the best feasible level of safety. But that such failures were not being anticipated was exemplified by the fact that airliners had flown for years under the "see and be seen" caveat, even though the crews *and* the industry knew that there was a potential for disaster. In fact, visual flight deviations through the uncontrolled airspace over the Grand Canyon occurred daily before June 30, 1956, and had for nearly a decade!

Despite an ocean of funds spent voluntarily by the airlines on the best training, maintenance, and other safety considerations, no one seemed able to look at the system as it was at the time and ask: What problems do we have here that might offer a high probability of causing a crash? "See and be seen" as a concept, the design of airliner windows, the IFR/VFR rules and their loopholes under positive control—all of these things should have set off alarms in the minds of the airline managers as well as the crews. They *knew* there were serious problems—hence the "concern . . . since 1948" as Pat Patterson put it—but the airlines and the crews continued to fly the same way in the same system. That, in itself, was a form of moral negligence.

The nation needed not only the centralized, coordinated control of an FAA and a positive air-traffic-control system, it needed as well a systematic approach to discovering and addressing potential human-performance foul-ups that might cause accidents, and it needed to make such discoveries *before* the CAB's Bureau of Accident Investigation had to piece together wrecked airplanes and deal with the grieving families of innocent victims. What it had, however, even with the new FAA, was a systematic means of *reacting* to human failures one by one, instead of working to prevent them.

104

United Air Lines President Patterson's attitude was solidly representative of the airline industry's thinking: Sure Grand Canyon resulted from a human failure, so we fix the system by taking human discretion out of the loop. Put "positive control" systems in place, and the problem will be solved.

Of course, only a few in the industry dared raise the point that unless *all* air traffic was positively controlled and separated, from takeoff to landing, the element of chance remained—the pilots would still have certain periods of time in which to operate in a potentially dangerous "see and be seen" environment.[10]

There had been a glimmer of a suggestion in the accident report on the Grand Canyon that perhaps this point was not entirely lost to the technological experts who did such an incredible job of detective work with the wreckage. Their conclusion that "the probable cause . . . was that the pilots did not see each other in time to avoid the collision" (the same conclusion Pat Patterson denounced in that subcommittee hearing) was followed by a listing of some of the possibilities of why they did not, or could not, see each other in time, and every item on the list identified potential human-factor problems and human-performance failures."[11]

10. Patterson himself made mention of this in the same subcommittee hearing on June 5, 1958. Talking about the proposal to regulate all traffic above 17,000 feet, he told the senators, "Let's be honest about this thing. You have got to get to seventeen thousand feet to be regulated. What is going to happen to you while you are going to seventeen thousand feet? . . . It's a pretty logical question."

The thrust of the argument, however, was that more positive control would be needed. Since it was somewhat apparent that there would be no way in the foreseeable future (with business and private and military aviation all zealously guarding their respective rights) to positively control all traffic at all times, Patterson was assiduously avoiding the question of whether improvement in the *ability* of aircrews to reliably maintain separation in visual flight environments could be useful. That point had to do with human capabilities, human factors, and that was foreign territory.

Patterson himself was a good friend of Professor McFarland, and familiar with his work. But like so many senior executives who looked at the subject of human factors in those days, his depth of understanding was limited to making sure that he had a good medical department in his company. In other words, human factors were presumed to be nothing more than the physiological problems of aircrews with which an airline-employed doctor might be concerned.

11. The Bureau of Accident Investigation board said: "The evidence suggests that [the failure to see each other in time] resulted from any one or a combination of the following factors: Intervening clouds reducing time for visual separation, visual limitations due to cockpit visibility, and preoccupation with normal cockpit duties, preoccupation with matters unrelated to cockpit duties such as attempting to provide the passengers with a more scenic view of the

The CAB's Bureau of Accident Investigation was staffed with an array of bright experts who had already gained considerable expertise in reconstructing the dynamics of accidents from little more than pieces of scrap metal (the initial process at the accident site is known as "kicking the tin"). Men such as Frank Taylor (who had joined the bureau in 1948) were to figure prominently in the field of airline safety through their meticulous investigative methods at the bureau and later at its successor, the National Transportation Safety Board.

But Taylor and his fellow investigators were not schooled in psychology or human factors. They were engineers, pilots, and technicians. They dealt in objective terms. They could show you exactly how the structure of a DC-7 wing could batter a hole in the metal skin of a Constellation, and even determine the angle and the velocity. They couldn't, however, reassemble the thought patterns of deceased pilots to find out why they might have become preoccupied, or why they let "intervening clouds" threaten their visibility. They could poke at the questions of cockpit visibility limits and suggest that there were human physiological limits to the pilots' vision. (This was the reason for including in the accident report charts portraying the difficulty of seeing out the cockpit windows of each airliner.) But other than dealing with such areas in an objective fashion, the investigators were unable (and unwilling) to go farther. That was too subjective to be handled with precision. You couldn't "get into a pilot's head," as the investigators (and the attorneys) used to say. (That attitude would still permeate the NTSB twenty-one years later when investigator Al Diehl began struggling to find the true cause of Jim Merryman's Downeast accident.)

Then too, there was the prejudice against looking for excuses. Those pilots were real men—dead ones to be sure—but real men nonetheless. They should not be accorded the indignity of having investigators hunt up excuses for their behavior. They had been ordered to see each other in time, and they had failed. It seemed as simple as that.

That list of contributing factors in the Grand Canyon accident report should have launched immediately an intensive study into just exactly *what* cockpit duties could become preoccupying, or

---

Grand Canyon area, physiological limits to human vision reducing the time opportunity to see and avoid the other aircraft, or insufficiency of air route traffic advisory information due to inadequacy of facilities and lack of personnel in air traffic control.

whether a pilot's obvious inability to see through "intervening clouds" meant that VFR flight (even when the airplane remained outside the clouds) was by definition unsafe if there were clouds in the same sky at the same altitude. Why didn't the questions arise: What were the motivations that would lead a flight captain to take risks in order to give his passengers a good view of the canyon? Was it excessive pride? Was it peer pressure? Or more ominous, was it company pressure?[12]

The only hope of preventing future crashes that might have their roots in the very same human frailties would be to examine and understand those human frailties—but in 1956 and 1958 the point was lost. There was too much to do to solve the basic equipment and regulatory problems. There was too much for the new FAA to do in bringing the air-traffic-control system up to the technological capabilities of the mid-twentieth century. The type of aggressive research and awareness that Ross McFarland was advocating—the hidden lessons of Grand Canyon concerning the human causes of human failures—would have to wait.

The fact that an NTSB investigator such as Al Diehl would have to go through an interoffice struggle some twenty years later in order to delve into the human causes of the Downeast Airlines crash would confirm the ridiculous truth that as late as 1979, those lessons were still being ignored—and airplanes were still crashing because of little understood human failures. With the destabilizing effect of airline deregulation, such ignorance could be lethal, and as always, it would be the trusting passengers (and their crews) who would pay the ultimate price.

12. In an airline as experienced and well run as United, the idea that company pressure could encourage an unsafe operation would have been (and is today) hard to believe, and in fact there was virtually no evidence of such a factor in the Grand Canyon tragedy. But did anyone really know? In the absence of a thorough human-performance investigation, any answer would have been speculation. As the next two decades would prove in crashes large and small, "pressure" could come from many sources even in a large, established company: self-imposed, casually suggested, or pressure by one's peers to conform to common but flawed practices (such as deferring maintenance write-ups during multiple-leg flights until the overnight station is reached). The potential is less at carriers like American, United, Delta, and other giants, but it does exist. Therefore, it must be guarded against.

Chapter 7

SEE NO EVIL, HEAR NO EVIL—
THE FAA IN ACTION

Why had the FAA ignored it? How on earth could the people at the Portland, Maine, FAA GADO (General Aviation District Office) have received so many tips, so many indications that Downeast Airlines was playing dangerous games with the rules, and yet have turned their collective heads over the years, pretending the problem didn't exist? Were they too busy to look into the problem, too intimidated, too friendly to the owner, or too trusting? Or was the system itself too overloaded and incompetent to be the public watchdog over commuter airlines? One thing was obvious: The "system" spawned by the Grand Canyon crash of 1956 and nurtured through the following two decades as the arbiter and defender of the flying public's safety, had failed in the Downeast situation!

Al Diehl, Ph.D., National Transportation Safety Board Human Factors Group chairman and investigator (for the Downeast crash) sat in his small, paper-strewn office at NTSB headquarters in Washington, D.C., and read through some of the letters again, shaking his head periodically in amazement. Lately he had been feeling a bit like the biblical David, indirectly attacking the Goliath of the FAA. The FAA, on the other hand, had been treating him like a minor irritant. Throughout the rather spartan eighth-floor offices at 800 Independence Avenue, there was a bit of an institutional inferiority complex where the FAA was concerned. The NTSB could recommend changes—but only the FAA could compel those changes by force of regulation and law. When the FAA chose to ignore the NTSB (as it had been doing with increasing frequency in the late seventies, issuing written rebuffs that dripped with contempt thinly veiled in withering disdain), it could do so with intimidating arrogance. The NTSB is supposedly a totally independent agency of the U.S. government, yet the NTSB staff members too often felt submerged within the bureaucratic power of the FAA. The fact that the NTSB's eighth floor was also the eighth floor of the FAA's Washington headquarters building only made matters psychologically worse.

It was a Friday in late August 1979, only three months since the Downeast Twin Otter had smashed into the trees at Rockland, Maine. The formal NTSB hearing on the crash was scheduled to get under way in just a few weeks in Boston. It was an event Diehl suspected might turn out to be a significant watershed in the advancement of human-factors and human-performance awareness. Certainly the revelations that Diehl's witnesses would offer would be galvanizing to the airline industry and the FAA. The hearing was going to make headlines.

In preparation, Al Diehl had been engaged in demanding detective work all summer, spending his days on the telephone and in the field, probing and questioning former Downeast employees as well as the friends and family of deceased Chief Pilot Jim Merryman, and digging up former Downeast employees from the early and mid-seventies. The nonstop work had produced a growing stack of evidence, but he was still struggling as hard and fast as possible to build a solid foundation under his original theory: that the root cause of the Downeast crash was inordinate, autocratic management pressure, which the FAA had failed to acknowledge and failed to address throughout eleven years of Downeast operations.

Diehl was also sure (along with his boss, Jerry Walhout, Steve Corrie, the investigator-in-charge of the Downeast accident, and several others at NTSB) that Downeast's owner, Robert Stenger, had learned nothing from the crash and was determined to keep operating with the same callous attitude that so many former Downeast employees claimed had been the norm for a decade. If that was true (and the evidence seemed damning), Diehl realized that the flying public was still in danger (though ridership at Downeast had dropped off dramatically since the crash). In effect, the NTSB was the only organization that could prevent a repeat accident, and the only one that could expose Downeast's operational methods for what they were: far more accident-prone than should be acceptable in a public utility, much less an airline carrying passengers for hire.

It was painfully obvious that the FAA, now on the defensive with its wagons circled against the implicit warcries of Diehl's probe into the inner dimensions of the accident, was not going to join its "adversary" at the NTSB by going after Downeast. It was too busy defending its GADO and the claim that Downeast was, and always had been, a legal and safe operation.

After all, it had papers to prove it. The FAA files in Portland were replete with neatly filled-out records of inspections by the principal operations inspector and others. The NTSB, however, was discovering that those papers were worthless.

What had really happened at Downeast was something the FAA was totally unprepared to deal with systemwide. A small air taxi had grown into a big-time commuter, led by a man whose management style was simply incompatible with an airline operation of any size, let alone one that held a major public trust to put safety ahead of profit no matter what.

But the FAA stands helpless and confused in the face of a Downeast-style management problem. "How do you regulate or enforce management style?" was a familiar FAA rejoinder to criticism that the FAA has the responsibility to keep companies with managements like Downeast's from operating in the airline business. "Besides," one Portland inspector continued, "someone like that will have all his records in order when we come around. If we can't prove anything is wrong, we can't take any action."

And so they didn't—which is where Robert Turner hit the wall of his official limitations.

The truth that Turner's bosses dearly wanted to keep hidden was that the Portland FAA office had known for nearly eleven years that Bob Stenger's operation paid little more than lip service to the rules. They had known, because Robert Turner among others had tried to get them to do something about it. Turner's bosses, however, had purposely turned their heads and ignored the entire situation, even when confronted with specific information from upset former employees. They didn't know how to deal with it, so they refused to try!

110

Turner had begun his tour of duty as principal operations inspector of Downeast by flunking the chief pilot, Hale Bartley, on a check ride. Stenger and Bartley were infuriated.

"I tell you what I'm gonna do, Turner; I'm gonna call my senator and congressman and have you removed as [principal operations inspector] right now!"

Whether Bob Stenger ever did make the calls, Turner never knew (though Stenger did complain to Turner's boss at Portland), but no one removed him from the job. In retrospect, Turner would wish they had removed him.

By 1974 Robert Turner had become sufficiently concerned over the way Downeast was operating to call for help from the FAA's Boston headquarters, which controlled the Portland GADO. He had come to realize that with the restrictions placed on him by the system and his boss, he wasn't going to be able to catch the airline "in the act" without help.

Every time he came to Rockland he was supposed to let Stenger know he was coming, since the FAA hierarchy didn't believe in surprise inspections as a normal method of watching a carrier.[1] With such warnings of an impending inspection, Downeast always had plenty of time to get its records in order, its pilots back into the books, and the entire operation on battle alert. It was no wonder Turner could find little wrong, considering the narrow scope of the things he was allowed to check.

Turner asked for a special inspection team (called a SWAP team), which was to descend on Downeast for several days and check everything. Such a team theoretically could bore in where Turner could not. He put in the request knowing that they would find major problems.

But they didn't. The SWAP team came and the SWAP team went, and Downeast was left with a clean bill of health. Turner knew better. He had no idea why the team had whitewashed the inspection—whether because of political influence or laziness—but he believed without question it was a whitewash.

Turner (his position somewhat tarnished in the GADO for call-

1. Certainly an FAA inspector can pull unannounced "ramp checks" and other spot checks if he desires, but those are meant to be supplementary. Descending unannounced on a carrier as a normal surveillance method is not condoned. Part of the reason has to do with the FAA's desire to be a helpmate to the carrier—to "promote" aviation. Part of it comes from the fact that too many unannounced inspections bring complaints from the carrier, and if those complaints go to Washington before they descend on the GADO, they can be embarrassing and dangerous to one's career.

ing down an inspection that had discovered nothing wrong) reverted to routine monitoring and administering of check rides.

Of course, Downeast wasn't his only headache. Turner had thirty-four Part 135 air-taxi operators for which he alone was principal operations inspector. Three of them, including Downeast, were scheduled commuters. There wasn't a great amount of time to waste hanging around Rockland and watching only Downeast, even if his boss had permitted him to—which he did not. To attempt an unannounced visit over several days invariably led to his being called on the carpet. To continually try to convince his GADO chief that the rumors about Downeast demanded better investigation also put his job in jeopardy.

The FAA hierarchy both in Portland and in New York did not want to hear about problems with "busted" minimums and overworked pilots unless there was hard proof. Otherwise it would appear that the FAA was harassing the air carrier, and that was contrary to the FAA caveat to "promote" as well as to regulate aviation.

By 1977 Bob Turner had finally realized that without support from above, and without catching Stenger and his airline redhanded on a host of ironclad, verifiable violations, he could do nothing but keep their paperwork and check rides in order.

It was a wild and undisciplined operation at Rockland, but even the pilots would lie to him to cover it up. Turner would confront a Downeast pilot with a training record showing a required groundschool course that Turner was convinced had never been taught, but the pilot would swear that the record was correct. How could he prove otherwise? There was no method for undercover investigation, nor would he have been allowed to try to interview the Downeast people clandestinely, away from the Downeast property. If he couldn't prove anything, his superiors certainly wouldn't support him. In fact, because of the "don't make waves" principle, he was not sure whether they would support him even if he *did* have proof! It was an impossible situation.

So, in the end Bob Turner did a very human thing. He gave up. Robert Stenger could operate Downeast however he chose. The FAA, at least in its present incarnation, would be unable and unwilling to interfere.

But then came the crash of Otter 68DE. Surely, Turner thought, his bosses would dig in and find the truth at last.

Turner's bosses, however, knew a potential scandal when they stubbed their toes on one. If the FAA attacked Downeast with a massive inspection and really did find the rotten operation that Al Diehl was discovering (if falsified company records, or routine rule breaking emerged from an investigation of *any* airline under their supervision), that would blacken the FAA officials' eyes, because it was *they* who had permitted the airline to keep flying. Consequently no major "close-watch" surveillance of Downeast was launched, and other than the same meaningless increased checks of pilot records and maintenance forms, Bob Stenger was left alone.

The GADO chief and his boss in Boston had more than just apparent incompetence to cover up. Beneath the veneer of righteous indignation that anyone would challenge the efficacy of their surveillance activities was a genuine fear that the NTSB, or the press and the public, might latch on to the fact that for so many years they had had reason to know the truth and had ignored it. It was the Portland GADO that had received complaints from employees and others about Stenger's methods years before—complaints full of details and dates and places—and that had refused to do anything at all about them because "the allegations were unproven," and "we get these sorts of unconfirmed rumors all the time."

The FAA's failure was worse than embarrassing, it was an irresponsible breach of public trust, but in the opinion of the FAA regional chiefs, it should be addressed quietly—in house. If the public discovered that the FAA had been warned and could have prevented the type of pressure that led Jim Merryman to such physical and emotional distraction—and ultimately led to the deaths of seventeen people—the FAA would be bloodied, Administrator Langhorne Bond would be furious, the Carter administration would be embarrassed, and the public's confidence in the FAA's ability to keep the system safe would be shaken.

Therefore, the Portland GADO's claims that Downeast was safe had to be supported at higher administrative levels. As a result, the NTSB, and Al Diehl in particular, were being effectively stonewalled by both Downeast *and* the very agency charged with responsibility for detecting and preventing the type of abuse of public trust that the former Downeast employees were documenting: the FAA!

And the allegations unearthed by Diehl's exhaustive detective work were incredible, and widespread. If he was dealing with disgruntled employees, he was dealing with an unprecedented number of them—all saying similar things. The former Downeast employees kept surprising him with their eagerness to tell their stories of marginal to downright illegal operations at Rockland over many years:

Dr. Al Diehl
NTSB
Washington, D.C.
August 24, 1979

I worked for Downeast Airlines from June 15, 1970 to May 26, 1972. On the day I was interviewed for the job . . . Mr. Stenger told me that the company's minimum descent altitude at the Rockland Airport was 300 feet MSL (above Mean Sea Level). He then asked me if that would be any problem. I said I didn't think it would be. Later, I found the [legal minimum descent altitude] to be 640 feet MSL. I realized then that if I had not answered his question in the way I did I might not have been hired.

Throughout the time I worked there, it was a policy of Mr. Stenger when reporting weather, to always report at least one mile visibility, even when the actual weather was less than ½ mile, so that approaching and departing Downeast planes always had [legal] minimums. . . .

It was Mr. Stenger's policy to overload the airplanes beyond their [legal, maximum] gross take-off weight. There was one plane that was overgross with its nine seats full, 300 pounds of bags in the nose to balance the load and [only] a half hour of fuel on board. As we all know one half hour of fuel is required just for reserve. The plane was used in that configuration routinely. . . .

I had a schedule that required me to be on duty 72 hours per week. . . .

If any pilot did not abide by any of the policies set forth by Mr. Stenger, he was fired. Each pilot was constantly reminded that his job was in jeopardy.

The above information is true and factual to the best of my knowledge.

[signed] Maurice A. Roundy

The same pattern was evident in a letter he had received back in July:

Dr. Al Diehl
National Transportation Safety Board
Bureau of Accident Investigation
Washington, D.C. 20594

Dear Sir:

From June, 1974, to Feb., 1976, I was employed as a captain with Downeast Airlines, Inc., and would like to relate to the Board, some of the questionable operating procedures dictated by the president of the airline, Mr. Robert Stenger.

During my interview with Mr. Stenger, company operating procedures were discussed including weather procedures to be used at the Rockland airport. Mr. Stenger told me that because of the installation for runway 3 at Rockland, our minimums were 200 ft. and a half mile visibility, but do not go too much below 300 feet indicated. [The legal minimums were 640 feet.] At this point, I asked Mr. Stenger if the company would be responsible for my salary if I were violated for "busting minimums." He indicated that he felt that would be a fair arrangement, if I were violated. Then I asked him if he would state such an agreement in writing, upon which, he declined, as did I. This discussion was held in the presence of the chief pilot.

During the period I was employed at Downeast, I witnessed numerous attempts at undermining a pilot's ability and judgment by pitting one pilot's skill against another, especially concerning weather operations.

On widespread IFR days, there was the problem of overgrossed aircraft because of the IFR fuel requirements. I had asked Mr. Stenger if I could reduce my passenger load to accommodate the required fuel load. He told me if I felt that strongly about it to leave luggage behind, but I would be the one to deal with the irate passengers. He also said that since no one else left luggage behind, why should I be the only one to do so.

Mr. Stenger generally showed a lack of support for any pilot who wanted to be legal.

On the positive side, Downeast's maintenance program is the best commuter maintenance program that I have ever been exposed to. Mr. Leo Gallant and his staff expressed a mature and conscientious attitude toward their work.

The statements I have made are true and factual to the best of my recollection.

Date: July 20, 1979   [signed] Joseph E. McGonagle

And still another from an employee who was there at the time of Merryman's death:

July 18, 1979

To Whom It May Concern

My name is Kurt Langseth. I worked for Downeast Airlines from June 1, 1977, to June 7, 1979. My responsibilities were charter and line flying.

I knew Jim Merryman for two years and considered him a good pilot and a sensitive individual. When he was promoted to chief pilot I felt that he was a puppet for Bob Stenger. On numerous occasions he had indicated to me that he would like to go back to the line because he felt that the boss was impossible to deal with. I also felt that Jim feared for his job. He knew that if he didn't go along with the boss that things could be rough. He once asked for a current set of [FAA required regulation manuals] and Jim told me that Mr. Stenger went thru the roof raving about the $30 that they would cost.

George Hines I knew only for a few months. I flew with him many times [under visual, clear weather conditions], and only once that I can remember [under IFR, instrument conditions]. George was very smooth and showed good ability to stay ahead of the aircraft. I remember thinking that he flew the [Twin Otter] smoother than I did when I started. The time I flew [under IFR, instrument conditions] with him to Boston he was behind the airplane when we started the approach. Approach [control] had given us a [specific instruction to be at a certain altitude at a specific point] which we did not make and we missed the [outer marker altitude] by about 500' altitude. We were also too fast to extend the landing gear [in the Piper Aztec], however, we were on the localizer. I did discuss the problem with George on the ground and he seemed to say no sweat, I'll get it next time. I feel George was undertrained for a Captain's position and did not know enough of the [Federal Air Regulations Part 135]. Knowledge of the Regulations was hard to get because there wasn't a set of applicable [Federal Air Regulation Manuals] made available to the pilots. In two years at Downeast Airlines, I never received any training, not flight, ground, or recurrent. I did receive some ground training on the [Twin Otter], but the training was not completed because we were needed to fly the line. Bob Stenger's attitude was "they don't need any training because they fly the line." I took check rides in [the Twin Otter], PA 31, Aztec [Cessna 182], and did not receive any training, just check rides. Ed Edwards did give me some training in June, 1977 in the Navajo, but there was absolutely no formal or informal flight or ground training.

The atmosphere at Downeast could change in a matter of seconds, depending on Mr. Stenger's mood. He has put most of the women at Downeast in tears and intimidated just about everyone including myself. If there was a common denominator, I think it would be money.

Mr. Stenger told me that "there was nothing wrong with shooting the localizer approach to Rockland at 300', 140 [feet] below the [minimum legal descent altitude]". . . .

Our chief mechanic told me that the boss had told him to reuse a tire that had made a hard landing and bent a motor mount and flap, even though he was not supposed to. What choice do you have when the boss condones this type of thing and even endorses it. Another example of this penny pinching attitude was when he told the weather people not to send up balloons, because they cost 90 [cents apiece]. There were some sort of [legal, regulatory] parameters for this, however, I don't know what they were.

The pressures that existed were not always spoken about. The example would be, flying over gross [weight]. You could sense the pressure from the boss and the dispatchers, why they were called dispatchers I don't know, because they didn't know anything about aircraft, regulations, or weather. Their weather reports lacked credibility, which accounts for some people taking extra fuel.

There was no excuse for anyone at Downeast to make any type of mistakes. The boss just insisted that all mistakes were not mistakes, but "intentional happenings aimed at [him] and Downeast Airlines." He told this to me a number of times. I too feared for my job, so much so that after about 9 months I had to confront him and ask if he planned to fire me. He said "no" with a laugh, but I realized that he was very prone to mood changes and I could be on my way the next day. Mr. Stenger said "if you want to work here, you do things my way."

I would like to close and state that some of Downeast Airlines problems were of an individual nature, however, over all the problems were created by the pressures, spoken and unspoken by Bob Stenger, and his unwillingness to follow the rules.

[signed] Kurt Langseth

The most frightening overall aspect emerging from the investigation, as Diehl knew well, was that Downeast was neither an isolated case, nor the worst of the commuters.

The number of accidents, and the number of deaths, in commuter-airline operations was historically far worse than that of the major airlines, but deregulation had blurred the lines between the two in the eyes of the public. People were climbing onto commuter carriers every day with the mistaken belief that they were statistically as safe as the large carriers. The constant drumbeat of fatal commuter-airline crashes through 1978 and 1979 had damaged that belief, but the NTSB was still not quite ready to conclude in public that the FAA was purposely ignoring the problem.

Al Diehl, five feet ten, of medium build, possessed of a slightly raspy voice and an animated, rapid-fire style of speaking (laced with fervor and sincerity), was an excellent interviewer. It had been his friendly informality and his directness that had elicited so much response in the Downeast case. His probe had kicked over the protective rock, and the real Downeast—a quintessential example of an unsupervised airline operation in myopic pursuit of profit—had crawled out from beneath, blinking and protesting at the heat and light of unwanted attention.

Though the NTSB usually spoke to the public (and the FAA) in formal reports and hearings, the day-to-day work of the board members and staff members alike was a constant flow and flux of conversations, informal meetings, interoffice memos, draft reports, bull sessions, and rumor exchanges, including interoffice politics. It was the very human and fluid process through which the recommendations and the reports of the NTSB were crystallized. But there were many rubber stamps around the eighth floor reading FOR OFFICIAL USE ONLY. They were there to protect that free exchange process from premature public scrutiny, because when the NTSB finally spoke as one, it had many things to consider.

Unfortunately, politics was too often one of the factors. In the summer of 1979, the Democratic administration had a large political stake in the success of the Airline Deregulation Act of 1978 sponsored by Democratic senators Robert Kennedy of Massachusetts and Howard Cannon of Nevada. It could be politically dangerous for the FAA to admit it was having trouble maintaining the level of safety that Congress had promised under deregulation. It could become a major political issue if the public perceived the initial processes of deregulation as putting severe and dangerous strains on the FAA's ability to watch over and regulate properly the commuter airlines of the nation—the Part 135 carriers.

Downeast, of course, was a commuter carrier. Even though its existence, its operating inadequacies, and its poor surveillance by the FAA predated the Deregulation Act, an NTSB accident report that spotlighted the FAA's failures at Downeast—failures that might be related in the public's mind to deregulation's pressures on safety—had to be thought out carefully. The NTSB staff could only recommend to the NTSB board members. It was the members (all political appointees) who signed the reports. Though the NTSB was by law independent of direct political control, no one on the eighth floor was naïve enough to think that the board members were unaware of the realities of a political appointee's longevity.

In addition, FAA Administrator Langhorne Bond, for all his safety crusades and his divisive "big stick" approach to the task of forcing commercial aviation to follow the rules his way, had put himself on record as having forced the agency to issue a long-awaited, long-recommended update and upgrading of the rules that governed commuter airlines (14 CFR 135).[2] Bond had prom-

2. The NTSB had sternly recommended in a special Air Taxi Safety Study adopted as far back as September 27, 1972, that the FAA should immediately "expedite the redrafting of FAR 135 in its entirety, recognizing that commuter air carrier operators are separate entities from the smaller air taxi charter operators." (That was in NTSB document A-72-171.) It had become apparent to the board even then (even before deregulation was a hot suggestion), that too many people were trusting their lives to "little airlines" called commuters to allow those commuters to be governed by rules that required very little of them in the way of maintenance, training, operational control, and especially FAA surveillance. The FAA assured the NTSB in that same year that it was already working on the redrafting and it should be ready shortly. Despite the assurances, nothing was done. For the next six years the NTSB kept prodding, and the FAA kept promising, and passengers kept perishing in an increasing number of commuter accidents—yet the FAA did nothing but play bureaucratic games. Finally, in December 1978, and as a direct result of Langhorne Bond's accession to the office of FAA administrator, the new Part 135 was adopted. The new rules, however, were not effectively implemented. As a result, the commuter carriers, such as Downeast, continued to operate much as they had before, and the FAA continued to pay lip service to monitoring them and forcing compliance with the new rules.

ised to do it, he had done it, and he was publicly proud of it. The fact that the rules had yet to make a dent in the growing fatality rate among commuter carriers was an embarrassing footnote to Bond and to the Carter administration—but it was going to have to come out. From the vantage point of the eighth floor, the NTSB appeared to be the only governmental agency that could make the point.

The thrust of the Downeast investigation, however, could be focused (by a subtle shift) more on Robert Stenger's style of operation than on the failure of the FAA to detect it and halt it. The temptation would be great to endorse that shift to minimize the broadside at the FAA. There was a rising level of interoffice pressure on Steve Corrie as IIC and specifically on Gerrit Walhout and Al Diehl, pressure to "stop fooling around with mumbo-jumbo and issue the Human Factors report." It was an indication that perhaps there was a sensitivity among the staff to the fact that the less depth to the investigation, the less politically embarrassing the Blue Cover report (the final, formal accident report issued by NTSB) might be.[3]

In the background, however, was the inescapable fact that a new role had been created for commuter airlines by the Deregulation Act. Even though the proponents of deregulation didn't want to admit that there were safety problems, the truth wouldn't permit a whitewash of the implications behind Downeast and Air New England—even if the board members had wanted to engage in one (and by and large the members of the NTSB were honorably mindful of the responsibilities to *stay* unmoved by political considerations).

Commuter airlines were filling the gaps left by the major airlines as the big boys terminated service to smaller communities. Indeed, Professor Alfred Kahn (considered the father of airline deregulation) had foreseen and advocated just that in his support of airline deregulation. The commuter carriers, according to the theory, could serve the interests of smaller communities with greater efficiency, greater responsiveness, and greater competitive spirit. Therefore, the public would be better served at lower fares.

But there was a legitimate issue of whether it was *as* safe to fly

3. In fairness, despite the desire to be "diplomatic" in its relationship with the FAA, the board members historically had shown little reluctance to blame the FAA when FAA inaction or inattention had slowed or blocked implementation of NTSB recommendations.

on the average commuter carrier as it had been to fly on the Uniteds, Braniffs, Frontiers, and other majors who used to serve those smaller communities. The subject was talked about in the prederegulation hearings on Capitol Hill and in the press, but the conclusions that came from those "discussions" were based on a flawed assumption: that the FAA would be "beefed up" to meet the challenge and keep them as safe.

The public need not worry. Commuters would be just as safe a method of getting from Cline's Corners to Bug Tussle as Eastern or Delta had been. However, since the public never really got involved in the deregulation debate, it simply inherited the glib conclusions without examination of their veracity.

No one had ever pointed out to the constituents that as good as they are, Twin Otters and light twin-engine aircraft do not possess as good a safety record as Boeing 727s and DC-9s. In fact, the type of aircraft that was pressed into service as air-taxi commuters in the early seventies (the Beech 99, the Piper Navajo, the Twin Otter) were never adequate for scheduled airline service! Some of these machines simply cannot do what they represent to the public that they can do. It is not possible, for instance, to load a Beech 99, a Piper Navajo, or a Cessna 402 completely full of people and baggage, fill the fuel tanks, and take off on a relatively short runway on a warm day with any significant margin of safety in the event of an engine loss. And engine losses on takeoff *do* occur. In addition, in many cases, twin-engine commuter "airliners" simply cannot stay level—hold altitude—on a single engine when heavyweight.[4]

The problem is not that the commuter-airline owners don't want

4. One reason Robert Stenger was at odds with his pilots so much of the time was that neither the Navajos nor the Otter could carry full fuel tanks and a full load of people and baggage from Boston to Rockland. Something had to be left behind. The pilots often tried to leave bags, but that caused problems with the passengers. Stenger felt that if anything, fuel should be left behind, and constantly harassed his pilots to keep their fuel reserves as low as possible. Unfortunately, too low a fuel reserve would mean too reduced a capability to go somewhere else if a pilot couldn't make it in to Rockland. The younger, less experienced, or more intimidated were constantly running into tight fuel situations with passengers aboard, unaware of their plight. In response to company pressure, Jim Merryman had instituted a personal policy in the Otter that he would not depart Boston with less fuel than it would take to make an approach to Rockland and return all the way to Boston. Stenger didn't like it, but he accepted it grudgingly. Few of the other Downeast pilots, however, were as assertive.

to buy larger aircraft that *can* perform with the safety margins of a Boeing or a Douglas; the sad fact is that as an industry (with some exceptions), seldom can they afford to buy such aircraft. The smaller and more thinly financed the carrier, the more meager its revenues, and the more pressure there is to buy used, less efficient, (and marginally performing) aircraft in which to carry trusting, paying passengers. But those same carriers were trotted out by Congress in 1978 as the future stars of the deregulated age: the carriers who would grow and mature into the backbone of the U.S. air-transport system for smaller communities.

No one explained to the public during those congressional debates what these carriers were really like, because officially not even the FAA really knew (though it had the responsibility and it could have known). No one spotlighted for the traveling public the fact that even the new Part 135 rules did not force the commuter carriers to adhere to the same standards of maintenance and training and operational control as the Pan Ams and TWAs, and Uniteds. No one ever adequately pointed out to the smaller communities and their citizens and mayors (who were being assured of continued air service) that airlines that don't make very much can't provide the same margins of safety in equipment, personnel, or practices as the Ozarks, Frontiers, U.S. Airs, or North Centrals (which ranked very close to the major Part 121 airlines).

Thanks to the admirable but unrealistic faith that Congress and Professor Kahn placed in the fine and dedicated—but human— people in the FAA (unrealistic faith that they could solve all problems and cure all ills in the vast, unknown Age of Deregulation), the American public was sold a pig in a poke. The congressional answers to the questions of what would become of safety in the brave new era of free-market airline flying were raw speculation laced with enthusiastic optimism unfettered by reality or historic experience. Nowhere was that more painfully apparent than to those in the FAA who watched their work loads soar out of sight (without accompanying increases in appropriations) as new and expanding airlines, both commuter (Part 135) and major (Part 121), began popping up like dandelions on a springtime hillside.

In that very germination of new competition, however, lay the seeds of potential disaster. While the commuter carriers were facing a massive task of metamorphosis—trying to upgrade their philosophy and professionalism as an industry to justify the public trust that had been thrust upon them overnight—they were losing legions of experienced pilots and other operational personnel to the expanding major carriers, who themselves were under intense pressure. The inexperienced, the unproven, and the excessively young airmen who would take the vacant positions were too often given too little training, too little guidance, and too little supervision to be safe and reliable.

What had been a delicately balanced national airline system (even if sluggish and inefficient) before deregulation, had suddenly become a confusing arena of constant change, and there were members of the industry who would (quite literally) not live through it. By mid-1979, the detractors of deregulation were already predicting the first "Deregulation Crash." (The proponents were just as adamant in their disdain of such predictions.) It was inevitable, however, that something tragic would happen, which would, at the very least, call into question the glib assurances of Professor Kahn's following. (Exactly where, how, and to which airline were questions no one wanted to field.) The deadly, explosive potential was there—and the fuse had been lit.

Chapter 8

A SKELETON IN THE AERONAUTICAL CLOSET

Like some sort of jet-age, ghostly apparition barely visible through the swirling ice crystals and airborne powdery snow swept from the ice-covered taxiway by the wind currents, the Boeing 737 rolled slowly, gingerly, past a runway maintenance crew huddled in their warm pickup—the cacophonous whine of turbine engines lending a surrealistic quality to the scene.

The roaring jet exhaust from another airliner in front of the 737, its form almost invisible now in the gray murk of a late winter afternoon, had been blowing loose snow toward the 737 as the two of them lumbered single file over the bumpy, ice-encrusted ribbon of concrete leading from the terminal ramp all the way to the end of the runway. It was taking far more time than normal today—the taxiways were extremely slippery and it would be dangerous to taxi with too much speed.

Periodically, the dark metal, clamshell-style doors at the tail end of each of the 737's engines would snap together, blocking the normal path of the jet exhaust and forcing it forward around the engines—reverse thrust—to help keep the jet from skidding. (Even with its brakes locked and the engines at idle, the compact airliner produced enough forward thrust to skid on very icy surfaces). Each time the thrust reversers were used, a cloud of loose snow and condensation billowed forward, engulfing the wings and the forward part of the body and nose of the jetliner. The men in the truck could see the process clearly, but the passengers sitting behind the lighted but nearly opaque windows in the warm passenger cabin could hear only the periodic change in the exhaust noise. Only a veteran crew member would recognize the sound as that of thrust reversers being used—or understand what that meant.

In the bitter cold the little Boeing, with its low-wing profile and its two underwing engines suspended just over the surface, looked almost as if it were hunching over, hunkering down against the frigid conditions. The maintenance crews had very carefully swept the loose snow from the wings and the body of the aircraft before it had left the gate, but with the hot exhaust of the reversed engines briefly melting the taxiway ice particles and blowing them over the cold metal of the wing's leading edge, it was no longer clean. The bright aircraft lights stabbing through the gloom highlighted the uneven coating of ice crystals now encrusting the front of the wing.

Finally, with the departure of the previous jet, it was the 737's turn.

The two pilots, receiving their takeoff clearance, maneuvered the slightly diminutive twin-jet onto the runway, made a final scan of their instruments, and advanced the thrust levers (throttles) to maximum power.

Back in the cabin the increasing acceleration pressed the sixty passengers back in their seats—noticeably. Some were reading or sleeping, some casually trying to see through the frosty windows, some talking, but most of them were oblivious to the impending rite of passage from groundbound vehicle to airborne flying machine.

"Eighty knots."

The call-out from the copilot was followed quickly by another. "Vee one."

It meant that they were now committed to takeoff. At that speed, they no longer had enough runway left in which to stop their 737, should they need to.

The indicated airspeed needles on each pilot's respective instrument panel hovered around 139 knots, the proper speed (for their particular weight and temperature conditions) for "rotation."

"Vee R."

With that call-out, the captain began a gentle pull on the control yoke in order to raise the nose of the aircraft, which would increase the pitch of the 737 and its wings, creating enough aerodynamic lift to cause the jetliner to leave the ground and begin a controlled climb.

At least, that was the normal reaction.

With his left hand putting routine back pressure on the yoke, the captain suddenly realized that the nose was coming up far faster than he had bargained for—faster than he had commanded—and now it was running away!

In three seconds the Boeing's main wheels had left the runway with its nose reared up nearly 18 degrees and still increasing at a startling rate, its captain shoving the yoke forward to stop the pitch-up.

The 737 was barely fifty feet in the air, but climbing abruptly, the deck angle now 21 degrees, the right wing dropping suddenly, uncontrollably, the airspeed around 139 knots.

With the copilot gripping his yoke in unison with the captain, both men rolled the control wheels (on top of the yoke) to the left, pressing the left rudder, shoving their respective yokes forward almost to the limit, trying desperately to recover before the airspeed began to decay.

But suddenly the yoke itself began to vibrate—shaken by a small electric motor designed to turn on whenever the aircraft was approaching a . . . stall![1]

1. As the wing of an aircraft increases its pitch—its angle of attack—while moving through the air, the amount of lift it generates increases, until a point is reached at which the angle of attack is so great that the airflow over the top of the wing begins to separate from the upper portion of the wing surface. When this point is reached, further increases in the angle of attack will *not* produce further increases in lift—exactly the opposite. The amount of lift produced begins to *drop* with further increases in the angle of attack. In some wings, this drop-off is abrupt, severe, and easily detected: The aircraft may shake and shudder as the disturbed, separated airflow burbles against the aircraft's tail. In swept-wing aircraft, the lift drops off more gently. The point at which this drop-off in lift becomes significant is called the stalling speed, and an aircraft that reaches that condition is said to be "stalled," or "stalling." If the angle of attack is not lowered by lowering the nose of the aircraft, the aircraft will lose altitude. In a Boeing 737 and most swept-wing jetliners, the drop-off in lift is not that abrupt or noticeable until the aircraft is well into a

The system was telling them that their 737 was stalling, but that was impossible! The minds of both pilots rebelled at the idea. The stall speed was 122 knots—it said so right on the data card—and their airspeed was 139 knots. There was no way they should be stalling—but the stall warning system had activated. Their response was automatic—jam the throttles full forward instantly and get the nose down—relax back pressure on the control yoke.

No more than ten seconds had passed. The amount of adrenaline instantly coursing through the veins of both pilots was significant. Time seemed to stretch—events were in a type of slow motion perception, so rapid were their thought processes and observations.

"What the hell?" The nose was not responding. The pilots activated the trim button, trimming the horizontal stabilizer to the "nose-down" position.

Get her forward . . . get that nose down . . . she's rolling right, left aileron . . . rudder . . . power . . . check airspeed . . . stick-shaker activated . . . Thoughts and impressions darted in microseconds through their minds as their instincts, honed through many years of commercial flying and training, guided their hands in urging the controls in the right direction, making the right moves.

In the cabin, the sudden pitch up and roll to the right had caused a few passengers to snap to immediate attention. They didn't know what was wrong, but *something* was definitely out of the ordinary. Boeing 737s didn't act like this on takeoff!

The Boeing began to respond, the nose pitching forward, lowering the body angle of the 737 to 15 degrees, then 10, picking up airspeed and rolling back to level flight. The stick-shaker had

---

dangerously stalled condition and has entered an increasing sink rate (rate of descent). It would be very dangerous for aircrews to wait for such physical indications to warn them that they had demanded too much of their aircraft. Since jetliners usually are exposed to maneuvering speeds and conditions that might be conducive to an accidental stall only when departing or arriving at an airport, and thus at low altitude, they are exposed to a stall and a high sink rate when they have the least altitude in which to recover. It is, then, very important that a pilot have some way of knowing that he might be getting too close to a stall before one actually occurs. This gives him time to add power and fly out of the stall in a high-powered turbojet airliner or lower the nose while adding power in more conventional aircraft, before he begins sinking toward the ground. Therefore, the manufacturers of swept-wing jetliners are required to install an artificial system that literally shakes the control yoke when the angle of attack of the aircraft gets dangerously close to one that would produce a stall.

stopped as well. The Boeing began to accelerate with the engines at maximum thrust: 145 knots . . . 150 . . . 165.

"Flaps up."

The copilot reached down and snapped the flap handle from the 5-degree position to 1 degree, then to full up.

And just as suddenly it was all back to normal.

As the two pilots tried to calm down and continue a normal climb-out with a machine that now seemed to be acting in every respect according to the book, their accelerated heart rates and adrenalized systems kept them focused on the puzzle of what had just occurred.

It was obvious that for some reason they didn't understand, they had nearly lost control. But what? They had no idea that there had been enough ice granules on the wing to change the flying characteristics of the entire airplane. They had no reason to suspect that their use of reverse thrust on a snowy runway had encrusted their wing's leading edge with frozen contaminants. If there were any notes on the subject in their flight manuals—if there had been any training by their airline—neither man could remember it.

Of course both of them knew the cardinal rule about cold-weather operations: *Never* attempt a takeoff with snow or ice adhering to any lifting surface or control surface of the aircraft. They were not guilty on that count. Their Boeing had been cleaned of snow at the gate, and they were certain that there had been no snow or ice on them when they pushed back from the gate area. Yet the wild gyrations had happened. There was no doubt about that. Could something have occurred between the gate and the runway? To the two veteran pilots, it was a deep puzzle.

To the company that built the 737, however, it was anything but a puzzle. Boeing was well aware that such incidents had happened before—numerous times. In fact, Boeing had been aware of the 737's vulnerability to small amounts of snow and ice contamination since the airplane's first flight-test series in the early seventies. Of course, the FAA-approved Operations Manual for the 737 prohibited flight with "snow or ice adhering to any lifting surface." Therefore, the reasoning went, the cure was simple. If pilots would obey the book, the bird would behave!

Nevertheless, over the early years of the seventies, incidents kept occurring (though none, thankfully, resulted in disaster). In most of the incidents, the flight crew was adamant that they had not violated the cardinal rule—they had *not* tried to fly with contaminated wings.

In 1974 Boeing issued an Operations Manual Bulletin to all its 737 customers disclosing and discussing the problem, and adding a "note" to the Adverse Weather section of the 737 manual. Very few airline flight departments and flight managers understood the significance of this strange malady called "pitch-up/roll-off," so very few pilots had their attention focused on the bulletin or even saw it. For some 737 operators who seldom flew in areas with snowy weather, the bulletin was just another notice to be filed away somewhere in the book. Many such bulletins arrived during the course of a year, some concerned with matters that seemed (to the busy executive or chief pilot) picayunish at best. The arrival of still another one would not be likely to set off alarm bells and galvanize the undivided attention of even the most professional of aircrews (unless, of course, that very airline had already suffered a well-publicized incident, or had a highly organized safety department dedicated to keeping up with such affairs). The typical reaction in far too many airlines was to distribute copies to all 737 pilots, who would eventually file it in their Operations Manual or stick it in their "brain bag" (flight case, map case, or pilot briefcase) with the fleeting thought, "If I ever need to know that, I'll pull it out and read it."

Strange as it seemed to Boeing, even though a bulletin had been issued, more incidents of wild takeoff excursions, "pitch ups" and "roll offs," kept occurring.

On February 23, 1979, after twenty-eight more incident reports, Boeing tried again with yet *another* bulletin: Operations Manual Bulletin 79–2:

Subject: PITCH-UP/ROLL-OFF AFTER TAKEOFF

Reason: To advise flight crews of a possible inadvertent pitch-up/roll-off after takeoff due to ice accumulation on leading edge devices.

THE FOLLOWING PROCEDURE AND/OR INFORMATION IS EFFECTIVE UPON RECEIPT

Background Information: Several operators have reported pitch-up/or roll-off after takeoff due to ice accumulation on leading edge devices. The incidents usually have occurred following the application of reverse thrust while taxiing on snow covered taxiways. In order to advise flight crews of this condition the following note is presently in the Adverse Weather section in the Operations Manual.

> NOTE: A buildup of ice on the leading edge devices may occur during ground operations involving use of reversers in light snow conditions. Snow is melted by the deflected engine gases and may re-freeze as clear ice upon contact with cold leading edge devices. This buildup, which is difficult to see, occurs in temperature conditions at or moderately below freezing. Crosswind conditions can cause the ice buildup to be asymmetrical, resulting in a tendency to roll at higher angles of attack during subsequent takeoffs.

Boeing was absolutely right—the note had indeed been in the Operations Manual since its first bulletin in 1974, but Adverse Weather was a back section of the manual, and the note had been all but buried there with little if any emphasis. The problem with such notes in operations manuals is that pilots, being human, cannot possibly know of or retain an encyclopedic memory of every item on every page and apply such information properly in every instance without initial and continuous training sessions.[2]

Operations manuals do not function in a vacuum. The wisdom and procedures contained on those carefully drafted pages must be

---

2. When an aircraft manufacturer such as Boeing builds a new airplane for commercial airline use, it compiles with incredible care and expertise a flight-operations manual for that particular aircraft, which is drafted in close cooperation with the Federal Aviation Administration's Part 121 (Airline Operation) experts, and subsequently approved by them. It is the bible for that plane and sets out the correct, proven, accepted, and safe methods of operation of the machine with the goal that if the aircraft is always operated in conformance with the manual, it will always be capable of safe flight. All the training and operations by the airlines must conform to the procedures and dictates in its various sections, such as Limitations, Normal Procedures, Emergency Procedures, Abnormal Procedures.

Nevertheless, an operations manual is a huge document containing hundreds of highly technical pages. Even the most dedicated and brilliant pilots cannot quote more than a few passages of such a manual by rote, though they may be intimately familiar with most of its content. When seemingly insignificant "notes" are inserted with no emphasis by the manufacturer, and when such notes do not become the focus of recurrent training classes or other eye-catching bulletins to the aircrews, they are ineffective at best, and invisible at worst. They become "cover your ass" memos, promulgated by the manufacturer to protect the manufacturer.

taught and emphasized dynamically and continuously if they are to take root and guide the actual flying of airplanes. An airline has to have someone, or some group, to ferret out the significance of any notes when they arrive, look into their meaning, and (if justified) take direct and affirmative action to, in effect, grab the pilots by the shoulders and focus their attention on the matter. The job doesn't end there, however. An airline has to incorporate these recommendations into *all* ground-training classes, and make certain that the check pilots check for knowledge of the matter "on the line."

These would be the normal functions of an airline's safety department—but precious few airlines have company-funded safety departments. The type of airline that would be operating Boeing 737s in the near future at cut-rate fares and cut-rate budgets would have little incentive and even less money to get involved in paying for a safety department.

Without such a structure in the majority of the airlines that flew Boeing 737s worldwide, Boeing's note on pitch-up/roll-off problems was far too subtle to be effective.

Of course there was another "practical" problem with such notes. In too many cases a manufacturer will put in a note simply to protect itself against legal liability. If an accident should occur because an aircrew failed to do something covered in such a note, the manufacturer has better grounds for saying in court, with great piety: "We warned them not to do that!" The fact that the note, the warning, might be a mere whisper lost in the thunderous roar of daily airline operations and practical considerations may be immaterial. It is far easier (and less risky) to say, "If it was there in the book, every pilot should have known about it!"[3]

Toward the bottom of the Boeing bulletin there were suggested guidelines on how to handle the problem—guidelines that had not

3. In fairness (as with Boeing and the 737), when the company does not consider the problem to be *that* significant a threat, and the FAA agrees, a note may be quite proper, even though marginally effective. Then too, Boeing does not run an airline, and does not have a complete grasp of the subjective, human dynamics and imprecise nature of the internal workings of a busy airline. In Boeing's world of strict engineering discipline and high precision, it becomes nearly inconceivable that an airline would receive one of its bulletins and fail to hammer its content into each and every affected employee. It is entirely reasonable that a manufacturer like Boeing would truly believe that the issuance of an Operations Manual Bulletin could solve the problem, though in reality the belief is very mistaken, especially in the case of newly expanded or newly created (upstart) airlines.

been in the Operations Manual note before but would now be included:

> Operating Instructions: Unless absolutely necessary, do not use reverse thrust while taxiing on snow covered taxiways. If reverse thrust has been used and clear ice accumulation is observed or suspected during taxi-in or taxi-out, have the leading edge devices inspected and ice removed, if present, prior to takeoff. If pitch up or roll off occurs after takeoff, *avoid over controlling.*

The airlines had had their note; now they had an addition, and yet the number of reported incidents continued to increase.

There were those in the industry in August 1979 who were convinced that however honorable Boeing's attempts to warn against this propensity had been, they were not enough. Chief among that small group were several aerodynamics experts at the Airworthiness Division of Her Majesty's Civil Aviation Authority in Great Britain. Headed by a veteran pilot and administrator with more than thirty years of experience (and an ex-RAF pilot), D. P. Davies, the British CAA had been watching the problem. Several airlines in the United Kingdom flew Boeing 737s. Both the incident reports and the Boeing bulletin of February 1979 sat on their desks, but they knew instinctively there would be more.

Chapter 9

OF NUTS AND BOLTS
AND HONORABLE MEN

As David Davies of the British CAA knew (as indeed all of aviation knows) the Boeing Commercial Airplane Company of Seattle, Washington, is perhaps the most accomplished and professional manufacturer of airliners on the planet. It has a justifiably proud reputation for producing solid, reliable, money-making aircraft, which last for decades. Its jetliners, such as the Boeing 707, 727, and 747, have made flying routine and trustworthy all over the world.

Boeing's deep and sincere dedication to safety is without question, if for no other reason than the obvious: Nothing else makes sense from a business standpoint. It doesn't take a genius to know that if you build airliners that crash on a regular basis, no airline is going to buy your airliners, and your company will become an endangered species—rapidly. Therefore, building safety into everything has always been the bedrock of Boeing's existence.

But Boeing is a human organization too, consisting largely of humans who are trained in the discipline of engineering, and proud of their work. At the very least, this background involves some tendency to downplay the lessons Professor Ross McFarland tried so hard to teach the industry in the course of his crusade to persuade engineers to take human beings into account in their designs.

Though Boeing has incorporated human-factor design groups and human-factor controls through every facet of its operation, the men and women in those human-factor positions fight a constant battle to overcome the professional prejudice of the typical aeronautical engineer who has difficulty understanding why he should bend his designs and calculations to fit the human element.

Submerged in their highly demanding technocracy, such folk quite often emerge from their calculations and pridefully present a new piece of equipment, only to run into what seems to them a nitpicking objection by a human-factors specialist. Their irritation at such interference with engineering perfection is very human and very understandable—and very dangerous. After all, the machine will eventually be operated by humans, with all their frailties. However much the engineer's baby may be a creation of technical perfection, when its gauges can be misinterpreted, its switches can be operated backward, or other elements of the design can invite a human screwup based on Murphy's Law, such problems cannot be ignored—even at the risk of bruising professional egos. Trying to get that single point across consumes the full-time talent of the human-factors specialists of the world.

An even darker threat, however, arises from company pride and the loyalty of good company people. If there's a problem with the product but the public-relations (and hence the sales) impact can be minimized within the bounds of safety, it will be. That, too, is human nature.

Of course, someone has to define "the bounds of safety," and in that subjective definition lurks the temptation to dodge, dissemble, conceal, and at worst, cover up.

In the late seventies, Boeing was becoming heavily dependent on the sale of more 737s, and on future expanded-capacity versions of the airplane, to stay financially sound. In trying to sell more 737s, the Boeing men were constantly engaged in a bidding war with a European consortium named Airbus Industrie. Airbus (part of Aérospatiale, Ltd.) has been charged with selling its airplanes for crazy prices, at or below cost, in order to keep the production lines moving—in part because it is government supported and can afford the losses. In such a high-stakes game, with the careers and security of tens of thousands of Boeing workers in the balance, Boeing could not be expected to be terribly eager to make too much of a public racket over an obscure glitch in the 737's flight characteristics—a glitch that had never caused more than a handful of quickened heartbeats in the previous ten years anyway. In fact, the fewer the potential customers who had their nose rubbed in that nasty little flight characteristic, the better.

It wasn't a cover-up—Boeing had not ignored the problem; it had simply been as quiet about it as possible. Besides, in Boeing's opinion, everything that ethically and legally should be done to address the pitch-up/roll-off propensity of the 737 *had* been done. Given the sales challenge and the company's growing need to fill the order books with 737 customers, it would be counterproductive at best to run around Europe and the world trumpeting the fact that such incidents had occurred and worse, were likely to occur, with any model of the 737.

Besides, the warnings were "in the book." Pilots should heed the warnings; it seemed that simple. Of course, to Boeing's engineers and test pilots, the bottom line was the Operation Manual's absolute prohibition against flying with snow or ice *adhering* to the wings. *Never* should the 737 be flown with snow and ice on its wings regardless of the cause. No ice, no pitch-up, no problem.

As David Davies and his CAA men in Great Britain knew, however, that attitude had an all-too-familiar ring to it. Another of the great American aircraft manufacturers, McDonnell Douglas of Long Beach, California, certainly as responsible and professional as Boeing, had nonetheless permitted a similar attitude and protectionist philosophy to metastasize into an act of life-threatening irresponsibility. It stemmed from a frenetic attempt at public-relations damage control and a drive by essentially decent people to save the public image of the troubled Douglas DC-10 in order to keep the sales alive for their company.

The DC-10's public-relations nightmare began with a single human mistake—this one involving a baggage handler for American Airlines in Detroit.

The man was trying to close the aft cargo door on Flight 96 to Los Angeles on the evening of June 12, 1972, but the locking lever on the DC-10's cargo door was resisting all efforts. The fellow was frustrated and disgusted with the recalcitrant handle. It should have been easy. Finally he put his knee into the job, and with gritted teeth forced it into place.

But it didn't look quite right. There was a little vent door in the much larger cargo door, and that little vent was designed to stay open until all the latches on the big door were secure. If the cargo door was not properly closed and latched, the little door wouldn't close, and the aircraft couldn't be pressurized. But the little door seemed slightly askew. Seated, but not right.

The baggage man thought it over for a second. He was under intense pressure to get the flight out of the gate, and it was he who was delaying things. Nevertheless, it didn't look right.

The baggage man departed for a minute and returned with a mechanic in tow. The mechanic took a cursory look and dismissed his worries. The door was closed.

The door, however, was not totally latched.

Flight 96 departed without trouble from Detroit and began its climb-out over a portion of Ontario. The flight appeared to be routine, and as the pressure outside the DC-10 decreased in relation to the controlled pressure of the passenger cabin, the forces pushing outward on all the doors of the jumbo built up to substantial proportions, just as they were supposed to.[1] The pressure on the

1. All modern jetliners are pressurized. As an airplane goes up and gains altitude, the atmospheric pressure outside its hull decreases. At 5,000 feet (the same altitude as that of Denver), there is effectively less oxygen available to breathe, though there is enough for most people. At 13,000 feet (the altitude of La Paz, Bolivia), the air is so rarified—thin—and the oxygen so diminished that a person who is used to the atmospheric pressure of, say, Los Angeles (sea level) will be short of breath, and perhaps a bit dizzy. Above 13,000 feet, unless a person is acclimated to—used to—living in extremely high levels (for example, an accomplished mountain climber or a resident of La Paz), there will be too little oxygen to the brain and a semiconscious or unconscious state will be the result *unless* that person breathes additional oxygen continuously. It would be a nightmare to attempt to get passengers on jetliner flights to put on oxygen masks at the beginning of each flight and breathe that way until arrival (not to mention causing a host of other physiological and engineering problems). Therefore, passenger jets must be able to provide their passengers with a breathable, pressurized atmosphere in the cabin that is, on the average, not much higher than that of Denver.

To do this, air under pressure is forced into the cabin at a controlled rate (like blowing up a balloon), and permitted to escape from the rear outflow valves at a controlled rate. Because the pressure differential (the difference between the air pressure inside and the air pressure outside) can rise to figures of 7 to 8 PSI (pounds per square inch), cabin pressure pushes outward with tremendous force on the doors and windows of the aircraft. For this reason, most jetliner doors, especially passenger cabin doors, are "plug-type." That is to say that when in place, they are slightly larger than the opening into which they fit. The more pressure applied to a "plug-type" door in flight, the more snugly it fits into its frame. Even if it comes unlatched, it won't open.

Some jetliner cargo doors, however, are outward opening for better access. These *can* be blown out if they are not securely latched to the surrounding door frame, and many different latching mechanisms have been designed to mate this type of door securely and to make sure that the DOOR OPEN warning light in the cockpit cannot go out unless that door is properly latched. In the

improperly closed door, however, finally reached the breaking point. The door assembly suddenly exploded outward, the escaping air sucking baggage from the cargo compartment (including a coffin), the rear floor of the passenger cabin collapsing as the pressurized atmosphere in the cabin pressed down on the floor with incredible force, trying to escape all at once through the gaping hole where the door had been. As the cabin floor buckled downward, creating a jagged hole in the rear of the cabin just in front of the galley, the torn metal severed some of the flight-control cables to the tail, setting up Captain Bryce McCormick and his crew with the battle of their careers: trying to coax the partially crippled and marginally controllable jumbo jet to what became a safe (if hairraising) landing back at Detroit's Metro Airport. The pilots, in one act of incredible airmanship, earned their salaries for several decades to come. The passengers escaped alive and uninjured.[2]

Douglas had designed the DC-10 in a hurry in the late sixties, responding to the need to meet Boeing's challenge with the 747, and more important, Lockheed's challenge with the three-engine TriStar, (the Lockheed 1011). During the incredibly complex design process a rather weak locking system for the aft cargo door slipped through with a human-factors bomb built into it. The door could be closed incorrectly. Therefore, Murphy's Law dictated that it was just a matter of time before it would be.

In addition, the huge floor area of the DC-10 did not have an adequate number of vents between the passenger section and the lower baggage compartment—vents that could let the tremendous cubic-footage volume of pressurized air equalize its pressure (in the event of a cargo-door blowout) by passing safely and almost instantly from the upper passenger area to the lower cargo area without taking the floor along with it. This was especially critical in the DC-10 design as it evolved, since Douglas (to save weight and money) decided to route the control cables for the tail surfaces (elevator and rudder) and the tail-mounted number-two engine (as well as the hydraulic-system lines) through the floor rather than the ceiling. If those cables were completely cut, the pilots would have no way to control the airliner, and a survivorless crash was all but certain. It was obvious to at least some people during the design

---

event an outward-opening type is only partially latched, the pressure, as it builds up to those tremendous values, may suddenly blow the door outward with an explosive force, causing an explosive decompression in the cabin.

2. Several passengers did receive minor injuries during the emergency evacuation, but no one was hurt in the cabin, though two of the stewardesses were nearly sucked out of the gaping hole in the rear cabin floor.

phase that if a cargo door should open explosively in flight, and the floor venting system was insufficient, the floor could collapse, taking the cables and the lives of the passengers along with it.[3]

Yet Douglas locked in the design, and the FAA "bought" it by certifying it—and thereby hangs a tale of inherent conflict of interest.

As far back as the early sixties it became apparent that there was no way the FAA's engineering experts could keep up with the latest state-of-the-art changes in aeronautical manufacturing so as to properly rule on each and every design that the manufacturers concocted. It would take an army of FAA men, and who would train them? The technology they would be called on to inspect would not exist at the time they might be in training. They would always be one technological step behind.

The only sensible method, it appeared, was to let the aeronautical engineers who were actually *designing* those latest developments inspect and rule on their safety themselves. Of course there had to be controls on that process, so the concept of the FAA designated engineering representative was created.

A DER would work for Boeing or Douglas or Lockheed (or any other aircraft manufacturer) in designing and building aircraft and their components, then change hats, taking upon himself the stern fiduciary responsibilities of an FAA inspector, and essentially inspect his own work. The potential conflict of interest was always obvious, but the program nevertheless worked with a surprising degree of reliability—especially when there was no adverse pressure.

When, however, the program was used in the building of general-aviation airplanes by manufacturers whose very existence might be threatened by the adverse ruling of one of its own employees acting as a DER, the pressure became incredible—and the system sometimes broke down. DERs sometimes let bad designs or marginal designs slip by to help their company. The major airline manufacturers (such as McDonnell Douglas), however, would in theory be far too professional to fall into the same trap.

3. This was known as the "size effect," which aircraft certification authorities in the Netherlands identified as potentially disastrous as early as 1969. The Dutch attempted to get the American FAA or the International Civil Aeronautics Organization or their counterpart agencies in other countries interested in their concerns, but no one wanted to listen—especially the manufacturers who shuddered at the thought of putting more blow-out panels or vent holes in the cabin floors of their jumbos. The disastrous nature of the "size effect" depended on doors coming off in flight, which was not supposed to happen.

But in the late sixties Douglas was in trouble. Sales and production of its main jetliner, the DC-8, were winding down, and with the Lockheed and Boeing jumbos nearing the market, Douglas had to act fast. The pressure on the McDonnell Douglas DERs was substantial. They were not blind. They could see who really signed their paychecks, and though most Douglas DERs were totally professional and responsible people, their futures were tied to Douglas. The effects of the heavy economic pressure were predictable: Questionable designs slipped through some Douglas DERs.

More startling to those who trusted the FAA's system to weed out the unsafe and the marginal, the pressure seemed to be more than internal.

The Los Angeles office of the FAA—the Western Region headquarters—was responsible for the certification of the airliner. Its overall boss, FAA Administrator John Shaffer, was also a Californian, and the McDonnell Douglas Company was based in that state. Even the president of the United States at that time, Richard Nixon, was a Californian. There was evidence to suggest that the designs of DC-10 were subjected to less scrutiny than they should have been as a result of concern that a California company might not get to the market fast enough with a California product unless the FAA helped out just a bit.

In the case of the not-so-sterling design of the cargo-door latching mechanism, Douglas (and the DERs) had full reason to know what might happen if the design was certificated: On May 29, 1970, while a preproduction DC-10 fuselage was being pressure-tested in a huge water tank, an improperly closed forward cargo door blew out explosively, and part of the floor collapsed.

Despite the immediate flurry of activity that resulted from the blowout in the test tank—hurried consultations and conferences between engineers of Douglas and the subcontractor, Convair, to decide what was wrong and how to cure it—only a series of "quick-fix" changes to make the latching mechanism "idiot-proof" were adopted instead of a substantive redesign. Douglas, against Convair's advice, kept the same basic latching mechanism, which would eventually and predictably be forced closed by the American Airlines baggage handler in Detroit, and come to grief minutes later over Windsor, Ontario.

Following the "Windsor" incident, as it was quickly labeled, the National Transportation Safety Board sent the FAA a list of urgent recommendations, chief among them the request that they take the proper steps to "fix" the DC-10 cargo-door latching mechanism and make it impossible for a baggage handler or anyone else to get the idea that the door was properly closed and latched when in fact it was not. (In the Windsor incident, even the DOOR OPEN warning light in the cockpit had been extinguished by the act of forcing the door lever closed. The pilots had been given no clue.)

Administrator John Shaffer's FAA was already looking into the problem. The chastened men of the FAA's Western Region office in Los Angeles—the team that had certified the DC-10 and let the poorly designed door-latching mechanism slip through in the first place—were already drafting an airworthiness directive (AD), which would have the force of law, requiring all U.S. carriers to immediately fix the problem through a series of modifications. That, of course, was the proper solution. Politics, however, as well as the product pride and loyalty of the McDonnell Douglas people, got in the way.

On Friday morning, June 16, 1972, the head of the FAA's Western Region, Arvin Basnight, received a shocking phone call from none other than the president of McDonnell Douglas Aircraft Company, Jack McGowan. McGowan told Basnight that on the previous day he and John Shaffer, the FAA administrator, had agreed over the phone that the FAA would stand down on the idea of issuing an embarrassing AD, and Douglas would instead take care of the problem quietly and internally through the issuance of a Douglas Alert Service Bulletin to the airlines then operating DC-10s in the United States. It was to be, McGowan told Basnight, a simple gentlemen's agreement between the company and the government to get the door fixed systemwide. It was the very essence of government-industry cooperation—and the very essence of political collusion, all to save the DC-10 from taking on any undue stigma.[4]

Basnight was shocked. The situation obviously required an AD, and his men were hard at work finalizing one at that very moment. In fact, one of his men had discovered that close to one hundred complaints from DC-10 operators about door-closing difficulties had been received—and suppressed from FAA scrutiny by Doug-

4. A certain amount of paranoia on the part of manufacturers over the potential of having the public perceive one of their creations as unsafe is not at all unusual—or unreasonable. In 1984 the Lockheed Electra, a technological bridge between the days of the pistons and the first pure jets, gained a terrible stigma due to multiple crashes caused, in part, by the effects of a malfunction of the turboprop engines on the technologically advanced rigid wing spar. Many passengers openly avoided the Electra long after the problem had been cured.

las. If that was their attitude, how could a gentlemen's agreement get immediate action, especially since the issuance of an Alert Service Bulletin was a request for totally voluntary action by the airlines. With an AD, they either complied or were grounded.

The veteran FAA man checked with his superiors in Washington, and though he was never told *not* to issue the AD, the message was loud and clear. In the language of professional bureaucrats, the oblique suggestion of a superior in a politically powerful position can have the force of law. Basnight, unsure, hesitated. By early the following week, Douglas had already issued its bulletin, and the FAA men in Los Angeles gave up. No AD was issued. The fate of future DC-10 passengers all over the world would be addressed solely by a voluntary compliance request in a Douglas Alert Service Bulletin. McDonnell Douglas, through some fancy political footwork and a gentlemen's agreement, had quashed the AD.

The Douglas bulletin, and several more that followed, attempted to fix the door with a series of changes that essentially made a bad situation worse. One of the changes involved the installation of a tiny, one-inch-diameter window in the door, designed to permit a baggage handler to view the locking mechanism to make sure it was *really* secure. That was, plainly and simply, a human-factors fiasco of the first magnitude. The window was too high to be seen without a stand, it was too small, and after a year in the system, few if any baggage handlers or mechanics had any idea what the mysterious little window was for, or how to use it.

The second change involved a small plate that was to be attached to the locking mechanism on the inside of the door. The presence of the plate would change the locking mechanism's mechanical advantage so that more than 440 pounds of pressure (rather than the previous 150) would have to be applied by anyone inadvertently trying to muscle the door closed before the latches were properly engaged. It still didn't correct the basic flaws in the design of the door, but it would eliminate the immediate problem: allowing someone to force the door handle to the closed position and put out the cockpit DOOR OPEN warning light with the latches only partially engaged. The problem, said Douglas, would not occur again. The NTSB's recommendation had been implemented.

But all the information on the fixes that were to prevent a recurrence of Windsor had been kept so quiet that the very individuals who would be closing the door over and over again in actual day-to-day airline service were never adequately informed. That fact would later become the stuff of scandals, but what occurred next, two months after the issuance of the Douglas bulletin, was beyond belief—and sealed the fate of 346 human beings.

Somewhere in the cavernous confines of the Douglas factory in Long Beach that fall of 1972, one of the Douglas inspectors who was charged with the responsibility for verifying the installation of that little metal plate inside the cargo doorframe did a very human thing: He lied. The inspector certified the plate's installation without ever actually verifying that it was installed. The little plate, however, had not been installed, and its installation was one of the key modifications recommended by Douglas to fix the cargo-door problem. Without that plate, any other baggage handler could overpower the door-latching mechanism just exactly as the American Airlines employee had done in the Windsor incident.

That particular DC-10 was delivered in December 1972, and the following year entered service with Turkish Airlines.

The FAA had been complicit in failing to issue an AD because Douglas was going to correct the problem itself. It was going to take care of the DC-10s then flying for various airlines (and thus out of Douglas's control) by issuing voluntary compliance recommendations to those customer airlines, and *assuming* they would make the changes.

But the "voluntary" nature of those bulletins concerned DC-10s that had left Long Beach and Douglas possession. It was inconceivable that DC-10s then on the Douglas production line would fail to get the same treatment. But that's exactly what had occurred.

On March 3, 1974, while climbing out of Paris on the way to London, Turkish Airline Flight 981, the very same DC-10, lost its pressurization, six passengers (sucked through the hole in the floor), and the integrity of its cabin floor when the cargo door, which had never received the required metal plate, blew out with explosive force. Predictably, the collapse of the cabin floor severed the control cables, though this time, all of them. The pilots were left with virtually no control over pitch and yaw, nor over the tail-mounted engine. One minute and seventeen seconds later the huge jumbo jet from Long Beach, traveling at nearly 500 miles per hour, slammed into the sturdy Ermenonville Forest northeast of Paris in a shallow trajectory, shredding in an incredible explosion of destruction the lives and bodies of the remaining 328 passengers and 12 crew members.

On the ramp back in Paris, a hardworking Turkish baggage handler had struggled mightily to close the balky cargo-door lever—and had finally succeeded.[5]

The response back in Washington, D.C., at the National Transportation Safety Board was shock. C. O. "Chuck" Miller, the chief of the NTSB's Bureau of Aviation Safety, began getting feelings of *déjà vu* from the first news of the accident. When he also heard that pieces of a small cargo door and six bodies had been found at a location miles distant from the main crash site, he knew instantly that history had repeated itself.

Even as he headed to Paris, Chuck Miller understood that Windsor had been just a dress rehearsal for the Paris tragedy.

But the NTSB had recommended immediate changes in that DC-10 door through the issuance of an AD after the Windsor incident. Surely that AD would have been issued. Wasn't it followed? Why hadn't it been done? Why didn't the NTSB know?

From Paris, where Miller's NTSB team (headed by one of his subordinates, Doug Dreifus) began participating in the French investigation (the French had invited the Americans as a courtesy because it was an American airliner, but they were *not* in need of American assistance, thank you), Chuck Miller phoned back to his staff people in Washington to find out what had happened with the AD.

The answer was deeply upsetting. No AD had been issued, and his staff had not found out. The NTSB recommendation from Windsor had been circumvented, and they had not discovered that fact.

To Chuck Miller it was more than a case of flawed follow-up. The entire reason for the NTSB's existence was to prevent accidents by making affirmative recommendations for change. It couldn't compel change, but it could keep pecking away at the FAA to take action—*if* there was an adequate follow-up program to track compliance with NTSB recommendations. Miller had set up such a program several years before, but his budget had been slashed and the program officially killed (over Miller's impassioned objections) by the new general manager of the NTSB, one Richard Spears. Miller had tried to keep the program going anyway, but as an "extra duty" assignment for his people, who were now com-

5. The entire story of Windsor, Ermenonville, and the DC-10 has been told with great accuracy and feeling by Moira Johnston in *The Last Nine Minutes,* which has been a substantial source of material verification for this chapter (William Morrow, 1976).

plaining that they had been too overworked, understaffed, and undersupported to carry out the follow-up in the case of the Windsor recommendations.

In Miller's view, Richard Spears was a marginally qualified political appointee who had been installed by the Nixon White House in 1972 to "keep an eye" on the NTSB (and presumably to keep the theoretically independent agency in lockstep with administration policy.)

Miller and his people had been working hard to upgrade the function of the NTSB. Historically, it had been merely the agency that discovered what caused a crash and made recommendations to the FAA (or other affected agencies) based on those discoveries. Miller wanted it to evolve into an agency that concentrated at least as much effort as it put into investigations on programs to prevent recurrences by aggressively monitoring and dogging action on NTSB proposals. Chuck Miller had spent years working as a human-factors and safety-engineering specialist for Chance Vought Corporation (later Ling Temco Vought, and part of the LTV conglomerate in Dallas) before joining the NTSB, but his ideas about preventive programs of "system safety," incorporating a profound understanding of human factors, were, sadly, ahead of the times even in 1974. It was ironic that Miller had wanted to have some of his people participate in FAA meetings monitoring the development of the DC-10 while it was still on the drawing board.

His immediate problem at the NTSB, however, was ignorance. The White House's inside man, General Manager Spears, seemed the antithesis of a professional accident investigator or aviation expert, and he had been systematically stripping Miller—and the presidentially appointed board members themselves—of their budget and the NTSB's autonomy. Miller and others suspected that Spears was simply one of a number of Nixon White House "agents" slipped into seemingly independent oversight agencies like the NTSB to blunt any criticism of the Nixon administration. There was little other explanation in Miller's view. The NTSB was having its budget cut back constantly, its manpower reduced, and its more vociferous members muzzled. Even members of the board themselves had been warned in highly improper fashion that their reappointment to office might be blocked if they were disloyal. So much for autonomy.

By the time of the Ermenonville tragedy, congressional hearings were already scheduled that would expose the White House manipulation, even as the Watergate scandal approached its denouement, and lead to passage of a new piece of legislation with

144

profound impact on the NTSB—good and bad—in the following decade: The Independent Safety Board Act of 1974.[6]

Ermenonville focused world attention on one major weakness of the DC-10 design: the vulnerability of the cabin floor to collapse from an explosive depressurization. That vulnerability, in turn, had been exposed (with murderous results) by another weakness, the Rube Goldberg door-latching mechanism. The question was, what other weaknesses lurked beneath that shiny metal skin? What other design or conceptual flaws were allowed to slide past the FAA's watchdogs? How many more parts had been inspected and signed off—but were not installed?

In the leading edge of the wing of each of the shiny new DC-10s that had been rolling steadily out of the Long Beach factory was another flawed system, which it would take the catastrophic failure of still another system to uncover. The leading-edge devices—the sliding surfaces that droop down from the front of a DC-10's wing to give it extra lift at slower airspeeds and thus allow it to fly more slowly on approach and departure—could become asymmetrical, leaving one wing with more lift than the other. The result could be loss of control, and at low altitude, that could mean disaster.

In order for such a split to occur, the main hydraulic system on one side (the LEDs were powered in and out by hydraulic pressure), *and* the backup hydraulic system (common to both wings), would have to fail simultaneously. That would seem unlikely, but Douglas had designed in one more vulnerability. The plumbing for both hydraulic systems ran through the leading edge of each wing.

6. The Safety Board Act, (which was strongly supported by the Air Line Pilots Association) helped gain tangible independence for the NTSB as an ombudsman agency with license to attack and critique any facet of government, but it also mandated the NTSB's involvement in transportation crashes, wrecks, and disasters of all types without providing the increase in money and manpower to do the job. The effect, ironically, was the opposite from what Miller had intended. Instead of beefing up the Aviation Safety function, it reduced it dramatically as Miller's people were pulled away to work on rail, sea, pipeline, and automotive mishaps.

Ermenonville would be Chuck Miller's last accident as an NTSB investigator. Suffering from a cardiac problem, Chuck Miller left the board in late 1974, and with him went one of the best and brightest hopes for early NTSB understanding of human-factors and human-performance failures. It would be the summer of 1979 before a glimmer of such understanding would emerge as an ember in the halls of the NTSB's Washington headquarters. More specifically, it would take the death of Jim Merryman, the dogged persistence of Al Diehl, Steve Corrie, and Gerrit Walhout, and the amazing revelations of the human-performance investigation into the Downeast crash to make the first real converts.

If any damage occurred that could sever one line, it could just as easily sever both.[7]

Boeing had designed the 747, 737, and 727 with internal locking devices to keep extended LEDs from retracting with the loss of hydraulic pressure. In addition, it made sure that whatever powered the left side powered the right side. Even if the hydraulic pressure does fail, the LEDs stay in place, and each wing produces the same amount of lift. (Similarly, Lockhead designed lockdowns on its L-1011.)

Douglas, however, somehow got by with a system that backed up the hydraulics with still more hydraulics, and had no locking mechanism to make sure that each wing kept the same amount of lift. It got by with it, that is, until the bloody afternoon of May 25, 1979.

By the spring of 1979 it had become apparent to McDonnell Douglas that some DC-10 operators were using the wrong method of removing the DC-10's wing-mounted engines from the wings for replacement or maintenance work. Instead of first removing the engine from the pylon (the metal structure that attaches the pod engines to the underside of the wings), some were using forklifts to hold the engine while the maintenance team detached the pylon from the wing. Then the team would lower the engine and the pylon as a unit on the arms of the forklift.

But that tended to crack the metal attachment bracket on the

7. The philosophy of building fail-safe airliners under the U.S. system includes the caveat that airliners should never be designed with a potential for unequal lift. If the left wing is producing a certain amount of lift, the right wing should produce exactly the same amount. Otherwise the airplane can become uncontrollable. Since modern, swept-wing jetliners have sophisticated panels that move back from the trailing edge of the wings (flaps) to enlarge the wing area and increase lift, and many have similar devices on the front of the wing (leading-edge devices), it would obviously be a problem to have the flaps and LEDs on the left wing come out while the ones on the right wing stayed in. In the parlance of pilots, that could "ruin your whole day." The left wing would produce more lift than the right wing, and the pilots would have a serious battle to keep the wings level by using the flight controls (ailerons and rudder).

On many jets, as with the DC-10, these lift-enhancing devices are operated by hydraulic pressure. As long as the pressure is normal, the LEDs and flaps will stay out. But what happens if hydraulic pressure is lost? Without some method of mechanically locking the devices in place, they could retract involuntarily and at different times, causing control problems.

Therefore it is quite important that whatever powers one side powers the other. It would be better to lose power to both sides simultaneously (as with a single hydraulic system and no locking devices) than to lose power to only one side.

top of the pylon, which actually mated to the underside of the wing. It seemed that the weight of the engine and pylon together could not be properly handled with a forklift.

So, as in thousands of earlier cases involving maintenance changes, Douglas issued another Service Bulletin, one of which went to American Airlines' Maintenance Base in Tulsa, Oklahoma. Somewhere in the beehive of activity, the bulletin's warnings got lost in the shuffle. American's mechanics kept using the same, speedier method with the forklift.

On Friday afternoon, May 25, 1979, as Downeast Chief Pilot James Merryman returned from Hyannis, Massachusetts, with the Twin Otter in which he would lose his life five days later, American Airlines Flight 191 rotated from Chicago's O'Hare runway 32-right, and the left wing-mounted engine tore loose. The rear casting had been cracked by American maintenance personnel in Tulsa utilizing the wrong method—a forklift—in removing the engine/pylon combination, EVEN THOUGH THE SERVICE BULLETIN HAD BEEN RECEIVED AND ACKNOWLEDGED.

The engine swung up over the wing, passed over the top surface with the damaged pylon cutting into the leading edge of the left wing, then fell to the runway behind the jumbo jet as it lifted from the runway surface.

O'Hare's tower controller keyed his microphone and asked, "Ah, American 191, do you want to return?"

The controller had seen the engine fall to the runway. The flight crew had not. The pilots can't see the wing-mounted engines from a DC-10 cockpit. The flight crew of American 191 did not realize that they had physically lost an engine—just that it had failed.

The departed left engine had contained the pump for hydraulic system number one, which operated the left wing's leading-edge devices and kept them extended. With that hydraulic pump and its system pressure gone, the left wing's leading-edge devices would stay extended with what is known as "trapped hydraulic pressure" as long as all the hydraulic lines were not cut. If the left engine's hydraulic-system lines to the LEDs were cut, the tail-mounted number-two engine's hydraulic pump (which powered the emergency hydraulic system) would keep the left wing's leading-edge flaps extended. But there had been yet another compromise in the design of the DC-10. The plumbing for both hydraulic systems had been routed too close to the leading edge. The pylon, as it ripped away, cut the lines of both hydraulic systems and spilled the "trapped hydraulic pressure."

Without a lockdown system to keep them extended, the leading-edge devices began retracting as Flight 191 gained some altitude in what should have been a routine, engine-out departure. Pilots are constantly trained to treat the loss of a critical engine on takeoff as a routine occurrence, though in fact it almost never happens. American's crews had also been taught to trade airspeed for altitude, a standard practice, so the crew of Flight 191 obediently let the airspeed bleed off to 159 knots as they gained altitude, rising to 325 feet above the airport surface.

One hundred fifty-nine knots was a safe speed for the right wing with its leading-edge devices still extended and powered down by the hydraulic system on the operating right engine (number-three engine). The left wing, however, was suddenly different. It had no extended leading-edge devices. It could not fly at 159 knots; 159 knots would not keep it level.

Suddenly, Flight 191's left wing was stalling, and the jetliner began a slow roll to the left, which steepened into a vertical bank and continued as the jumbo came down with thunderous force in a vacant lot, upside down, doing over 165 miles per hour, its crew fighting to correct a situation that made no sense to them. Two hundred fifty-eight passengers with not a clue as to why the ground had suddenly begun rushing up at them, 13 crew members, and 2 nearby residents in the wrong place at the wrong time died in the massive impact. The improbable, once again, had occurred.[8]

8. The deaths of 273 members of the human family in the impact of American Flight 191 was followed by the worst possible public-relations disaster for McDonnell Douglas.

Three days after the crash, on May 28, 1979, the FAA Western Region issued an AD requiring visual inspections of each wing-engine pylon. The next day, twenty-four hours before the crash of the Downeast Twin Otter in Rockland, Maine, the Western Region issued a second AD requiring further inspections. During all the flurry of ADs and inspections, the DC-10s kept flying. By June 4 the target of the inspections had narrowed to the engine-pylon removal sequence, further inspections of certain Model DC-10s were directed, and cracks began showing up. On June 6, 1979, FAA Administrator Langhorne Bond grounded all DC-10s, citing as justification the suspicion that the DC-10 might not be capable of safe flight. Specifically, Bond's Emergency Order of Suspension said in part: "The Administrator has reason to believe that the Model DC-10 series aircraft may not meet the requirements of Section 603(a) of the Federal Aviation Act for a Type Certificate in that it may not be of proper design, material, specification, construction, and performance for safe operation, or meet the minimum standards, rules, and regulations prescribed by the Administrator."

On July 13, 1979, Bond lifted the suspension, but the damage had been done. The airlines and the manufacturer had lost untold millions. In the interim, some of the finest minds in commercial aviation and manufacturing were

A flawed design, passed by FAA designated engineering representatives and FAA inspectors under political pressure, exposed by a procedural flaw brought too quietly to the attention of the airline, whose personnel had ignored it and whose supervisors had failed to catch it, had formed a destructive chain, which ended in death and space-age rubble strewn along a patch of Chicago real estate. American 191 was, above all else, a very human disaster. It resulted from the inherent propensity of humans to fail, to take shortcuts under pressure, to be unduly influenced by company loyalties, and to do the wrong things for the best of reasons. Even though there was no pilot error involved, there was plenty of human error. It was a series of mechanical flaws brought about by multiple human failures, and it spotlights the fact that the understanding of (and allowance for) human performance—human nature—in *all* aspects of the airline system is vital.

There are times when all the best intentions of the best people working for the best of airline manufacturers are not enough to overcome the desire to protect the company interests. Permitting the aircraft manufacturers to decide what is a vital safety matter and what is not can be a bit like letting the fox guard the chicken coop—the temptations can overpower the noblest of intentions. Only the strong and continuous—almost brooding—presence of an impartial and nonpolitical FAA can come close to guaranteeing that the temptations won't win out.

By the same token, establishing a system of designated engineering representatives—letting the industry monitor itself—without maintaining the strongest of outside FAA controls to enforce its honesty is as potentially flawed a concept as trusting a salesman for a critical evaluation of his own product.

The principle is as valid throughout the airline industry. Wherever airlines large or small are allowed to police themselves without adequate FAA supervision, surveillance, and inspection, the temptation to cut corners and backslide on safety precautions—the temptation to interpret the rules in the least costly manner—counteracts the best of intentions. The greater the economic pressure on an airline to cut costs, the greater the backsliding. The less experienced the personnel (as in new and rapidly expanding carriers), the less extensive the safety margin—especially where pilots are concerned. By August 1979 the chaotic results of too much expansion and too little surveillance were becoming critical, and in the case of one upstart carrier in Miami, painfully obvious.

---

brought together on an emergency basis to evaluate the DC-10 and determine whether or not it was safe to continue in service—something that should have been done during the design phase in 1968 and 1969.

Chapter 10

FILLING THE SQUARES

Al Koleno was astounded! He had known before joining Air Florida that he would have his work cut out for him, but this was incredible. If he were still an FAA inspector, he mused, he would have no choice. He would have to shut the airline down.

Just one month before (in July 1979) Koleno had been comfortably engaged in the business of being an FAA air carrier inspector for well-established Piedmont Airlines in Winston-Salem, North Carolina. Suddenly one morning, like a bolt out of the blue, Dick Skully, the vice-president-operations for Air Florida, had called and asked him to fly down to Miami for an interview. It seemed Air Florida needed a chief of training for the pilots.

Skully was an old FAA friend of Koleno's, a veteran FAA man who had risen from assistant administrator of the FAA Academy in Oklahoma City to become one of the most important FAA chiefs in Washington. But Dick Skully, an administrative type all his career, had received a better offer in 1978. Skully had not enjoyed the best of all possible relations with the FAA rank and file, so the offer came at a propitious time. He promptly retired from federal service and took on the rather exciting job of operations chief for

Air Florida, an aggressive little upstart carrier in Miami. The, rapidly expanding airline was being led from the woods of intrastate regionalism to the status of big-time interstate carrier by one of the most politically adept executives in aviation, former Braniff president C. Edward Acker.[1]

The planned growth curve for what had been a tiny in-state Florida commuter carrier was breathtaking. Suddenly Acker and the firm's founder, Eli Timoner, a wealthy Floridian, had inflated the company to the status of a full Part 121 carrier with Boeing 737s and Douglas DC-9s carrying passengers hither and yon just like the big boys, eyeing much greater expansion under the alluring temptations of the Airline Deregulation Act of 1978. To Air Florida, the fruits of deregulation—the newly available routes—hung there in the breeze of optimistic rhetoric about free-market competition, ripe and tempting. The thought that some of that fruit might be unpalatable—financially poisonous, in fact—never stemmed the rush to expand, despite the operational problems and training deficiencies that would sow the seeds of a bitter harvest.

Skully had warned Koleno that Air Florida needed help in organizing the ground-training function, but he had not prepared the veteran air carrier inspector for what he would find in his first excursion to the other side of the fence: from inspector to potential inspectee. Koleno would be charged with bringing the training program up to the point where it could pass an FAA inspection— something it already should have been able to do.

But as Al Koleno sat in his comfortable new twelve-by-twelve-foot office in the rented building complex on Thirty-sixth Street near a corner of Miami International Airport, he knew he had joined a carrier in deep trouble. The box in front of him contained the training records of all Air Florida's pilots.

A box.

Not a file cabinet, filled with properly indexed and updated forms tracking what training the pilots then flying the line had received and when, but a pitiful, half-filled, dogeared cardboard file box. A ridiculous repository containing a mishmash of incomplete training forms and scribbled records so far from legally sufficient that they would have turned the stomach of any responsible FAA air carrier inspector.

1. Acker, who had executed a timely departure from Braniff in the face of a South American ticket scandal, had spent a couple of years outside the airline business before returning as head of the Miami-based company.

Of course Koleno, at heart, was still an uncompromising air carrier inspector, and his stomach was indeed turned.

The problem was growth: rapid, urgent growth of the tiny air carrier's route system into new areas of Florida and surrounding states with a growing number of personnel located in different places. There were tremendous demands on the attention of its small (and somewhat inexperienced) middle-management force, which was suddenly inundated by unforeseen and sometimes trivial details, the inevitable effects of inflating a little company—or a little airline—into a big one.

Acker and Timoner, chairman and president respectively, were both rather autocratic (though Eli Timoner had a reputation as an empathetic and somewhat "softhearted" employer). Acker, a Dallas financial expert by training, had learned the labor-intensive business of airline management at the side of Harding Lawrence of Braniff, while Timoner had always been his own boss. Both men were possessed of forceful styles and short tempers.

Since the airline had a small executive force, authority at Air Florida flowed directly from Acker and Timoner, and accountability flowed back in the same direction. There was immediate hell to pay for creating any problem that came to their attention, and that included the sins of getting flights out late, or causing delays in the system. Air Florida held a "stand-up briefing" every morning in the conference room adjacent to Al Koleno's office, and it was usually a "pin the blame on the donkey" session for whatever had gone wrong the previous day.

To that extent, all the other executives—including Skully (despite his former bigwig status at the FAA)—were running scared. Though no one would accuse either Chairman Acker or President Timoner of being blasé where safety was concerned, few in the organization wanted to march into their offices and tell them the real truth of Air Florida's position in August 1979: It was expanding too fast to be *as* safe a carrier as the public had an absolute right to expect. Too many things were left hanging—like the box full of licks and promises sitting in front of a crusty old ACI like Al Koleno.

As Koleno had already discovered, the story went far deeper than nonexistent training records. If he were still an FAA man, he could probably shut them down then and there, just on that evidence alone. But the pitiful contents of that box represented just the latest problem. Since he had arrived in mid-August, the shock of how disorganized and unprepared this company was deepened daily.

His arrival had been tragicomic. Koleno had been greeted on the doorstep of his new employer by VP Skully, only to be interrupted by an Air Florida pilot.

"Do you gentlemen know where the recurrent training class is this morning?"

Skully, the operations chief, looked puzzled.

"Ah, no, I don't."

As they stood there and mulled it over under the hot morning sun of a muggy Miami summer day, it became apparent that Dick Skully did not even know what room such a class would be in if it was being held, or who might be there to teach it, for how long, or concerning what aircraft.

Koleno spent the first two hours of his employment with Air Florida trying to find someone to teach the course the pilot was looking for, and finding a room to teach it in. No one at the headquarters of the airline had any idea who might be in charge of such things.

"Well, Al, that's why we need you, I guess," Skully had said, somewhat embarrassed.

The rather disturbing truth imposed itself in stages on Koleno, who couldn't keep from being amazed. After all, Air Florida was not a paper airline just getting ready to start operations (or a 135 carrier trying to upgrade); this was an operating airline under Part 121 with authority to carry passengers at the same level of trust granted to United or American, yet no one was in charge of something as vital as aircrew training!

In fact, no one had been specifically assigned to any aspect of training, including the scheduling of training courses or the tracking of training requirements. Chief Pilot Dave Mulligan had been in the habit of simply asking pilots at random if they would volunteer to teach whatever ground course needed to be taught. No background necessary, no preparation necessary, no training aids readily available, and seldom a classroom to use.

But the squares had to be filled. The FAA required recurrent (or upgrade, or transition) training, so that's what they were going to do—even if the pilots were self-taught.

In fact, even the training for the airline's stewardesses in vital emergency-evacuation procedures and other required subjects was taught by whatever stewardess might want to teach a course that day—there were no paid, full-time instructors.

In truth, Air Florida had hired an excellent cadre of pilots as it rushed to expand. Its standards were high (three-thousand-hour minimum flight time at one point), and many of those who came on board had air force or navy experience flying thousands of hours as pilots of heavy turbojet or turboprop aircraft. That sort of excellent preparation for an airline career would qualify them to transition to any Boeing or Douglas airliner.

And there were pilots with extensive academic and engineering backgrounds, such as Captain Jim Marquis who held a master's degree in aeronautical engineering and had worked for the Chance Vought Corporation for years before coming to the airlines. (Marquis usually taught the courses on aircraft performance.) High-quality people like Marquis were the backbone of the Air Florida pilot corps, as well as the backbone of the pitiful aircrew training "program," which was barely functioning when Al Koleno arrived in Miami.

Air Florida's flights arrived and departed without incident during that period because of the extensive capability and experience the pilots brought with them when they joined the company, not because of training acquired at Air Florida. For the most part, they were professional pilots who had enough pride in their professionalism to police themselves.

But there were some glaring exceptions.

In the drive to expand, Timoner and Acker had acquired a small Key West commuter airline called Air Sunshine in early 1979, and as part of the deal agreed to take the Air Sunshine pilots—provided they came up to Air Florida "standards." Unfortunately, there was sufficient confusion over what Air Florida standards really were (and enough pressure on Dick Skully's Flight Operations to take on the Air Sunshine pilots) that some very marginal people slipped into Air Florida cockpits among the fourteen out of thirty-two Air Sunshine pilots who were brought on board. They may have been good commuter pilots of piston/propeller-driven aircraft (up to and including the DC-3), but as a group they had had little or no exposure to turbojet airliners such as the Boeing 737 and the Douglas DC-9, and little military experience.

The act of hiring Air Sunshine pilots should not have presented a problem. It should not have compromised anyone's safety.

Any group of conscientious, professional pilots with decent ability, regardless of their commuter background, could have been trained into a higher class of airliner and have become excellent air-carrier captains—eventually. That sort of metamorphosis, however, requires time, and more important, it requires training—continuous, professional, dedicated, carefully structured, and expensive training. There are no acceptable substitutes for that

154

training if such people are to be turned loose to fly around as airline pilots, entrusted with live passengers who have every right to expect to stay that way.

Above all else, the Air Sunshine people had needed extensive training in how to be responsible pilots for a growing Part 121 airline. They needed serious instruction on how to fly an airliner as an integrated, cooperative crew, rather than permitting the captain to operate the aircraft as if he were alone in a single-engine airplane (too often making his own decisions with a fool's confidence and ignoring his copilot).

They needed indoctrination, serious and extensive, into the weighty responsibilities of being an airline captain responsible for hundreds of lives. (Air Florida was soon to be operating three-hundred-passenger DC-10s to Europe.)

And most important, they needed a well-organized, professionally polished training department to teach them in no uncertain terms that to fly for Air Florida meant to adhere faithfully to a higher ethic than just company loyalty. They should have been taught that the company expected them to obey the Federal Air Regulations in all areas and at all times—and especially to follow the rules.

As Downeast and many others had proved too many times, commuter carriers were not the best of environments in which a young pilot could learn the principle of rigid compliance with the rules, and Air Sunshine's pilots (with some notable exceptions) had been steeped in the realities of financially pressured commuter operations. In that world, all too often, pilots "carry the aircraft on their backs" and make their own rules to get the job done.

And they should have been taught to follow standard procedures from carefully written, standardized flight manuals (which, of course, Air Florida did not possess). They needed all this as a bare minimum, but the pilots from Air Sunshine didn't get it—because in the first half of 1979 Air Florida had no training department equal to the task.

Had they joined the ranks of a major Part 121 carrier, such as United or Eastern, in a similar shotgun wedding, the Air Sunshine pilots *would* have received the proper training. They would have entered a carefully structured world of ground-school classes, qualification check rides, and professional instructors, and would have begun their careers in the flight engineer's seat of (most likely) a Boeing 727. After a sufficiently lengthy period (and once their seniority number permitted advancement), they would have progressed individually to the position of first officer/copilot. By that time, the professionalism drummed into them by the training, the safety-awareness functions of their pilot union, and the peer pressure of the seasoned professionals around them would have molded them into the type of professional pilot who belongs in the cockpit of a turbojet airliner; or the process would have identified them as incompetent and dismissed them.

Being hired by a major carrier and integrated into its pilot force is, in effect, a natural, evolutionary method of weeding out most of the "cowboys" and other pilots who simply can't make the transition from being a seat-of-the-pants pilot to mature and responsible airline pilot.

But there's more to the process. Had they been acquired by a major, established airline, the Air Sunshine people would have been exposed to a vast array of good and bad weather and problem flight conditions. They would have seen how a captain handles icy runways at Washington National, in-flight ice forming on the wings in a holding pattern over JFK, or the use of engine anti-icing systems to keep his engine instruments accurate. They would have experienced firsthand the process by which a captain decides conditions are unsafe and the flight will not depart, period—despite the fact that his company may be on the financial ropes and need the revenue from that flight. They would have witnessed the transition that a responsible and experienced aircrew makes when it begins a takeoff roll—realizing the seriousness of the act of takeoff, understanding that if anything is significantly wrong or suspect, their airliner full of trusting passengers must not be launched into the air.[2]

The Air Sunshine pilots would have experienced all those things and more with a major Part 121 carrier (prior to deregulation) before ever having the opportunity to fly as captain, and they needed such experience to make the transition. At Air Florida, what they got was little more than what the law required, because the airline was too busy expanding to concentrate on the finer points of aircrew training.

Yes, they had a cadre of highly qualified pilots of substantial experience and impeccable ethics who were involved in their flight training. And yes, they did have an official company attitude that promoted strict adherence to the rules. Certainly they were

2. The go/no-go decision on takeoff can be one of the most critical moments in flight. Sometimes it is far safer for the pilots of a large airliner to continue a takeoff than to abort it (while the opposite response might be best for a smaller aircraft).

checked out for the most part in accordance with FAA rules and procedures. They were technically legal—but legal was not enough!

What the largely inexperienced commuter pilots from the little Key West airline did *not* get (a thoroughly professional training course, a proper indoctrination to Part 121 responsibilities, and the seasoning, maturing process of experience through gradual advancement-under-scrutiny) prevented a sure, safe transition from the habits of pilots for small-sized commuters to the ethics of the turbojet airline pilot.

Some made it anyway, through intelligence, native ability, and extensive help from their peers. Some, however, did not. It would be one of the latter group who would bring the entire airline to the brink of ruin.

There were those who knew as early as 1979 that some of the Air Sunshine group were strangers in a strange land, brought on board with no guide, and speaking, in effect, a different language. There were those who lobbied the management to do something about it but who were ignored. The goal was rapid expansion, and nothing could be allowed to retard that goal.

The Air Florida economic master plan was not to be disturbed by mere glitches in bringing the training department up to speed or in meeting all the FAA-imposed requirements, even though by summer of 1979 Air Florida had just hired more than a hundred new pilots in addition to the Air Sunshine group, most of them highly qualified, but all of them requiring considerable training. Much of that training would have to be done by other airlines in their flight simulators and in airplanes under contract to Air Florida, which was where Al Koleno had come in, so to speak.

At Piedmont, in his capacity as an FAA ACI, Koleno had given numerous type-rating check rides to Air Florida pilots whose company was paying Piedmont to provide upgrade training in Piedmont's equipment.[3] Apparently Skully had become aware of

3. In order to operate a Boeing 737 or a DC-9 as captain in airline service or otherwise, a pilot must have what is known as a type rating in that class and category of aircraft. After a prescribed ground school in the systems of the aircraft, the type-rating candidate has to have (typically) time in a flight simulator with a check ride, and a check ride administered by an FAA inspector in the actual aircraft in flight. The FAA man giving the check ride must also be qualified and type-rated as a pilot in that aircraft, so he'll have a firm idea of what he's expecting the examinee to do. The type-rating system is another

Koleno's whereabouts through feedback from some of the newly rated pilots, who liked the FAA man's style. Koleno had not been fully aware, however, that Piedmont, during the same time period, had been having some troubles with Air Florida.

When Air Florida's pilots were sent to the training facilities of other airlines to take their required flight-simulator training, the normal practice was to take fellow Air Florida instructors and check pilots along as well to administer the training and check rides. In those cases, the host airline providing the training facilities wouldn't be concerned with what operations manuals the Air Florida people were using, or the exact form of checklist, or their Air Florida flight procedures. The Floridians simply came, bought time in the flight simulators, and departed.

When, however, Air Florida pilots needed such training at another airline and could not bring their own Air Florida personnel along, the host airline had to provide the instructors and check pilots. For that host airline to train the Air Florida pilots, they needed Air Florida operations manuals, checklists, and procedures. It would make no sense (not to mention being illegal) for Piedmont, or United, or American to attempt to train Air Florida pilots with Piedmont, United, or American procedures. Each Part 121 airline writes or adapts its own checklists, manuals, and procedures for each aircraft it operates, and those manuals and procedures are specifically approved by the FAA air carrier inspectors assigned to monitor that airline. Such books and procedures are always slightly different among the different carriers even for the same airliner.

The problem was that Air Florida didn't have any standard operations manuals. They had told their Miami FAA inspectors that they did (just as they certified that they had a working ground-training program), but that was a paper promise. In reality, Air Florida's airplanes had been purchased and leased from so many different operators that no two airplanes seemed to have exactly the same instrumentation, and the operations manuals that came with each airplane were slightly different. The airline had been too busy expanding to formulate a single version for everyone. In-

method of assuring that pilots who cannot handle the demands of a large turbojet airliner do not end up in command of one. Naturally, if it is administered lackadaisically by the FAA (or administered lackadaisically by a carrier's own check pilots under FAA approval—the Designated Check Airman Program), marginal pilots can (and do) slip through, obtaining type ratings that can become licenses to kill.

158

stead, the pilots were using a hodgepodge of the basic Boeing operations manual pieced together with a wild assortment of changes and added looseleaf pages that were not necessarily compatible. This meant that checklists were not exactly the same between manuals, nor were the procedures by which the airplanes were to be flown! So when Piedmont or United wanted to know what Air Florida's procedures were, in too many cases no one at Air Florida could be sure, because no one had taken the responsibility for formulating an "Air Florida procedure" to begin with. They had been too busy growing.

That, of course, was why they needed Koleno, as he now recognized. But Al Koleno kept coming back to the same chilling realization: This was not a start-up carrier. This airline was already being trusted by the general public to know what the hell it was doing.

And for that matter where in the name of regulatory enforcement were his old counterparts in the Miami office of the FAA? Where was the principal operations inspector for Air Florida, or the principal maintenance inspector?

As the first week in September 1979 came to a close, Koleno realized that the FAA men in charge of keeping Air Florida on the straight and narrow path of compliance simply hadn't the slightest idea what was going on. There was no excuse for their not finding the hideously incomplete training records (most of the training had been done in accordance with the rules, but the records couldn't prove it), no excuse for the FAA's missing the fact that Air Florida maintenance was in a terribly confused and disorganized state, working out of borrowed hangars and the corners of ramps. Nevertheless, the local FAA Air Carrier District Office sat there day after day and let the airline get by with incredible errors.

Suddenly, it became all too clear why a former senior official of the FAA could be such an effective vice-president of operations of a small carrier growing frantically under the freedom of deregulation: The man knew the ropes. Who would know better how to handle the FAA than the FAA, which was what Skully personified. The ins and the outs, the methods of influence, and the right strings to pull were simple tools of the trade for such an old hand at the game. Then, too, the local air carrier inspectors must have felt a bit of glory hobnobbing with such a high FAA official, despite the fact that his official status was a thing of the past. Men of influence remain men of influence even after the fact, especially in the eyes of those who stand in awe of power.

In addition, as Koleno was discovering, the warm, buddy-buddy relationship between the local FAA men assigned to Air Florida and the Air Florida brass was a well-nurtured affair. The resulting objectivity of the FAA's surveillance of Air Florida was, therefore, suspect by definition, and it was that surveillance and enforcement by the FAA that was supposed to be protector of the flying public—the bulwark against a lowered threshold of airline safety in the wake of deregulation's new freedoms and economic pressures.

But as Koleno knew, and as Al Diehl of the NTSB was coming to know, the role of the FAA (according to the assurances so glibly enunciated by Congress in 1978 as it threw airline regulation to the wind) was not being fulfilled. Even before deregulation the FAA had failed to catch the problems at Downeast Airlines, and now, in the very throes of the classic case of deregulatory expansionistic fever infecting the starry-eyed entrepreneurs of Air Florida, the FAA was failing to catch other problems with as serious a potential for public harm.

By September 11, 1979, a young, extremely self-confident former DC-3 captain for Air Sunshine, Larry Wheaton, had completed his required Air Florida training, and had begun flying the line as copilot on the Boeing 737. Wheaton had passed his check rides but underwhelmed his ground instructors as overly cocky. It was hard to tell Wheaton anything—he seemed to feel he knew it all. After all, he had been captain on a DC-3.

Wheaton had never flown turbojets before joining Air Florida, but had served as DC-9 copilot for 471 flight hours after being brought into the fold in October 1978. Now he had transitioned to the Boeing. Larry Wheaton knew little of crew-concept flying. He knew little of life as a Part 121 airline pilot. He did know that he wanted to get back to the left seat—to be a captain once again. He could fly the Boeing 737 as a copilot; therefore he could fly it as a captain. It never occurred to him that maybe, just maybe, there was a difference between being a commuter pilot-in-command and a Part 121 airline captain. It never occurred to him that he was not ready—and without a change in attitude might never be. As Wheaton settled into the right seat of the 737, Air Florida was making plans for service to even more cities, and Larry Wheaton would get his chance at the left seat all too soon.

Also by September 11, 1979, Al Koleno had come to understand with frightening certainty how much work would be needed to bring Air Florida up to where it was supposed to have been all along. In doing so, however, he would soon be butting his head against an unexpected wall of resistance. They had done things their own way so far and it had worked; why should they have to change? Though he didn't realize it, that very day in Boston, an NTSB hearing was beginning that would spotlight a similar— though more extreme—attitude by another organization that had held the public trust—and dropped it. The formal hearing into the May 30, 1979, crash of Downeast Airlines, the hearing that Al Diehl knew would galvanize a host of opinions about commuter carriers and titillate the press, was finally under way.

Chapter 11

KICKING OVER THE ROCK

"[Captain Fenske], with regard to pressure on pilots, were you ever pressured to make overgross [weight] takeoffs?"

The young commuter captain, formerly with Downeast, now with Air Wisconsin, considered the question for a minute.

National Transportation Safety Board Human Factors Group Chairman Alan Diehl, the man who had asked the question, knew approximately what the answer would be. He had been documenting this story for months with letters, phone calls, and interviews. Now, at last, the story of what life was really like at Downeast was spilling into the public arena.

Captain Jim Fenske, a star witness for the NTSB, the pilot who had refused to fly Otter 68DE the day before Jim Merryman's death because of the ratcheting noise in number-two engine, looked Diehl in the eye.

"Yes, sir."

"Would you describe the situation?"

Fenske, his hands clasped in his lap, leaning slightly forward toward the microphone, thought for a split second.

"I had left a few bags behind in Boston on one particular day. . . . Upon arrival in Rockland, I was brought into Mr.

162

Stenger senior's office and asked why I had left these bags behind. He felt I didn't need the extra fuel . . . and that it didn't make much difference whether the airplane was at the [maximum legal weight of] seven thousand pounds [or] seven thousand twenty-three pounds, which [represents the average weight of one bag added]."

The example was mild compared to many of the allegations collected by Diehl, but it was now on the record. Few people outside the ranks of commuter pilots would understand the significance of his answer. Overloading an Otter or a Navajo even to the extent of one bag might, under certain circumstances, spell the difference between a safe return or disaster in the event of an engine failure.

In the audience of the hearing room (a nondescript part of the Federal Building in Cambridge, Massachusetts) sat Robert Stenger senior and Robert "Rocky" Stenger junior, watching and occasionally conversing in whispers. Rocky Stenger had already testified, stating with wide-eyed innocence that he did not recall conversations and incidents cited in letters sent to the NTSB from former pilots, nor did he recall George Hines or Jim Merryman telling him on the phone from Boston that there was anything wrong with the aircraft. Rocky's performance had been nondamaging, and a good beginning for Downeast. Stenger and Downeast were represented by a Boston law firm, and their attorney would get a turn at each of the witnesses who had disparaging things to say about the airline.[1]

1. Formal hearings are usually held by the NTSB when a high death toll or sensitive issues are involved and the evidence may be contradictory. The use of a public hearing to develop sworn testimony can be helpful, but the hearing itself decides nothing. The evidence developed from the witnesses in an NTSB hearing simply provides more material for the board to consider in making its final decisions on what caused and contributed to the accident, and what should be done to prevent recurrences.

Even though the hearing phase—if one is held—is in no way a trial, sensational revelations pertaining to the accident and its contributing causes often provide the press with headline copy. Since these are public hearings in almost all cases, radio and TV as well as the print media are allowed in to cover it. The NTSB investigators are available to question witnesses, as are attorneys for other involved parties, such as the carrier, the manufacturer of the ill-fated aircraft, the insurance carrier, and the Air Line Pilots Association if an ALPA airline is involved (which Downeast was not). Other interested parties may be admitted to observer status at the discretion of the board but not allowed to participate in questioning of witnesses. An NTSB hearing is not conducted by the normal "rules of evidence" of a formal courtroom, whether federal or state. The format is more loosely controlled, and the fairness depends in good

Diehl decided to press more deeply into the subject of over-weight takeoffs, numerous examples of which had been reported to him by many former Downeast pilots.

"[Captain Fenske], did [Robert Stenger] ever tell you that he would pay a fine if you were ever violated for [overweight take-offs]?"

Fenske smiled slightly at the memory.

"Yes, he told me—well, my answer to him was that my [pilot license and ratings] were on the line if I ever got [ramp checked by the FAA]. [His] response was 'How often have you got ramped? How often is the FAA around?' . . . And I was advised that he had paid Jim Merryman's fine and that he would do so for me."

Diehl continued, a long list of topics in his hand.

"Do you ever remember pressure to take off with less than legal takeoff [weather] minimums?"

Again Fenske replied with a slight smile.

"Yes. The policy was that the pilots would determine their weather for a particular trip, and there was always someone there to question them as to their ability or reasoning for doing something.

"Mr. Stenger said to me one time that the only agent he could trust to get the flights out when the weather was bad was his son, Rocky Stenger, and the reason was that [Rocky] would call his father . . . if a pilot said that the weather was such that he couldn't go. . . . Pressure would be applied from [Bob Stenger senior]. Most of the other people, providing Mr. Stenger was not around, would accept what you had to say."

Al Diehl looked up from his list again.

"Specifically, were you ever pressured to launch in a Navajo in a thunderstorm environment?"

"Yes, sir."

"Would you describe that incident, [Captain Fenske]?"

Jim Fenske shifted slightly in his seat, examining his memory before looking up at Diehl again.

"I had a late afternoon flight in the summer, I believe it was, of

---

measure on the chairman of the hearing, a National Transportation Safety Board member. Sitting in the hearing chairman's seat during the Downeast hearing was board member Francis McAdams, who had been a board member for many years. Though McAdams had never shown great enthusiasm for deep investigation into human-performance causes of accidents, he was beginning to come around and was willing to permit deeper questioning on ancillary subject matter in the Downeast accident than in past hearings.

1977. I elected not to cancel the flight, but to delay the flight, due to a line of thunderstorms moving from—at that time they were just to the west of Portland moving east across my flight path en route to Boston. . . . I was then asked in Mr. Stenger's office by Mr. Stenger why I was delaying the flight. I explained the situation and went back out to the front behind the counter. He advised me that he had made some telephone calls to Flight Service and to Brunswick Approach.

"I was called back into his office and asked when I intended on leaving . . . and I explained that when the line of thunderstorms was far enough to the east where I could get around, then . . . I would go.

"He told me that Brunswick Approach was [looking at] the thunderstorms on radar, but that there were some holes, and they would guide me through the holes. I elected not to go through the holes that Brunswick Approach had offered to put me through, according to Mr. Stenger.

"The flight went out forty minutes late. And from that day on, for about a month and a half, I was constantly harassed as to any kind of thunderstorm activity; for instance 'They're reporting thunderstorms in San Francisco; are you going to go today?'"

"Who did this harassment come from?" Diehl asked.

"Mr. Stenger."

"Just Mr. Stenger senior?"

"Yes, sir."

"Did that Navajo have weather radar on board?"

"No, sir, it did not."

Diehl looked at his list again, keeping his expressions under tight control. Whatever he might feel in reaction to the answers, however bizarre the situation being described from the stand, he must not react to it openly. A grimace or shake of the head would be unprofessional, and would probably evoke an immediate rebuke from Francis McAdams, the chairman, who seemed on the edge of tolerance at this line of questioning to begin with.

"[Captain Fenske], did you ever observe anybody requiring pilots to make multiple approaches when landing at Rockland?"

"I myself have been told to do that, yes."

"What is the effect on Instrument Flight Rule fuel reserves if you make multiple approaches?"

"Well . . . you're supposed to have fuel for one approach and a [missed approach climb-out], and then proceed to an alternate [which will leave you with] forty-five minutes reserve [fuel]. Not multiple approaches."

"Were you ever directed to go to a certain alternate?"

"Yes, sir."

Fenske described a flight in the Otter during the summer of 1978 in which he was unable to see the runway at Rockland during his first approach due to a fogbank that sat only over the airport (the rest of the Rockland area was sunny). Robert Stenger came on the company radio and reported improving weather, urging Fenske to try a second approach, though Fenske had not loaded on enough fuel in Boston to cover a second approach. He tried a second time anyway, and was again unable to land. Once again Stenger came on the radio advising him of improved visibility to elicit a third attempt.

"My first officer and myself decided to go on to Portland. We were then advised to go to Augusta, and I had decided at this point that I would not go to Augusta due to my fuel situation and the weather at Augusta at the time, and Portland having an ILS with two hundred [feet and a half-mile visibility] for minimums."

"Who advised you to go to Augusta?"

"Mr. Stenger. He was still talking on the radio when I shut it off. I don't know what he was talking about after that. I just shut him off. . . . The whole coast was starting to sock in with fog. . . . My first officer looked down and saw Wiscasset Airport wide open. . . . I called Brunswick Approach. . . . They cleared me [to land at Wiscasset]. Upon landing at Wiscasset, I was advised by Rocky Stenger [by telephone] . . . that there was no ground transportation [for the passengers], that I was supposed to [fly] on to Portland now because that was where the ground transportation had gone. I got a telephone call from Mr. Stenger asking why I [would] not go on to Portland, and I explained to him that given the situation and my fuel . . . I would just stay with Wiscasset; at which time I was told that I would do things his way or I will not do them at all."

"Who told you that?"

"Mr. Stenger."

Fenske told of being sent down to Wiscasset the next day by rental car to fly the Otter empty back to Rockland. The Otter had only the same six-hundred pounds of fuel on board with which Fenske had landed the day before. The weather was still lousy. Fenske wanted more fuel, but no jet fuel was available at Wiscasset, so he prepared to put some 100-octane aviation gasoline on board (an acceptable, though abnormal, procedure).

"The phone [at Wiscasset Airport] rang within five minutes. I was ordered not to put one hundred-octane on that airplane, and that it was Visual Flight Rule weather at Rockland, and to get [the Otter] back now, just like that. I told him I would not move the airplane unless I added fuel—one hundred-octane, which was all they had—and the telephone conversation ended.

"The phone rang again. . . . It was Mr. Stenger once again. . . . He was sitting on his boat in Camden, [ten miles north of Rockland], which was VFR [Visual Flight Rule weather], and that I should get the airplane up to [Rockland] immediately. I told him I would not do it.

"I went out and added one-hundred-octane to the airplane and flew it back to Rockland, which was a minimums approach to get in. I was advised by the manager of the airline that Mr. Stenger wanted to speak to me, to call him immediately on his boat. I did so, and I was advised that if I did not do what I was told to do that I'd be finished."

Diehl moved in again.

"Captain, let me make sure that I understand this. Mr. Stenger was on his yacht, and he told you it was VFR in Rockland, and [you had been told by someone else] that it was marginal IFR [Instrument Flight Rule weather] at best?"

"Yes, sir. They were [reporting] three quarters of a mile."

"And a low ceiling?"

"Yes, sir. When I arrived at Rockland, I was [fourth in line] for the approach. There were three Downeast Navajos holding over Sprucehead for clearance to get in."

"What happened after that? Were you ever reminded of this incident?"

Fenske chuckled.

"Oh, yes. My job was threatened right there on the spot, that I would either do things the way he wanted them done, this was his airline; if I didn't want to conform to those standards, to pack my things and get out. I met with him the next morning . . . and explained my situation and why I did what I had done, and that was the end of it. There was no apology, or no 'I understand that,' or anything. It was just the end of the conversation."

Al Diehl was coming to the end of his list of subjects on which to question Jim Fenske. A host of other witnesses was waiting, including Robert Stenger himself. The press corps in the back had been hastily scribbling and recording the proceedings, and it was a sure bet that the stories they would file this afternoon would border on the sensational.

"[Captain], did you ever find that disagreements with the management of Downeast affected your flying ability?"

"Oh, yes . . . I had my job threatened four times, that I can recollect, in the three years that I had been there . . . and I would find myself in an aircraft [under instrument conditions], instead of paying attention to what I was doing, trying to figure out a way to cover my behind, or to come up with an explanation as to why I exercised certain judgment in a situation. And I would have to just—I'd have to sort of push it away, just to concentrate on what I was doing."

"So did you find it distracting then?"

"Oh, very."

"When Mr. Stenger returned from vacation, was there usually a period of more or less intense arguments?"

Fenske again fixed Diehl with a steady gaze before replying.

"The day after was generally a battlefield."

He paused briefly before continuing.

"Everyone would upon knowledge of him returning, everyone would get together with themselves and try to get their stories straight as to how they were going to explain what they did in the three-week or month absence or two-day absence while he was gone, because everyone knew that they were going to be called in at one time or another for something."

"And when was Mr. Stenger scheduled to return with reference to the accident?"

"The day of the accident, I guess."

Chairman McAdams leaned over and addressed Fenske.

"Did you, or to your knowledge did any of the other pilots ever complain in writing to Mr. Stenger about the conditions you've described?"

"No, sir."

"Did you ever have occasion to talk to the FAA with respect to the operation of Downeast Airlines?"

"Yes, sir. The opportunity sure was there, but no one would do it with their job on the line that I know of."

There it was.

He had believed all along that they weren't telling him the story—weren't leveling with him. The reason had been no secret, but to FAA Principal Operations Inspector Robert Turner, sitting in the audience waiting his turn on the stand, it was good to hear it read into the record. What could he have done without help, without corroboration?

168

Al Diehl ended his questioning of Fenske, and Downeast's attorney, Lloyd Starrett (of Foley, Hoag and Eliot in Boston) moved in to attempt to punch holes in the superstructure of Fenske's damaging testimony.

Starrett began by asking Jim Fenske questions regarding the amount of training he had received at Downeast. The subject was one that Al Diehl had researched extensively. There were excellent records available to the FAA that indicated all the Downeast pilots had been trained on the ground and in the air according to the regulations. But many of the pilots had said flatly that no such training had ever been done, or very little of it at best. It was a very touchy and vulnerable area for Downeast, and the NTSB technical team sat dumbfounded as Lloyd Starrett walked boldly into that minefield, apparently unaware of the trip wires all around.

"Mr. Chairman, may I show the witness a couple of documents that may help to refresh his recollection?"

Chairman McAdams looked at the papers in the attorney's hands.

"What are they?"

"They are training records, sir. This is a training record of his training."

McAdams frowned. He had not seen these before.

"Of Mr. Fenske's training?"

"Captain Fenske's training in 1976, that shows a total of eighty point four hours of training in that year."

McAdams nodded, and the official records from Downeast's files—the same ones shown to the FAA—were handed to Fenske, who had already testified that over his entire period of employment of three years he had received no more than 2.6 hours of training. There was a glaring discrepancy, and something was very wrong. At its simplest, either Jim Fenske was lying, or Downeast was lying, although there was always the possibility of a failure to communicate. "Training" might not mean the same to one as it did to the other. Starrett thought he had caught the young pilot in an act of factual assassination. Here was a disgruntled ex-employee with a monetary dispute with his former employer (Stenger had refused to distribute Fenske's profit-sharing account to him) now making things up to retaliate. Here were official records. The opportunity to "refresh his recollection" was a tried and true courtroom technique used to confront a witness with hard evidence of his falsehood, scare him out of his smugness and his story, and induce him to change his testimony to what the attorney believed was the truth.

But Jim Fenske was not impressed. He had seen these "official" records before, and he knew what they represented. He also knew that he was only one of many Downeast pilots who had told of seeing records of training that they had never, in fact, received.

"Does that refresh your recollection, Captain?" Starrett began, watching for a crack in the facade.

"No. This is—no. This is my handwriting."

"That is your handwriting?"

"That's correct."

"But it's not training?"

Starrett seemed puzzled. Fenske looked him in the eye.

"No. These are hours that I kept for myself as I rode in the right seat of the airplane."

Starrett fished for another document, handing it to Fenske.

"I show you this record, which is dated September twenty-fifth, 1978. Is that your handwriting?"

Again Fenske looked briefly at the paper, then at Starrett.

"No, sir."

"Do you know what that represents?"

"I guess training. Yes, training record."

"Do you know, Captain Fenske, where other records such as these are? Do you know where yours are?"

"Yes, sir."

"Where?"

Fenske smiled slightly. The confrontation had failed, and Starrett was searching.

"I have my training records." he said flatly.

Starrett and Mike Pangia, the FAA's attorney, got involved in a discussion over making copies of the Downeast records. When Starrett resumed his questioning, he decided to stay the hell away from the subject of training records. The witness had just contradicted the accuracy of Downeast's records and claimed to have the true ones himself, which might prove that only 2.6 hours had been received rather than 80.4. The Downeast attorney smelled the minefield now, and was tiptoeing gingerly out of it.

"[Captain Fenske], do you recall how many check rides you were given by Captain Edwards?"

"Three, sir, to my knowledge."

"Do you recall how many check rides you were given by the FAA in Portland?"

"Three."

"In fact, Captain Fenske, you had to be given several check rides to pass at one time, didn't you?"

"No, sir."

"No?"

"No."

Again a dead end. Starrett abandoned the check-ride issue and continued probing other aspects of Fenske's testimony, getting nowhere. The knockout blow to Fenske's credibility had not occurred.

"When you left Downeast, did you have any bad feelings toward the company?"

"Yes, sir."

"Why?"

"Just the general environment."

"In fact, did you have any monetary disputes or do you now have any monetary disputes with Downeast?"

"Yes, sir."

Starrett consulted his notes again.

"Do you recall having an argument with Mr. Audie [the Boston station manager for Downeast] in Boston sometime after the accident?"

"Yes, sir."

"What were the circumstances of that argument?"

Fenske sighed and adjusted himself in his chair. That was quite a question. He wondered if the attorney had any idea what kind of Pandora's box he had just asked Fenske to throw open.

"I flew the first flight the morning after the accident. Upon arrival in Boston, I found the ramp agent, Joe [Falzarano] in tears, physical tears, and very upset. And I pulled him aside and I asked him what the problem was, and he started to explain that he overheard a conversation and he doesn't know what he should do, if anything. And I told him that he should talk to the FAA or the NTSB or someone about it."

Fenske was referring to Falzarano's overhearing of the conversation between Jim Merryman and George Hines at the Boston counter before their fatal flight. Hines had talked to Rocky on the phone, as had Merryman, and they had been advised to bring the aircraft home. Neither wanted to try Rockland, but Hines was dead set against it. Falzarano had heard most of the exchange, and had chatted with both men briefly. When news of the accident reached him late that night, he had been thrown into instant agony over what to do—and whether he should have done something that night.

"I believe it was the Monday following the accident, I had gone down to Boston, and through rumor I was told that Conrad [Audie] had told his people to keep their mouths shut. I approached . . . Mr. Audie with this question, if this were in fact true. His reply to me was that they're dead, and to not make any trouble because I was leaving the company [to take a job with Air Wisconsin], and not to cause trouble for the people left behind.

"Upon getting back to Rockland, Mr. Stenger advised me that Mr. Audie had called him and recommended me to be fired from the company with two days left of service before my notice was up, because I was going to cause trouble over this accident. Mr. Stenger also advised me that if I had anything I felt was important to say to the [NTSB] to go ahead and do so. He . . . told me that I was disloyal for leaving the company . . . and that myself and Kurt Langseth should have been killed in the aircraft."

Starrett paused for a moment, looking at Fenske.

"Anything else?"

"That was enough. I walked out. And I did not fly to Boston. I was pulled off the line [to keep me away] from Conrad Audie."

Lloyd Starrett finally gave up and sat down. Downeast might be denying everything detrimental that Jim Fenske had said, but Fenske's testimony was damning—if for no other reason than what the press would inevitably do with it. He was going to have his work cut out for him defending Bob Stenger's interests, and the trial in the court of public opinion was just beginning. Tomorrow's headlines would surely set off a holocaust.

Chapter 12

THE OLD TOOTHLESS TIGER ON A CIVIL SERVICE LEASH

Mike Pangia, in his early thirties and already possessed of a fine reputation as a sharp aviation lawyer working for the FAA, arranged his notes and prepared to take over the questioning of Jim Fenske. Pangia's job was to make sure the FAA was "protected" on the record—challenging or clarifying any testimony that seemed to indicate that the FAA might have failed to do its job at any point. Some of what he would need to clarify might help Downeast, since it was axiomatic that if Downeast had really been a rotten operation, the FAA's failure to discover it and do something about it would amount to a significant failure in the public eye.

Pangia had been making voluminous notes, and he began guiding Fenske back through them one by one, including the question of visibility at Rockland. Rockland always seemed to be at three quarters of a mile visibility, even though the pilots were unable to find the runway on approach because of fog that was far below three quarters. But the visibility was measured from the Downeast terminal, over a mile from the approach, and the fog at Rockland could be quite different in the space of one mile. This had not been brought out.

"Now, [Captain Fenske], did pilots of Downeast Airlines have the knowledge—were they taught where that three-quarter-mile visibility marker was in relation to the airport?"

"All I was ever told was that it was an individual's house set up on the hill."

"North of the airport?"

"Yes."

"Okay, and as a pilot, if you were making an approach to, say, runway three, and you were to receive a visibility value of three quarters of a mile at Rockland, you would suspect or at least expect that the visibility on the approach to runway three might be different than that?"

Fenske nodded. "Yes. We ran into that quite a bit. The airport on a number of occasions could be VFR, the approach be solid [fog all the way down] to the trees."

Pangia continued on to several other subjects, finally getting to a central issue.

"Now, were you ever pressured by the company to make approaches to airports under minimum weather conditions . . . in other words, go below the [minimums], even though you [cannot see] the airport environment?"

"No, sir, I was personally not."

"Do you know of anyone who had received any such pressure?"

"No, sir."

Francis McAdams interrupted.

"Does the term 'company minimums' mean anything to you?"

Fenske looked back up at McAdams.

"It's a term I heard used, but if the question is, did Mr. Stenger use 'company minimums' with me, he did not."

McAdams continued.

"Does the term 'cement-plant approach' mean anything to you?"

"Yes, sir. There's a Martin Marietta cement plant in Thomaston, Maine, which is off the approach end of runway thirteen. When I first went to work for the company, I had heard the pilots talk about coming up the St. George River below the fog deck, then getting in sight of the cement plant and making a right turn, something about a red house, and start a descent down to runway thirteen. That was my interpretation of 'cement-plant approach.'"

The chairman, McAdams, paused a second, fixing the witness with a stern gaze.

"Do you know of anyone who used the so-called 'cement-plant approach' below minimums?"

Fenske replied evenly. "No, sir, I do not."

To attorney Starrett the last few exchanges had taken some of the sting out of Fenske's statements, though the testimony yet to come, and the letters and other statements entered in the NTSB record, would go much further than Fenske on such subjects as the cement-plant approach and company minimums. Of course those last answers only helped Fenske's credibility and helped destroy the idea that he was nothing but a disgruntled employee with a made-up tale. If Jim Fenske was going to throw the truth to the wind to "get" Robert Stenger, he certainly wouldn't hedge on questions about cement-plant approaches and company minimums.

To Robert Stenger senior, however, nothing Fenske said had been moderate. Fenske, in his opinion, was obviously "out to get" Downeast and him, as were the other pilots who were about to testify—pilots like Richard Mau, the young captain who had canceled a flight in Boston because of concern over thunderstorms, and who took the stand next, followed the next day by Joseph McGonagle, a corporate pilot who had flown for Downeast between June 1974, and February 1976.

Step by step, corroborating statements kept coming out. McGonagle, who had written one of the early letters Diehl had received, was asked about the discrepancies in the pilot training records:

"It was common knowledge, and a procedure that was commonly used at Downeast, to record all sorts of time in the training folder that was not given to the pilot." McGonagle went on to outline the severe discrepancies between the Downeast training records on him, and what he had actually received. It was getting to be a familiar story, though this time it was the FAA's Mike Pangia and not Downeast's Lloyd Starrett who brought up the subject.

Mike Pangia also asked McGonagle whether the level of compliance with the rules changed at Downeast whenever the FAA was known to be on the way:

"Oh, you betcha . . . you want to see a clean operation is when the FAA's around."

"And as soon as they leave, it might go back to the ordinary?"

"It goes right back the way it used to be."

Richard Arnold, a highly qualified former military pilot who had stayed with Downeast only briefly in 1974 before leaving in disgust at the loose procedures, told of his impressions:

"I thought it was a highly pressurized atmosphere, from the day I walked in the door until I left. I thought that pilots were asked to do things that, if not in violation of FARs, were at least questionable in the light of safety.

"I also thought it was a fairly amateur operation. I come from a background of flying [North American] Sabreliners in Europe, and I was flying VIPs, including ambassadorial staff, and so on and so forth. That was my previous flying assignment.

"If I were to compare the two things, one was a professional operation and the other was amateur, as far as weather, as far as suborning the pilots to violate regulations. I had never been subject to that type of aviation pressure before in over twenty years of flying."

And finally, on the second day of the hearing, Robert Stenger senior got his turn.

"[Mr. Stenger], did you ever state to anyone anything to the effect that you felt the wrong crew was aboard the aircraft that crashed?"

Mike Pangia of the FAA had asked the question. Robert Stenger was leaning back in the witness chair, an expression of disgust— almost a scowl—on his face. He considered the entire proceeding to be more or less a setup, especially since he was not allowed to testify on the first day when most of the allegations were made and the press was the most interested. He just could not understand why an owner of an airline would be grilled, castigated, or criticized for trying to be in tight control of everything that went on.

Robert Stenger glared at Mike Pangia and answered.

"I stated to someone—I stated that to Captain Fenske from a statement that was made to me, yes, when he was in a discussion of other things that was being brought up about the accident, and him saying different things about the company, and I just mentioned to him when things got in conversation on it that other pilots in the company had made that statement that 'the wrong crew was on the flight.'

"And that's when he comes back to me and stated—well, I think his exact words is 'That's nothing. Half of Downeast hoped that you and your wife was on it.'"

Pangia pursued a number of other subjects with Stenger, who had denied that any of the allegations of pressure or harassment of the pilots was true.

"[Mr. Stenger], what is your explanation as to this general impression of many employees that they weren't getting along with the company . . . ?"

"It seems to be mostly in the pilots, pilot group. We don't have a turnover in the maintenance or the office of the reservationists.

176

They mostly seem to be the pilots. I don't know if it's because I used to be one, as far as making a living, or that it's—and I can see through some of the things that they try to pull.

"This is one problem that—like in regards to Captain Mau refusing to fly the flight when Rocky was at the airport; he don't really know if they're trying to pull his leg or if they know what they're talking about, because so many times he's found out that they'd take advantage of him because he don't know about being a pilot and what they can do or what they can't do, or if they're using their skills to do something else maybe that they prefer to do instead of fly an airplane back to Rockland or to Boston."

Pangia tried again.

"Well, what explanation do you have for the general impression . . . that there were some . . . pressures to violate Federal Aviation Regulations with respect to weather minimums, weight limitations, and their general impression that that would be condoned because of business exigencies or business pressures?"

Stenger shifted forward in his seat.

"I don't—I deny those. Every one is different insofar as one saying I told him to take off at two-hundred, and this one say three-hundred, the next one say three-hundred-and-fifty. And none of them is consistent. Every one of these that's [got] one of these reports in here, as far as I'm concerned, has got a grudge against the company. Most of these that are here have been discharged by me for reasons that I felt were justifiable.

"I don't think anyone is scared of their job, as some say, if they do what they're supposed to. . . ."

Mike Pangia consulted his note pad a moment, then continued.

"What is your explanation about some of the impressions . . . that there was some question as to whether the pilot for Downeast Airlines was in fact the person who makes the final determination as to the flight?"

"It depends on the pilot giving me the facts, if I'm involved in it."

"Well, the regulations state that the pilot-in-command is the final authority. . . . Isn't that correct?"

"That's what it states, yes."

"And is that the philosophy that Downeast Airlines should follow in all cases?"

"That is the philosophy, yes. But if he—that's his prerogative."

Robert Stenger paused a second, then shifted forward, as if trying to drive home the underlying explanation—the key to the entire problem.

177

"If he wants to cancel the flight due to his discretion, then as far as I'm concerned it better be a legitimate reason, not just because he wants to do it. And everyone else is willing to fly and other companies are still flying, and he decides not to, then I don't feel that he's using his best judgment to his fullest capacity, either."

"And you would feel in those situations that your judgment, rather than the pilot's judgment, should overrule?"

"Pertains to the pilot."

Pangia looked puzzled.

"Beg your pardon?"

Stenger shot back. "It pertains to the pilot involved."

"And I understand from your testimony that there [were] circumstances where you let pilots go because they would not fly when you thought they should. Do you remember that testimony?"

"Yes."

"Can you, [Mr. Stenger], give us occasions when you did let pilots go when you thought they should fly when they made a decision as pilot-in-command that they should not?"

Stenger brought up the name of Richard Mau, the dismissed pilot who had called Al Diehl the day after his firing, ready to talk about pressures at Rockland. He fumbled with the pages of the exhibits for a second, finding a copy of Mau's letter regarding his decision to ground a Downeast flight in Boston because of threatening thunderstorms ahead.

Mike Pangia moved in again to the witness.

"Now, he made a decision as pilot-in-command that he did not want to fly because of severe weather conditions forecast for the area, to wit, Sigmets, which are potentially dangerous to all aircraft, convective Sigmets meaning thunderstorms.[1] Did you feel that your decision should overrule Captain Mau's in that particular situation?"

Robert Stenger was furious over the incident, and his being questioned again about it. The pilot had disobeyed him, and he had fired the offending man. What more was there to it?

Stenger stared straight at Pangia.

"Seeing that Sigmet didn't involve the area he was flying in, yes."

"Do you recall . . ."

1. "Sigmet" is weather shorthand for *sig*nificant *met*erological information. Sigmets are issued on an alert basis to call the attention of pilots to important weather developments in their area of the nation. Each Sigmet is numbered and distributed over various frequencies, by Air Traffic Control Centers and approach/departure control facilities, and by teletype.

Stenger cut him off. "As far as this particular incident here, Captain Mau was asked to fly; he refused. That's all there was to it. He was bused to Rockland. He wasn't fired."

Pangia pressed on. "Was he reprimanded?"

"He was reprimanded to the point that it was decided by the other pilots involved with Downeast that Captain Mau"—Stenger pronounced the "Captain" with distaste—"that refused to fly single-pilot autopilot, which he was hired for, and which we acknowledge, "okay, if you don't want to do it, we'll give you [your] copilot, but drop it right there. Don't try to harass everyone else that's willing to do it, and call a single pilot other than a dual pilot concept unsafe."[2]

The answer didn't make much sense to Pangia. In fact Richard Mau had been "reprimanded" to the point that he had been effectively fired, but Stenger didn't want to use the words. In any event, in his opinion he had the right to fire the man without having to explain it to anyone. He had been in the business twenty years—he had pulled himself up by the bootstraps to become a wealthy man, and he had never needed anyone's permission to run his show his way. The anger and disgust he felt over the negative attention he was getting were all too obvious to the curious people who watched him stalk from the hearing room at the end of the day—men and women from various branches of government and industry as well as the press who had heard the disparaging testimony about Downeast and wondered how, if all were true, could anyone be so callous?

Who was this Robert Stenger senior anyway? Who was this man who could, according to the allegations of some of the witnesses, urge, cajole, browbeat, and force his employees to go against their better judgment—imperil the safety of the flying public—then climb on the stand and deny that any of the accusations were true?

There are those who would compare Bob Stenger with the residents of the Rockland area who epitomize the fierce independence and hardworking nature of people along the rugged coast of Maine: the fishermen and lobstermen, farmers and sailors, living a sometimes hardscrabble existence, learning the value of a dollar

2. The use of a single pilot in instrument conditions is permitted under Part 135 operations as long as the aircraft has an operable autopilot and several other conditions are met. It is an easy exemption to abuse, however, and when the single pilot is marginally qualified or overloaded, it can rapidly deteriorate into an unsafe operation.

early in life, and striving for stability—valuing most highly the absence of change.

Change comes hard to such folks. The rapid, kaleidoscopic life of a big-city dweller is foreign and disturbing to someone who measures his yearly needs by the seasons, and who is comfortable with making money only six months a year, wintering over on savings the other half.

Stenger, to a certain extent, was such an individual—with one major exception. Bob Stenger's flint edge, the hardness—some would say coldness—of his character and his attitude toward others, was atypical.

The people of Downeast Maine may be hard to get to know, and even tougher to best in a business deal, but underneath they tend to be generous and caring (if private) people, the milk of human kindness running full just beneath the hard crust of a wintry facade. They also tend to be family oriented, drawing together in tough times, circling the wagons around home and kin, community, region, and state.

The Stengers were indeed family oriented, and they knew how to build a wall of defensiveness around their clan, but there the similarities end. Those who have known Bob Stenger most of his adult life claim to understand him, and to respect what he has accomplished through very hard work, but few would characterize him as warm. To the large circle of people—employees and acquaintances—who professed to detest him, the principal reason was his tight fisted nature with money and his often loud and abusive, argumentative style of defending what he perceived as his own interests. Trust and business friendships came hard to Robert Stenger senior, and his method of dealing with people made him few friends and many who might describe themselves as enemies.

But Bob Stenger was not a monster. He had not set out to contribute to the death or injury of any passenger or employee. He had no intention of driving Jim Merryman to desperation—or to his death. It seemed all he wanted was to make as much money as he possibly could—his way. And, as hard as it was to discern, beneath the flint-edged visage Robert Stenger showed to the world, the man was not totally devoid of caring.

At first glance he seemed not to know how to care, how to trust himself to others without fear of being burned. Whatever had transpired in his background to develop such an ingrained suspicion of people, the former U.S. Army sergeant simply could not trust others or understand the effect of the constant verbal abuse he heaped on those around him. But there were wild contradictions in his behavior if one wanted to condemn him as uncaring. This was a man who would give an unexpected gift of money to an employee or associate; then, almost as if he were afraid to be perceived as soft, unmercifully chew out the same individual in front of his peers. This was a man who could put incredible pressure on the majority of his pilots, yet pull his punches on one of the younger men because he felt somewhat protective of him. (Of course, he was also an adversary to be taken seriously if he decided someone was against him.)

And that, too, is the key to his "management" of Downeast.

He had never learned how to be a manager. He had never gained the training, the qualifications, or the professional managerial balance to handle people in an organization of increasing complexity—and certainly not in a professional airline. Downeast was basically a Mom and Pop operation that got out of hand. Suddenly the little seat-of-the-pants fixed-base air-taxi operation Stenger had started and could run himself turned into a gold mine—a sure-thing route to Boston that could make him millions. To fly it, though, his new service needed a lot more than licenses and airplanes. It needed professional management (though to make money it needed only Bob Stenger).

To uphold the public trust—to be a responsible, safe, professional airline with a detached dedication to ethical compliance with the rules—a Mom and Pop operation simply wouldn't do. Overnight Stenger was wrestling with personnel problems that would not respond to his methodology of threats and intimidation underlying ironfisted control. He had none of the people-managing skills to handle it, but who was going to tell him?

To that extent, it was less Robert Stenger's fault than the fault of the air-transportation system—a system that allowed an unqualified management to reign free and uncontrolled over an enterprise the public trusted to be safe.

Then too, there were the shortcuts. The shortest distance between two points—today's costs and tomorrow's profits—seemed to be the only course that made sense to Bob Stenger. The end of maintaining service for Rockland by making a profit justified whatever means he had to use to get the flights in and out. In his view, economic considerations were inseparable from takeoff and landing decisions and weather minimums.

After all, he was providing "good" scheduled service, service schedules far better than the Rockland area had had with Northeast Airlines years before. (Of course, it is significant that Stenger measured "good" service in terms of frequency of flights and predictability of arrivals, rather than safety of the passengers carried.) If he was to continue to provide that "good" service, then the company had to remain profitable. Ironfisted control was the only method he knew, and just as he would be entitled to exercise such control on any fishing boat under his captaincy, or in a Mom and Pop grocery store or filling station, Stenger felt he had the right to exercise it in his airline business to the same degree. "This is my airline!" he had snapped at Jim Fenske. As he pointed out constantly, he had been in the air-taxi business for twenty years at Rockland. Who knew the conditions better than he? Why would anyone, FAA, NTSB, the press, or anyone else, have the audacity to question his judgment on such matters as how low you could really go and still be safe on an approach to Rockland? If his chief pilot screwed it up and killed a bunch of passengers (and wrecked his new airplane, the acquisition of which had been such a frightening change to him in the first place), it was simply because Jim Merryman had somehow failed to follow Robert Stenger's instructions—his way of doing things. Why couldn't they see that? Why all the hullabaloo?

As Robert Stenger left the Federal Building in Cambridge, his resentment over those who were "out to get him"—those who (in his view) felt themselves intellectually superior and resented his success, whether the Fenskes, the NTSB people, or the press—showed clearly. Those adversaries would not prevail. Nobody, by God, was going to run Robert Stenger out of business or tell him how to run his business. Nobody!

The tragedy, of course, was that Robert Stenger couldn't understand, would never accept, that the nature of the business in which he had chosen to engage had changed in the very moment that he undertook to carry the public as a scheduled airline, regardless of size. He could not understand that the resulting responsibilities to a higher ethic transcended all his instincts about making profit first, and invalidated all his training about the benefits of a penurious nature—training acquired in the school of hard knocks.

The problem had nothing to do with the public, the press, the NTSB, or the pilots begrudging Bob Stenger his financial success, or resenting his contentious, cantankerous nature. The problem was that he had no right to impose his autocratic management on a scheduled airline—even one with a single, solitary route. Nor did he have the right to encourage, permit, or even to look the other way and ignore, any illicit or corner-cutting operations in a company the public had a legal right to trust.

Because Downeast was a licensed, regulated, scheduled airline to the general public, its primary duty was the highest standards of safety—even at the expense of reduced profits. Nothing, however, in Downeast's methods over eleven years indicated that there had ever been a company belief in such a duty. It was Bob Stenger's company and he would run it the way he saw fit. Period.

And run it he did, on into the fall of 1979 and the spring of 1980, unfettered by any intense FAA scrutiny, his licenses and certificates still in force regardless of the allegations and the evidence that at the very least made a case for change in Downeast's attitudes.

To be sure, Downeast's charter business evaporated because of the press coverage. What had been a steady "you call—we haul" operation to the nearby islands of the well-heeled shrank to a trickle. But that involved pure air-taxi work, which had always been represented to the public by the NTSB and FAA as a greater risk than scheduled airline service.

Despite greatly reduced passenger loads, however, the operation of the airline portion of Downeast—the Boston-Rockland route—continued. Bob Stenger's one-man control of the operation continued. The foggy weather, poor visibility, and challenging flight conditions of fall and winter continued. The meaningless emphasis on checking the written records of pilots and maintenance crews by the FAA's new principal operations inspector (Robert Turner had taken medical retirement because of hypertension) continued. As Al Diehl, Steve Corrie, and the other members of the NTSB investigative team worked on preparation of the final accident report (and as Downeast's law firm fielded the damage suits for injury and death), Downeast apparently kept on going as if there had been no shattered Twin Otter in the trees of Otter Point, as if no retinue of pilots and former employees had faced the glaring lights of a public forum in Cambridge and told their tales of life at Rockland.

It was as if the system had not worked—could not work—and it was very frustrating to those who had been brave enough to step forward and testify. (Many of the former Downeast pilots who spoke up had truly expected repercussions, believing that Bob Stenger had friends and political acquaintances throughout aviation who could damage their flying careers wherever they chose to go.) It almost seemed that in the final analysis, when faced with management problems and few (if any) hard-core Federal Aviation

Regulation violations, the FAA was nothing but an old, toothless tiger on a civil-service leash, cowering in the corner, fearful of even growling at an offending airline, lest the poor beast have its leash yanked by its owner in Washington. Its image as a proud and fearless guardian of the airline passenger's safety had been tarnished badly by the reality of its cowering—its ineffectiveness—but since the Downeast situation was essentially regional, the damage nationally was small.

The effects of deregulation on the FAA's limited capabilities—the effects of unfettered expansionistic fever on the industry—had become an irresistible force. The certainty of human failure in any given situation—especially in an unsupervised, technically demanding environment—had always been an immovable object. The combination of the two guaranteed a vastly increased potential for avoidable air accidents caused by human foul-ups—preventable human foul-ups made worse by poor management practices undiscovered and unaddressed by the FAA.

The NTSB was struggling to minimize that potential with accident reports and commuter studies spotlighting the weak areas. But the FAA was being protectionist and slow to combat the very decline in air safety Congress had promised would never occur. There were situations festering all over the country of which the FAA (and the NTSB) knew little, or about which they could do little. The rapidly expanding world of Air Florida, the opportunities available to an airline like Air Illinois, the wide world of profitable challenges facing even United, American, and Eastern were changing the equations in a delicately balanced system, and few wanted to admit it.

The old problems of human-performance failures were still there—still unaddressed. The new problems were taking on greater significance. Massive changes in the system were needed, but changes were not politically popular, and were not yet possible. But as always, just as in the mid-fifties, the price of delaying needed changes in the system would be human lives. The reason for delay never really matters. The effect does.

The tragedies waiting in the wings would not be limited to commuter aviation. The lessons of the Downeast disaster and the FAA's weaknesses affected all carriers, great and small.

Chapter 13

BEYOND PILOT ERROR

John Merryman sat at his dining-room table in Brunswick and opened one of the copies of the NTSB "Blue Cover" report, which Al Diehl had been thoughtful enough to send. Almost exactly a year had passed since that horrible evening of May 30, 1979, when the news of his brother's death had invaded their home. The wounds were still raw.

Yet the report had to be read. The encounter was the point of convergence of two dissimilar worlds—the highly technical work of the NTSB finally touching the highly emotional environment of those left behind.

Sharon Merryman opened a copy as well, and the two began reading.

The National Transportation Safety Board determines that the probable cause of the accident was the failure of the flightcrew to arrest the aircraft's descent at the minimum descent altitude for the non-precision approach, without the runway environment in sight, for unknown reasons.

Although the Safety Board was unable to determine conclusively the reason(s) for the flightcrew's deviation from standard instrument approach procedures, it is believed that inordinate management pressures, the first officer's marginal instrument proficiency, the captain's inadequate supervision of the flight, inadequate crew training and procedures, and the captain's chronic fatigue were all factors in the accident.[1]

The report went through the litany of technical descriptions and findings and, exactly halfway through, repeated the charges against Downeast's practices. Even though the Safety Board had the power to rule conclusively on whether the problems did, in fact, exist, caution had won out. The heading was: "Alleged Company Unsafe Practices."[2]

1. These quotations are from the NTSB Aircraft Accident Report, DOWN-EAST AIRLINES, INC., NTSB-AAR-80-5, Technical Report Documentation Page, Item 16. Anyone may obtain a copy through the Documents Section of the NTSB in Washington, D.C.
2. "During the course of the investigation and public hearing, fourteen former Downeast pilots and several other employees provided written statements and/or sworn testimony which were critical of the Downeast president's management practices and policies as they related to safety.
  A brief summary of these alleged practices and policies includes the following:

  1. Establishing "company minimums" between 200 to 350 feet, which is below the legal FAA minimums for the Knox County Regional Airport.
  2. Using unapproved instrument approaches.
  3. Avoiding the mandatory procedure turn (which was previously required for the NDB approach to Knox County Regional Airport).
  4. Ignoring takeoff and landing visibility minimums.
  5. Directing pilots to make repeated instrument approaches and to "get lower" during adverse weather conditions.
  6. Directing pilots to go to a particular alternate airport solely on the basis of ground transportation availability, regardless of the reported weather conditions.
  7. Pressuring pilots not to carry "extra" fuel, especially IFR reserve requirements.
  8. Pressuring pilots into flying over gross weight limits and reportedly permitting ground personnel to overload aircraft and provide pilots with knowingly inaccurate baggage weights and counts.
  9. Failing to provide pilots with current training manuals and company operating manuals.
  10. Discouraging the training officers or chief pilots from providing adequate flight training by suggesting that training is unnecessary.
  11. Permitting grossly exaggerated or inaccurate flight and ground training rec-

Though the Board would not come right out and validate the allegations, it at least chronicled them and incorporated them into the list of the things that contributed to the disastrous end of Twin Otter 68DE. That, in itself, was at once a victory for the NTSB members who had worked so hard to bring out the human-performance failures of the crash (including the human failings of management), and for those who had testified.

On the underlying question of the FAA's involvement, however, the NTSB pulled no punches:

> The Safety Board believes that the FAA's surveillance of Downeast Airlines' operations practices should have detected, and caused to be corrected, the deficiencies discovered during the Safety Board's investigation of this accident. The FAA also should have acted when it was informed by a Downeast captain of questionable company practices. The Safety Board realizes that the same FAA operations inspector responsible for surveillance of this company

---

ords to be presented to FAA inspectors.

12. Offering to pay fines of pilots who received violations and suggesting that FAA enforcement actions were unlikely.
13. Ridiculing pilots in front of others and suggesting that pilots who were unable to land when others had landed were less skilled or were cowardly.
14. Failing to report incidents as required by 14 CFR 135.57 and 135.59.
15. Using an aircraft with a history of propeller feathering problems in 14 CFR 135 passenger operations.
16. Pressuring pilots into flying aircraft with known mechanical defects contrary to the 14 CFR 135 requirement, or contrary to good operating practices.
17. Threatening a pilot for cancelling a revenue flight because of a mechanical defect which had occurred away from Downeast maintenance facilities (e.g., landing gear problems at Boston) and generally insisting that aircraft, if "flyable," always be brought back to Rockland.
18. Firing a pilot for cancelling a revenue flight which in his judgment could not be conducted safely because of weather conditions.
19. Firing a pilot for de-icing an aircraft without prior approval.
20. Providing only minimal training to mechanics on equipment with which they were unfamiliar (e.g., DHC-6 aircraft).
21. Permitting unsupervised weather-observer trainees to make and transmit observations and the use of uncertified personnel to make weather observations.
22. Discouraging weather observers from using balloons because of the expense.
23. Intimidating weather observers with regard to their observations." (NTSB Aircraft Accident Report, DOWNEAST AIRLINES, INC., NTSB-AAR-80-5, pages 18–20).

was also responsible for about 23 other Part 135 operators in the New England area. The size, and more particularly, the distant locations of these operators would have created a heavy workload and, therefore, made it difficult to accomplish these inspections adequately. Nevertheless, the detection and correction of operations such as the one uncovered during this investigation are vital to safe operations in the commuter/air taxi industry, particularly with the advent of deregulation and the introduction of larger, more sophisticated aircraft into the industry.[3]

The report went on to list its twenty-one findings and its safety recommendations to the FAA.

It was all there. Jim Merryman had "bought" a substantial amount of the blame, but the underlying human causes had made it far more accurate than a simple accusation of pilot error.

John Merryman closed the report and sat for a moment, deep in thought. Bob Stenger's operation had been laid bare for the world to see. Unfortunately, it was too late. Too late for his brother, Jim, and too late to create any direct action to bring Downeast in line. Robert Stenger had sold his Boston-Rockland route to Bar Harbor Airlines just the month before. The problem had been solved by default.

On July 22, 1980, another report was issued by the Safety Board, this one a bombshell critique of the FAA's foot-dragging in the area of commuter-airline safety from the beginnings of the industry in the early sixties to the present. The Downeast investigation had been a large motivating factor in the report, which examined a host of other commuter crashes to reach the conclusion that the only path to commuter-airline safety was through tough enforcement and surveillance action by the FAA.

But first, the FAA had to admit that commuters were airlines, shouldering the full responsibilities of large Part 121 airlines. Therefore, the rules for Part 135 commuters must be revised to require standards for commuters as close as possible to the standards required of the large airlines.

In the final analysis, it was the FAA and only the FAA that could make the system safe, or keep it safe. Looking to the indus-

3. Ibid., page 28.

try for voluntary compliance would never be enough (though the majority of the industry, commuters and majors alike, usually have maintained the highest standards voluntarily). There would always be people willing to put profit before safety.[4]

The board ripped the FAA apart for taking six years to rewrite Part 135, and then for failing to put any substantive effort into really implementing it. The study cited an earlier NTSB report all the way back in 1971 in which a host of safety recommendations had been made to the FAA on the same subject: Realize commuter airlines are airlines and upgrade their standards. Most of the recommendations had been effectively ignored during the years when the commuter death rate remained more than two and a half times greater than that of the major carriers.

Then deregulation had been cranked into the equation, along with the pious words of Congress stating clearly that it did not intend the changes in economic regulation to be accomplished at the expense of safety. "The FAA," said the study, "was slow to recognize the commuter airline industry as an air transportation industry or to devote the necessary manpower, resources, and assistance to overseeing its activity." The overloaded FAA work force was also cited, the impossibility of individual inspectors adequately handling work loads that included twenty and thirty Part 135 carriers being stated in print for the first time. The fact that airline-surveillance functions were being required of general aviation FAA men (who were not as a group trained in the philosophy of handling airline operations) was also spotlighted, as was the statement of one FAA District Office supervisor who told the board flatly that proper surveillance of just two commuter airlines in his district would take the full-time attention of one inspector.

In fact the board was documenting a problem almost as serious as aircrew fatigue—inspector fatigue. Inspector burnout, disillusionment, disgust, and retirement were rising in direct response to the incredible pressures that the deregulated environment was placing on their shoulders, yet manpower was being reduced in-

4. "The basis to sound commuter airline safety must come from a coordinated program which includes the implementation of 14 CFR 135 [the rewritten and long-delayed Part 135 rules], Federal Aviation Administration surveillance and enforcement efforts, and a strong safety-oriented posture by commuter airlines. Coupled with this program must be the permanent recognition by the Federal Aviation Administration of the commuter industry as an airline industry rather than as a segment of general aviation." (NTSB-AAS-80-1, July 22, 1980, Technical Report Documentation Page, Item 15)

stead of increased. The "quality" of what little surveillance already existed in the case of dedicated FAA inspectors was headed downhill in every region.

> [Several] FAA [General Aviation District Office] inspectors have testified at several Safety Board accident investigation public hearings that to accomplish GADO workload requirements this position requires, loose interpretation of what comprises adequate surveillance [was necessary]. [Another stated that] "surveillance is obviously causing us to leave other things out," [and another official said], "The FAA took the position that deregulation wouldn't affect manpower, but it *has* affected our manpower."

Flight- and duty-time problems for the aircrews were documented: the ability of a company to fly its commuter pilots up to one hundred hours per month (actual block-to-block flight time) while the major carriers, usually limited by union contract, would not allow more than seventy-five to eighty-five hours—though commuter flying was usually far more tiring, involving shorter legs in worse weather, using less sophisticated aircraft and sometimes primitive instrumentation. The report talked about the ludicrous loophole that let a Captain George Parmenter fly a crushing flight schedule at age sixty-one when he would not have been allowed to fly a single leg anywhere as captain on a major carrier after his sixtieth birthday. The lack of heavyweight, single-engine flight training for commuters was mentioned, along with the less stringent and barely enforced rules requiring minimal qualifications for flight-operations managers and other members of the commuter carrier's management.

And on the central issue of the Downeast case, the influence of management and the direct connection between economic pressure and lowered safety, the report listed the areas of major concern, which could affect even the biggest and the best of the commuters, or for that matter the majors:

> The field survey revealed that about 65% of the commuter managers believe that there is a relationship between safety and financial and economic posture. According to the survey, financial problems could lead to: (1) disregarding procedures and regulations; (2) discouraging pilots from listing maintenance deficiencies which might ground an aircraft; (3) increasing the likelihood of placing extra passengers or cargo on an aircraft that was at the

maximum gross weight; (4) reducing the spare parts inventory, which would affect the maintenance program; and (5) reducing the quality of pilot training programs.

The number of commuter airline managers who encourage or tolerate regulatory noncompliance and poor operating practices seem to be few. . . . The Safety Board believes that the FAA has the means to bring about management improvements in the companies, and that company managers, pilots, and employees must demand uncompromised standards of safety within their respective airlines. Since management's philosophy is a subjective topic, it is difficult for government regulators to assess the adequacy of the management of an individual company before the company is certified, especially if the minimal requirements are satisfied and an acceptable management structure exists. Thus it is incumbent upon the FAA to emphasize continuing surveillance [of] the commuter airline industry to develop sound management structures which meet the expansion needs of individual commuter airlines so as to maintain a high level of management integrity regarding the safety aspects of operational, training, and maintenance programs.

Those are fine words—a caveat that the FAA (so some truly believed) would surely heed. But in the FAA as a group and in the minds of the individual chiefs and inspectors, another viewpoint prevailed: We cannot regulate management style. . . . We have no *way* of regulating management style. . . . We are *helpless* to regulate management style.

At the very moment that the commuter-safety report was issued, those same problems of economically suppressed safety considerations were eating away at the foundations of three carriers in particular—one a major-come-lately, the other two established commuters-gone-big-time. The same influences were also at work in the established majors, causing corner cutting in training, reduced inventories in maintenance, the dispatch of airliners with lower and lower reserve fuel loads, and hundreds of other permutations. It was never one single item, it was the trend that would put each of those three in the headlines in the coming few years, widowing wives, orphaning children, and wiping out entire families in crashes that could have been avoided had the FAA known what was really going on (and had the FAA had some method of preventing such problems). Ultimately it was to be the management styles—the oversights, the mistakes, the influences, the abdications of responsibility—that would lead to more aerial disasters. People problems. Human-performance problems. Subjective problems. All of them far more in control of the system than the FAA.

On any airliner, on any day, in any airport, most any passenger you might have asked in the summer of 1980 would have given the same reason for trusting his or her life to that carrier: "If they weren't safe, the FAA wouldn't allow them to fly!" But seldom have so many people been so profoundly deluded. The FAA has no such control. In terms of the vital influence of management, the airlines are on the honor system.

The true bombshell that rolled out on the floor during the Downeast hearings back in September 1979 was hardly noticed. It simply fell into that arena with a thud, and because the press didn't see it and the NTSB sidestepped its darker implications most adroitly, it lay there gathering dust, but ticking toward detonation nevertheless.

The revelation that said it all didn't reach the written record until most of the Downeast hearing participants had gone home.

It was the third day, the thirteenth of September (before the Air New England hearing got under way in the same hearing room). As the headlines about inordinate management pressure on pilots and the other allegations ricocheted around New England and the national wires like a battlefield crossfire (with Robert Stenger in the middle), an FAA chief from the Boston area by the name of John Roach sat on the witness stand and calmly shot down the claim that the FAA could protect the flying public.

John B. Roach was both a veteran air-force pilot/officer (a retired full colonel) and a veteran FAA man with eleven years of service. He had just taken over the position of FAA General Aviation and Air Carrier Branch chief in the New England Region, headquartered in Burlington, Massachusetts, when Otter 68DE was destroyed. Now, after two days in the audience at the Downeast hearings, he had been called in to testify before the NTSB on general aspects of commuter-airline enforcement and surveillance. He had not intended to make headlines, and in fact he did not. What he had to say would escape the notice of most of the media, and its significance would remain hidden to all but a few at the NTSB.

Bill Hendricks of the NTSB team had just asked him what could be done when information about management pressures on pilots and similar problems were brought to the attention of an over-worked FAA inspector in an FAA General Aviation District Of-fice—such as Robert Turner's office in Portland.

Colonel Roach smiled and waded into the question.

"As a field-office chief, I have had pilots, mechanics, flight atten-dants, come in and report things that did not fall within the pur-view of the FAA. As a result of that, the only thing you can do is approach the operator and discuss the item.

"Now, the [complaining] individual in most cases wishes to re-main anonymous, and I can understand that. If the individual is real serious and has a very serious problem, even though it is not covered by the Federal Aviation Rules, we would like that in writ-ing. There are several reasons.

"First of all, for the last two days I sat and listened to a discus-sion where the operator [Downeast] was allegedly coercing and ha-rassing pilots and trying to entice them to break minimums on instrument approaches. There is no rule which says 'Thou shalt not harass pilots.' Okay?

"The only thing we can do—and this has always been an FAA policy, is to approach the management people and try to ask them to refrain from this type of an attitude . . . one of the reasons being that you will find the pilot who's weak enough to go ahead and violate the rules. And the pressure from management people may be just the thing that causes that kind of thing.

"It is our policy that we will discuss that with the management people . . . that is as much as we can do. If the [management] individual sits there and says, 'No, I'm still going to tell the pilots to break minimums,' we discontinue to talk to him. It's very inef-fective, but there's no other way to approach it. As long as the pilot does not break minimums, then there is no safety involve-ment, there is no hazard to people, there is no violation of Federal Aviation Regulations.

"The pilot always has—as we say, the pilot in command is the final authority. And if he elects to exercise that authority, fine, then he has provided a level of safety that we desire. If the com-pany should turn around and fire the individual for that particular [exercise of his authority] . . . there's no way we can protect him and ensure that he gets his job back. . . ."

The answer was incredible, though quite accurate. What it meant was that even if Robert Turner had uncovered full, detailed evidence of all the allegations of pressure and coercion at Downeast—even if he had sworn testimony in hand—unless he could actually prove a specific violation, the Federal Aviation Administration of the United States of America, with all its statutory authority and its solemn mandate to protect the flying public, was virtually powerless to do anything but come hat in hand to Robert Stenger and beg him to please mend his ways. That a manager like Robert Stenger would have been unmoved by such a plea was obvious.

The implications were staggering. Here was the Deregulation Act of 1978 blurring the lines between Part 135 commuters and full-fledged Part 121 air carriers—forcing the public to regard Part 135 carriers as an integral part of the air-transportation system—and the governmental agency charged with keeping it all safe was virtually unable to act in the face of any major management problem that might adversely affect an airline, unless actual violations could be proven.

That, of course, begged the question of whether a Downeast-style management problem could even be detected by FAA "surveillance." What's more, the problem by definition could not be limited to commuters. No management problem of any scope would be within the "purview of the FAA" under the present system—including the problems that had finally driven retired FAA Air Carrier Inspector Al Koleno to quit Air Florida: a management unwilling to spend time or money on the finer points of safety because of the more important demands of remaining profitable.

Al Koleno jumped ship in June 1980, just as the NTSB report on Downeast came out and the commuter-safety report was being readied. He would be replaced by another former FAA man from the confused Miami office, Ed Cook, who would try to build his training program around the wishes of management.

Significant damage had already been done at Air Florida, however. The influence of rapid expansion, the history of its effect on the growing carrier, had loaded and cocked a gun. In the winter of 1981–1982, it would go off.

Chapter 14

A CERTAIN MOMENTUM

Something was not right. It was not obvious—in fact Joe Stiley wasn't sure exactly what was bugging him. But something about this crew and the way they were going about their job of getting the jetliner ready to leave the gate wasn't the way it should be.

Joe had opened one eye to survey the interior of the stuffy green and blue cabin of the Boeing. He had tried to relax. He had even considered a nap, but another passenger had come back from the front part of the cabin to talk to an acquaintance immediately behind Stiley. The man stood with his arm half on the seat back and half on Joe's head, talking in a loud voice and jostling the seat, and Joe Stiley couldn't keep his eyes closed. Sleep apparently was not going to save him from the irritation of the moment, any more than the homogenized, happy-talk in-flight magazine—*Sunshine* — which had failed to divert his attention. He was mentally restless and he knew it, but delays like this could approach the ragged edge of intolerable.

Joe's administrative assistant, Patricia Felch (Nikki to her friends and employer), an attractive young brunette with sparkling eyes and a quick smile, sat next to him in the window seat, engrossed in her own thoughts. Well, not actually in the next seat—the middle seat between them was empty. That would give them room to work on the way southbound. Joe's suit coat sat folded in the middle seat, his airline tickets stuffed in an inside pocket along with his glasses.

The window over Nikki's left shoulder had finally fogged over. The wintry view of heavy snowfall and freezing temperatures, wisps of steam and spray from the de-icing operations going on around the airport on various planes, could only be imagined now.

That, in fact, was one of the irritants. Joe Stiley recalled that only one side had been done—at least on the second go-around. It was like waiting for the other shoe to fall, waiting for the noisy-as-hell sound of fluid under pressure to hit the fuselage once again. Nikki and Joe were sitting just aft of the left wing on the left side—seats 18A and C—which is where the de-icing had started over an hour earlier. No warning from the pilots, no warning from the cabin crew. Just a sudden cascade of cacophonous noise inches from their heads, hitting with such force that it caused them to jump, and streams of reddish-colored liquid flowing down the windows—colored ethylene glycol de-icing fluid no doubt, mixed with hot water, washing off the ice and snow and trying to leave a protective film of antifreeze. Joe Stiley knew the routine, and he also knew such protection probably wouldn't be effective for more than a half hour. After that, the constant snowfall could start building up again on the airplane's fuselage and wings, and any melting water would freeze into ice.

Damnit! They had been sitting there at the gate, in the airplane, for over two hours already. The airport had closed down for snow removal just before their noon arrival—and here they remained.

Joe was a veteran flyer. He made this trip down south as often as once a week, and usually on this airline. The other choice to Tampa—an Eastern flight—wouldn't get him there until late. That would mean a 1:00 A.M. arrival at the hotel and a groggy following day at the GTE office. He'd rather arrive early on this flight, have time for a brief late-afternoon meeting, then a good night's sleep.

The de-icing had begun again—the second de-icing—about a half hour ago—this time with perfectly clear liquid. Joe had noticed that. Possibly hot water with no de-icing fluid. He wondered why. That had been on the right side. They hadn't done the left side yet. Surely they would repeat the process before push-back. This snowfall was nasty, and no pilot would want to take off with snow or ice adhering to the wings of his aircraft—not to mention the fact that to do so was illegal.

This was all familiar territory to Joe. He too was a pilot. Not an airline pilot, but a flight instructor and private pilot with a commercial ticket and more than ten thousand hours flight time. Quite a few of his students had gone on to join the airlines, but he had not followed them. Joe had never piloted a Boeing jetliner, but he had been around them—having worked for Boeing years before as a guidance-system engineer on the Minuteman missile project. For a while he had flown back and forth to Europe on Pan Am as a flight engineer—testing a Boeing inertial-guidance system. That had been fun, but as an electronics engineer with a bachelor's in physics and an MBA, his chief love was computer technology. It was an ordered world—a procedural world. With such training, when something was out of sync, he noticed it.

And something was definitely out of sync around this operation.

As he sat there, trying to ignore the fellow still standing and yapping behind him, he recalled a small thing he had seen just before boarding the aircraft: There were no footprints in the snow.

Just before he and Nikki had stepped into the cabin, he had noticed. Just a little fact, but unusual and, well, not normal. There were footprints around the baggage compartment, and tire tracks all over the place—but no footprints around the wing tips and the engines and the tail. Strange.

A raucous group of men (he thought he had heard the company name of Fairchild) had come on board just before the last de-icing sequence and sat down in various rows behind them. A number of them were kidding with the stewardess, trying to talk her into free drinks, and generally laughing and talking among themselves. Apparently some sort of enjoyable business trip, which was going to be more of a holiday.

The noise of an aircraft cabin door being closed, and the accompanying flux in the cabin pressure caught his attention, followed by the muffled noise of a gasoline engine and a gentle lurch—as if they were getting ready to push back.

Joe looked up the aisle at the cockpit door, which was open. Those two guys had been sitting there since he had come on board, snug and secure in their warm cockpit, wearing short-sleeve uniform shirts with the striped epaulets that identified one as captain (four stripes), the other as first officer or copilot (three stripes). Joe had glanced in as he had boarded. A Boeing cockpit and two pilots. Nothing unusual—except that the door was still open.

One of the stewardesses looked familiar—he was sure he had flown with her before. She had been in and out of the cockpit during the last two hours. Now she was coming down the aisle, apparently in preparation for departure. But where was the de-icing truck?

There must be a warm hangar with de-icing facilities inside awaiting this jetliner. Surely they were going to tug it over there and clean it up. Joe was convinced that the airplane had too much snow and ice on it now for takeoff.

Irritations had been accumulating all day just like the snow outside. A 10:00 A.M. meeting with the president of his GTE division (Joe Stiley was director of program management for Subscriber Equipment) had been preceded by the news that Huntsville, Alabama, his first stop on this business trip, was snowed in and the airport closed. So they would have to go to Tampa, the second stop, first. Nikki had changed all the arrangements.

The meeting itself had been difficult. The drive through growing snowdrifts was even more annoying, especially since the weather bureau had been mistaken in assuring the city of a slackening snowfall. Whatever it was going to do later, just before the noon hour, Eastern standard time, January 13, 1982, the weather was working hard to attain the classification of "blizzard."

The hassle at the ticket counter had not helped Joe Stiley's mood. The airline shared space with American Airlines, whose agents couldn't help him. Twenty minutes passed as he and Nikki Felch stood there waiting. Finally, an agent from the airline had shown up and processed their tickets—explaining that the aircraft had made it in, was on the ground, and would depart when the airport reopened. Twenty minutes! This particular airline had been expanding rapidly, but in his experience it wasn't usually that disorganized.

The sound of turbine engines starting up caught their ears. First one, then the second. The noise changed with a "whooshing" sound, indicating reverse thrust. The power came up a little, the aircraft rocked back and forth a bit, then the engine noise fell, and died. They were still at the gate.

The captain came on the public-address system within a minute, explaining that the ramp was slick and the tug they had been given could not get enough traction to push them out. A larger tug had been ordered. When they got out, he said, they would be number three or four for takeoff. The airport was just now reopening. Joe Stiley looked at his watch. It was exactly 3:30 P.M.

Bert Hamilton's fellow executives from Fairchild may have been having a good old time, but he was not. This delay was getting to him, too. Bert had looked at the ceiling speakers during the captain's explanation—as if by reflex—pulling his attention away from the book he had been reading, a paperback tome with a positive-thinking theme. He had been working hard to avoid boredom—and to escape the slight twinge of apprehension that he'd been feeling all day.

The apprehension had nothing to do with the weather, although the heavy snowstorm, which had really cut loose about 10:00 A.M., had given him a hard time driving the thirty miles to the airport with his boss. Just five miles from the airport he had momentarily lost control of his car coming down an icy hill and had almost slid into oncoming traffic. That had been enough excitement for one day. Basically, he just did not want to go on this trip. He was a purchasing agent for Fairchild and had been with them only some ninety days. He didn't really know any of these fellows, and he couldn't understand why this trip was necessary. They were going to Tampa to do a vendor survey on a government contract. Why on earth would the company pick up the buyers and take them along as well? That made no sense to him at all—but then, he didn't make policy.

And then there was the matter of this particular airline. Bert Hamilton never minded flying the big carriers, but these smaller ones—the new-entrant carriers and upstarts that hadn't had years of stable service—bothered him. He had always felt that way. He would rather stay off the less established airlines, and wanted nothing to do with commuters and their tiny airplanes. Deregulation was acceptable in principle to a good Republican like him, but he wasn't too sure whether he liked the real, in-the-flesh results.

The other Fairchild employees were sitting ahead of him in rows 19 and 20. He was in 21D, the aisle seat on the right side, and the very last row in the airplane. He felt more comfortable in the back part of the cabin—probably because of that old theory about being safer in the tail end of an airliner. That was silly, though. That old theory was from piston and prop days. These days, with jet airliners, he thought it an unshakable truth that if you crashed, you died, no matter where you were sitting.

Bert had looked up periodically from his book and taken note of goings on in the cabin. A young woman with her baby and husband was sitting four rows ahead. They looked awfully young. He noticed the open cockpit door, and he noticed the flight attendant, Kelly Duncan. For some reason she seemed a bit of an outsider among the crew members. The other two flight attendants seemed to be talking to each other and the pilots, but no one was talking to this gal—and she was taking the brunt of the passenger irritation over her supervisor's relayed refusal: The airline would not serve drinks despite the delay. Bert felt sorry for her. She acted out of place.

Bert Hamilton was not a pilot, though he was a retired, twenty-one-year air-force veteran. He had been an enlisted man in the supply area all his career, not flying. Yet he now flew often enough that this should all be routine. Unfortunately, today it didn't *feel* routine. There was still that unexplained, nagging, apprehension.

Eastern Captain Larry Jones, a forty-two-year-old veteran airline pilot, took his brightly colored Boeing 727 out of the electronic clutches of the autopilot and began hand-flying it through the beginning of the descent. The weather ahead was awful and it would be a minimums approach at their destination airport. There was a snowfall in progress, runway braking factors were probably poor, and it would be a very tight approach. As he knew well, it is at such times that airline crews earn their pay. This is when experience counts.

This time the tug was big enough. The Boeing jetliner began moving backward out of the gate, over the frozen puddles of water and the slick de-icing fluid that had spilled on the ground—the glycol—which had prevented the smaller tug from gaining any traction. The American Airlines ground crew had been caught off guard by the problem—and by the captain's attempt to use reverse thrust to help the smaller tug. Using reverse thrust to get out of the gate like that was specifically prohibited in American's procedures. The tug driver had told the pilots that over the interphone, but they used reverse anyway. Maybe this airline's procedures were different.

One of the American ground crewmen had inspected the engine inlets after the engines were shut down, and they seemed clear of ice or snow. With the reverse thrust applied, there had been an awful lot of snow and ice and water droplets blowing back in their faces, however, as well as swirling around the leading edges of the wings—and that without much more than idle power on the two Pratt & Whitney turbojets.

As they moved backward, the tug driver could see the snow that had already accumulated on the fuselage and the wings. It seemed ironic that they could move so much air with those jet engines and yet the ground crews had to struggle with brooms and de-icing trucks to clear the stuff off the aircraft in such weather.

200

He keyed his microphone button, talking to the cockpit crew as he steered the tug and tried to ignore the biting cold.

"Bet those vacuum cleaners would do wonders as a snow melter."

One of the pilots replied in his earphone, "Sure do!"

A bit farther out onto the ramp he keyed the mike again.

"You can start engines if you want. I don't know whether you've got 'em running or not."

With all the noise from the surrounding machinery on the ramp, the swirling wind gusts carrying wisps of snow and ice crystals, and the noise from his tug's engine, he couldn't be sure.

The same voice came back in his earphone.

"I'll tell you what; I'm gonna wait till you disconnect before I start them up so I can get the [thrust reverser] buckets closed."

The tension in the airport's control tower was high. The conditions could hardly be worse. The airport was just reopening with the snowplows finishing up on the main runway, and a long line of flights was out there unseen in the distance, in holding patterns, waiting to get in. There was just as much of a crowd of snow-dusted planes at the various gates, waiting to get out. The airport was too tight on real estate to permit a gate-hold procedure.

At the bigger airports around the country, ground control could keep the aircraft at the gates until it was almost time for their departure slot. That way they could be de-iced, pushed back, and then get to the end of the runway with no snow or ice problems and without wasting fuel and time sitting on taxiways. That was elsewhere. At *this* crowded facility there were too few gates crammed together in a complex of tight ramps. With inbound flights arriving now, there was no way the outbound flights could be held in their gates.

Approach control was already sequencing the incoming airliners, each one of which would be handed off to the tower controller for final landing clearance only a few miles out. It didn't take a genius to realize that the next few hours were going to be a real trial.

Rudolph West was manning ground control, and he already had a snarl of traffic, with several jets trying to push out at once. The visibility around the airport was so poor he couldn't see the run-up area at the end of the runway—and they had no ground-position radar. If anybody wanted to get out of sequence down there after they got out of his view and into the lineup for takeoff, it was going to be a hassle.

In the cockpit the engine start sequence was under-way again, as first number two, then number one came up to speed. The tug crew had disconnected and were waving them off—indicating it was clear to taxi. The first officer, former air-force F-15 fighter pilot Roger Pettit, picked up his microphone and dialed in the ground-control frequency.

West keyed his ground-control mike again. A New York Air DC-9 (all New York Air flights were called "Apple" in air-traffic radio transmissions, along with the flight number) was trying to get by the small Boeing on push-back from Gate 5. West asked the Apple flight if he could get around the Boeing, whose radio name was Palm 90.

One of the pilots of the Boeing, Palm 90, keyed his mike button.
"Ground, Palm 90. We're ready to taxi out of his way."
"Okay, Palm 90, roger, just pull up over, behind that, ah, TWA and hold right there. You'll be falling in line behind a, oh, Apple DC-nine."
"Palm . . . 90."

As American Airlines Flight 624 sat on the ramp, waiting to taxi to the gate Palm 90 was vacating, the pilots were noticing the snow on Palm 90's wings and fuselage. The buildup seemed excessive. But then again, maybe it was dry snow that would blow off.

Now Joe Stiley was worried. The low-grade irritation had accelerated to genuine concern. The damned engines had started again, this time the jetliner was taxiing, and unless he was mistaken, they were headed for the runway without any more de-icing. All he could make out through the opaque window were the lights of the terminal, and they were now sliding out of sight. Their gate—Gate 5—was far behind.

And there was something else. A couple of seemingly unrelated facts finally had connected—two tumblers falling into place and unlocking a realization. No tracks in the snow, and the pilots sitting in their cockpit.

Those guys had never left the airplane! They had never done a walkaround! For over two hours with snow falling and de-icing operations going on, vehicles roaring all around and God knows what else was happening out there, neither of those two pilots had left his warm cockpit to go outside and personally look at his aircraft to see whether it was free of ice and snow, and whether it was flyable.

Oh, the ground people probably took a look. But not the pilots. Joe was sure of that. A walkaround by the cockpit crew might not be a requirement, but it was unprofessional not to do one in his opinion. Even a good private pilot would have gone out and looked.

Suddenly, for the first time since he had been flying on this carrier, Joe Stiley was ready to get off one of its airplanes—*this* airplane—right now!

Unfortunately, he no longer had that choice.

Bert Hamilton was relieved to be moving. Maybe they could get the takeoff behind them so he could relax. He always felt uneasy during takeoffs and landings. He assumed the fellows up front knew what they were doing, but it still bothered him. He felt the Boeing lurch over the packed snow, saw the flight attendants pulling out their demonstration oxygen masks and seat belts, getting ready to give their cabin announcement, and realized that they had stopped taxiing again after only a short distance. Now what?

First Officer Roger Pettit was surveying the frigid scene in front of the Palm 90's cockpit. He had never had much experience in winter weather. Oh, some, yes, along with the required training. But never regular operations in this stuff. It wasn't what his airline normally flew around in, and it was a bit disconcerting—especially with this captain.

The Apple DC-9 had finally taxied in front of them. They were to follow all the way to the runway.

"Boy, this is shitty; it's probably the shittiest snow I've seen!"

The sound of a takeoff warning horn sounded in the cockpit as the captain advanced the throttles to begin taxiing behind the Apple. He had decided to keep the flaps up until they were ready for takeoff. With the flaps up, the warning horn sounded every time the throttles were advanced. The system was designed to prevent pilots from mistakenly taking off without takeoff flaps selected. All part of the preflight routine.

Back in the cabin, Pettit could hear the girls beginning their cabin PA announcement. The captain finally responded to his pronouncement about the snow, saying he wished they could go over to a hangar to be de-iced.

"Yeah. Definitely." That would be a good idea, Pettit agreed. A hangar. Not a messy gate with nearly an hour to go before takeoff. Unfortunately, there was no procedure for de-icing in a hangar.

Pettit had reason to worry about this captain. Though he had

*flown small observation aircraft while in the army during the Viet-
nam War, his background was mainly civilian, commuter flying. He
didn't have Pettit's strict air-force training and indoctrination in procedures and checklists, and it showed. It was hard to tell him anything, so great was his self-assurance. Of course, for that matter he,
Pettit, didn't have much training in multiple-crew operations. He
had always been a fighter jock. Single-pilot operation in which you
do everything yourself. Aircraft commander by default.*

*One thing was certain, though. In an airline environment, the
captain was king. He had to live with this particular guy all month—
they were going to be flying together through the end of January. If
he got irritated enough with this captain's easygoing nature to challenge him on anything, it could be a frosty cockpit for the next two
weeks.*

*Pettit addressed the captain again: "It's been a while since we've
been de-iced."*

*The man with the four stripes and the primary responsibility for
the safety of Palm 90's passengers apparently agreed. "Think I'll go
home and play . . ." he said.*

*Pettit caught sight of a Cessna Citation business jet, a small twin-
engine machine sitting low in the snow on its stubby landing gear.*

"That Citation over there, that guy's about ankle deep in it."

*Both of them chuckled at the sight, as the senior stewardess,
twenty-three-year-old Donna Adams, stuck her head in the cockpit.*

*"Hello, Donna." Petit acknowledged her as she wedged in be-
tween the two seats.*

"I love it out here!"

*"It's fun," Pettit replied, watching the Apple DC-9 just ahead for
movement. It was 3:41 P.M.*

"I love it!" she repeated.

Roger Olian had left St. Elizabeth's Hospital early. He was one
of the building maintenance men—a sheet-metal worker. He enjoyed his job, but the snowstorm had prompted most of the city's
employers—including the government—to send their people home
early. A massive traffic jam was in the offing as it was. Maybe they
could relieve it a bit by an early shutdown.

That had been at noon, but Olian had not left until well after
2:00 P.M. For the next hour he was stuck in traffic, moving at a
crawl toward the suburb where he lived. Finally it looked as if the
traffic was moving a bit faster as the arterial approached a bridge
ahead. The time was 3:47 P.M.

Priscilla Tirado was fascinated by airliners and flying. She didn't know much about the subject, but she was curious and excited— and eager to fly. Beside her in the still-stuffy cabin of Palm 90 was her baby son, Jason, all of nine weeks old—her first child—and her husband, José, a native of Spain. The family was on its way to a new job waiting for José in the Tampa area. Priscilla's father was a well-known newspaperman in the same city. She was looking forward to being close to home again.

Priscilla's window next to seat 17F was fogged over, which was upsetting because she wanted to see out. Every few minutes she could hear the muffled roar of other airliners as they took off. She hoped it would be their turn soon.

Palm 90's pilots were right behind the Apple DC-9, letting the hot exhaust gases from the DC-9's two tail-mounted engines blow on their aircraft, possibly helping blow off the snow and ice. The captain was apparently straining to see out his left window, looking at the left wing.

"Gonna get your wing now."

"Did they get yours? Can you see your wing tip over 'er?" Pettit asked.

"I got a little on mine." The captain again.

"A little . . ." Pettit didn't seem too sure about this procedure, but it sounded sensible. "This one's got about a quarter to half an inch on it all the way."

Both pilots resumed their monitoring of the Apple DC-9 just ahead.

Pettit had spotted icicles.

"Look how the ice is just hanging on his, ah, back, back there, see that?"

No response.

"It's impressive that these big old planes get in here with the weather this bad, you know, it's impressive."

Still no response from the captain, who seemed lost in thought.

"It never ceases to amaze me when we break out of the clouds, there's the runway anyway, d'care how many times we do it. God, we did good!"

Pettit laughed and the captain joined in. The first officer finally had his attention, so he pointed at the DC-9 again.

"See all those icicles on the back there and everything?"

"Yeah," the captain acknowledged at last.

"He's getting excited there, he got his flaps down, he thinks he's getting close." Pettit laughed again, this time alone. He let a few seconds pass, apparently running his eyes over the gauges, including the engine pressure ratio (EPR) gauges on the center panel—the instruments that measure the amount of thrust put out by the engines.

"See this difference in that left engine and right one?"

Pettit had reason to be suspicious. The engines were both at idle, but the EPR indication on one of them seemed to be fluctuating when it should not have been.

"Yeah," the captain replied.

"Don't know why that's different"—Pettit paused, but the response didn't come—"'less it's his hot air going into that right one. That must be it . . . from his exhaust."

First Officer Pettit fell silent for a second. That couldn't be the reason. "It was doing that [at the gate] awhile ago but, ah . . ."

The voice of ground controller Rudolph West cut into his concern.

"Okay, Palm 90, cross runway three and if there's space, then monitor the tower on [119.1]; don't call him, he'll call you."

Pettit keyed the mike to acknowledge: "Palm 90."

On the overhead panel, just above and slightly left of his head, a small toggle switch labeled ENGINE ANTI-ICE *remained in the* OFF *position. Pettit had verified that it was* OFF *during the After Start checklist. The switch should have been* ON.

Learjet captain Roger Hough, a veteran corporate pilot with more than sixteen thousand hours of flight time, was appalled at the severity of the snowfall. He had been preparing his airplane for departure at the north end of the airport, but he was not sure he wanted to go when the airport reopened. Now, apparently, it had reopened, and he would have to make a decision. Hough strained to see the first in a salvo of departing jetliners launching off runway 36 into the murky skies.

Tower controller Ronald Montague took the strip with Palm 90's identification on it from ground controller West. Palm 90 would go out right after the Apple "9." The flow was working out pretty well. As long as no one took too long getting off the runway, it should be just a few more minutes for this guy. Behind Montague, controller Stan Gromelski was getting ready to take over Montague's position.

Joe Stiley was becoming acutely aware of the sounds and movements of Palm 90 as he sat within the belly of the Boeing—essentially trapped. He thought about the snow on the lifting surfaces and the snow and ice on the runway. He thought about the long period of time that had passed since the last de-icing, and the fact that only one side had been done. And he tried to accept the proposition that those fellows up front had received many, many hours of highly refined, technical training in this machine, and probably had more than enough experience to be thoroughly aware of what they were doing. Surely there were logical, professional, *safe* explanations for all the telltale signs that were worrying him. Surely the FAA would never have certified this captain without a thorough knowledge of cold-weather procedures. He thought about that—but again his mind came back to the growing list of worries over this flight, and the fact that there was no way out. He and Nikki simply did not have the realistic option of saying, "Thanks, but I think we'll take a later flight—on another airline," and quietly stepping out. You couldn't simply open the door and step out onto a frozen taxiway eight feet below.

Roger Pettit's eyes were filled with the images of ice and snow. Snow blowing, snow falling, snow and more snow accumulating—especially on this airplane.

"Boy, this is a, this is a losing battle here on trying to de-ice those things. It gives you a false feeling of security—that's all it does."

The captain had heard him this time.

"That, ah, satisfies the feds," he responded simply.

"Yeah." Pettit mulled over the way things should be. "As good and crisp as the air is and no heavier than we are, I'd . . ."

The captain interrupted him.

"Right there is where the icing truck, they oughta have two of them, you pull right . . . like cattle, like cows . . . right . . . right in between these things and then . . ."

"Get your position back," Pettit finished for him, but it was the wrong ending. The captain continued.

"Now you're cleared for takeoff."

Pettit agreed. "Yeah, and you taxi through kinda like a car wash or something."

"Yeah . . . hit that thing with about eight billion gallons of glycol."

The captain continued, apparently remembering one of only seven times he'd flown in such weather.

"In Minneapolis, the truck they were de-icing us with . . . the heater didn't work on it; the fucking glycol was freezing the moment it hit."

"Especially cold metal like that."

"Yeah."

The Apple DC-9 moved up a few more yards and Palm 90 followed—creeping inexorably toward the takeoff point.

Stanley Gromelski, standing with the tower controller microphone in his hand and peering into the terminal radarscope in the darkened interior of the control tower, was balancing the "picture" in his mind at high speed. He had just taken over as relief for Ron Montague. Things were really moving now. The digital-clock readout showed 20:56 Zulu (Greenwich mean time), 3:56 P.M. local. He had just cleared Eastern 133 for takeoff, and the Boeing 727 was pushing up his power, beginning his takeoff roll somewhere out there, unseen by Gromelski on runway 36. The winds were zero-one-zero at one-zero, basically out of the north at ten knots. The New York Air DC-9 was next—that would be Apple 58—then Palm 90. Trans World 556 was approaching the outer marker inbound for the same runway for landing, and behind him—though not in contact with Gromelski yet—was Eastern's 1451. Both would be Boeing 727s.

Gromelski keyed the microphone.

"Apple 58, taxi into position and hold; be ready for an immediate."

He knew Eastern 133 would be on his way down the runway before Apple could get into the same spot and stop. That was "position and hold." Taxi onto the runway, line up, then stop and wait. TWA would be on top of them in a few minutes. He would clear Apple to roll as soon as he was sure Eastern 133 was airborne.

"Position and hold, Apple 58." The acknowledgment from the New York Air flight registered in his mind.

"TWA 556 is inside Oxonn." That was the outer marker. TWA was less than five miles now south of runway 36, inbound at about 150 miles per hour. The blip showed clearly on his radarscope, each sweep of the radar beam producing a tiny flare of phosphorescent light—a "hit" in radar/air controller terminology—moving each time closer to the airport.

"TWA 556, roger, runway three six." Gromelski confirmed.

The wind was kicking up wisps of snow in front of the tower cab, and the overall visibility was rotten—3,000 feet out on the approach end of the runway according to the south transmissometer, 1,800 feet on the north one. He could see Eastern 133 rotating now as the 727 rolled past, northbound on runway 36. Within a few seconds the silver and blue paint scheme would disappear in the distance. That's why the radarscope was so important on such a terrible day.

Gromelski checked the wind again, and keyed the mike to give the inbound TWA the latest. "Trans World 556, the wind is zero one zero at one zero, you're cleared to land runway three six, visual range is three thousand, touchdown is at, ah, rollout is one thousand eight hundred."

There was no response from TWA. The crew was obviously very busy with the approach. The 727 was now some three miles south of the runway and clearly visible on his radarscope. He had to get Apple rolling, and it should be in position by now. It would be such a help if he had a remote TV camera down at the end of 36. He had to guess at where some of the fellows were.

"Apple 58 cleared for takeoff; traffic's three [miles] south for the runway."

"Apple 58 takeoff," the New York Air pilot responded, undoubtedly pushing up the throttles at the same time. It would take the DC-9 about forty seconds to get to rotate speed—and TWA by that time would be one and a half miles distant from the end of the runway. Apple would have rolled at least 2,500 feet by then—a half mile—so Gromelski's required separation of two miles between arriving and departing aircraft would just be made. He was trying as hard as he could to get these birds in and out, and sometimes the rules had to be stretched a bit.

"Eastern 133, contact departure control." Gromelski had seen 133's progress off runway 36 on radar. It was time to make him a ward of Departure Control—thus the radio switchover.

"Okay, sir, good day." The Eastern pilot's friendly salutation.

"Good day," Gromelski replied.

Now he had only the departing Apple, the inbound TWA, and Palm 90 to worry about. Wait. There was also that inbound Eastern behind the TWA. He hadn't called in yet.

Pettit and his captain in the cockpit of Palm 90 were finally at the number-one position, watching the Apple DC-9 disappear down the runway into the murk, the roar of the two turbojet engines still audible in their cockpit. "I think we get to go here in a minute," Pettit remarked. He knew there was a TWA on short final.

He read through a couple more checklist items before coming to the power setting.

"EPR all the way two oh four." The setting would be for full power, 2.04 on the EPR gauges—the same gauges that had been fluctuating earlier.

"Indicated airspeed [markers] are [at] thirty-eight, forty, fortyfour," Pettit recited as he positioned the little movable markers on the airspeed indicator on his panel to correspond to 138, 140, and 144 knots of airspeed. When the airspeed needle reached each of those points, he would call out "Vee one," "Vee R," and "Vee two" in sequence.[1]

"[Tower], TWA 556, [are we] cleared to land?"

Gromelski had already cleared the TWA, but had received no response. Apparently they had not heard the call.

"Five-five-six *is* cleared to land," he confirmed. "The wind is zero one zero at one zero."

"Cleared to land, TWA 556."

The Trans World 727 was almost over the approach lights. It would pass Palm 90's place on the taxiway momentarily and Gromelski could get Palm 90 moving into takeoff position.

Finally the inbound Eastern flight checked in.

"Eastern 1451 by the marker."

Gromelski had watched him coming, of course, on radar. This was his first radio contact with the inbound Eastern since Approach Control had called to say he was coming. "Eastern 1451, runway three six."

"Fourteen-fifty-one." The acknowledgment from Eastern.

This one was going to be tight. Eastern was less than five miles

1. "Vee one," or V1 is the commit speed. Theoretically, on a perfectly dry runway, the aircraft can reach V1 and either continue the takeoff with an engine out or stop safely within the remaining runway if an engine is lost and the captain decides to abort the takeoff. "Vee R," or VR, is the rotate speed at which the nose of the aircraft is lifted and the aircraft is flown off the runway. "Vee two," or V2, is the safety speed at which the aircraft can continue the climb-out and clear all obstacles with an engine out. In practice, a normal rotation would still allow enough acceleration time for the 737 to reach V2 speed plus ten knots by lift-off. (In most cases on the Boeing 737 and 727, V1 and VR end up being the same airspeed.)

south, and TWA was over the approach lights, seconds from touchdown. He had to decide whether there was time to get Palm 90 off between them. Was there time for TWA to slow down and taxi off the runway, get Palm 90 into position, cleared for takeoff, and then get him off the runway, before Eastern 1451 touched down? He doubted he could keep them two miles apart. He was going to be pressing this one beyond the limits if he tried it.

"Apple 58, contact Departure Control."

"Fifty-eight, so long."

Another distraction out of the way.

Pettit had asked the captain if there was anything special he should do to compensate for the slushy runway they were dealing with. Slushy runways could be very dangerous, retarding the acceleration of an airplane and using up runway. Was there any special technique the left-seater wanted him to use? It was Pettit's leg—his turn to fly the twin jet.

The captain responded simply. "Unless you got anything special you'd like to do."

Pettit thought it over. "Unless just take off the nosewheel early like a soft-field takeoff or something . . ." No response from the left seat. "I'll take the nosewheel off and then we'll let it fly off." There was no such procedure for large turbojet airliners, but Pettit seemed nervous about the conditions.

Still no response—not even that infernal whistling the captain had been doing for the last half hour. Periodically he'd just sit there, looking out the window and whistling.

Pettit began reviewing the departure, probably as much for his own peace of mind as the captain's.

"Be out the three two six [degree radial], climbing to five [thousand feet], I'll pull [the power] back to about one point five five [EPR], supposed to be about one point six [EPR, noise abatement procedure], depending on how scared we are."

Both of them laughed at that, as the voice of tower controller Stanley Gromelski cut through the cockpit. Gromelski had made his decision, with TWA still over the threshold of the runway and Eastern 1451 bearing down on it, and none of them visible from the tower.

"Palm 90, taxi into position and hold, be ready for an immediate [departure]."

And Pettit responded.

"Palm 90, position and hold."

Air Florida Flight 90, a Boeing 737, began moving toward the takeoff position to await clearance. The time was exactly 3:59 P.M., Eastern standard time. The runway was number 36 at Washington National Airport, located along the Potomac River adjacent to downtown Washington, D.C.

The Air Florida flight was under the command of thirty-four-year-old Larry Michael Wheaton, a former Air Sunshine pilot from Florida.

Ahead of the brightly colored 737, less than the length of the longest runway at nearby Washington Dulles Airport, was the Fourteenth Street bridge—covered with bumper-to-bumper traffic.

The de Havilland DHC-6 Twin Otter, a popular and reliable commuter airliner especially suited to short-runway, clear-weather operations

NTSB investigator Alan Diehl, on the morning of May 31, 1979, taking documentary photos inside the wrecked cabin of Twin Otter 68DE

Jim and Jimmy Merryman ice fishing in February 1979, three months before the last flight of Twin Otter 68DE

Jimmy Merryman standing before the wreckage of the Downeast Twin Otter on the morning of May 31, 1979, after all the victims had been removed from the scene

Robert Stenger, Sr.,
president and owner
of Downeast Airlines,
at the NTSB hearing
in Cambridge,
Massachusetts, in
September 1979

(Portland Press Herald)

FAA inspector Robert
Turner, now retired,
served as principal
operations inspector
for Downeast Airlines
during most of the
1970s.

The smashed cars and damaged boom truck hit by the aft fuselage and landing gear of Air Florida Flight 90 sit sprawled in the snow on the Fourteenth Street Bridge. The photo above is from the Virginia side looking north.

The fragmented tail section of Air Florida's 737 was lifted from the icy Potomac nearly a week after the crash. It had been the opposite side of this section to which Joe Stiley, Patricia Felch, and Priscilla Tirado clung as they watched Roger Olian trying to reach them.

The cockpit top crown section on the bridge after being removed from the river

Joseph Stiley

Captain Jim Marquis

The Hawker-Siddeley 748, a rugged and highly reliable twin turboprop mini-airliner manufactured by British Aerospace

A structural member from the fuselage of Air Illinois Flight 710 embedded in a tree by the impact

Chapter 15

ALONG FOR THE RIDE

"Eastern 1451, keep it at reduced speed; traffic's going to depart. Trans World 556, left turn off [the runway]; if you can, turn at the next taxiway, it would be appreciated . . . nothing's been plowed."

Stan Gromelski had set up a very tight situation here and he knew it. Palm 90 had better get into position fast. Eastern 1451 was bearing down on them rapidly, and TWA, he assumed, was just touching down. He might have to give Eastern a go-around if this didn't work.

The public-address speakers in the cabin of Palm 90 came alive with the voice of one of the pilots.

"Ladies and gentlemen, we have just been cleared for takeoff; flight attendants, please be seated."

Joe Stiley took a deep breath and frowned. He glanced at Nikki Felch who was glancing at him, wondering why her boss seemed so uptight.

In seat 21D Bert Hamilton adjusted his seat belt and waited for the acceleration.

Stewardess Kelly Duncan finished her final walk-through and sat down in the rear stewardess jump seat. The flight attendant who would have been her companion in the rear cabin was Marilyn Nichols, but Marilyn had decided to sit in the front jump seat with her old friend Donna Adams. Kelly was alone in the back.

Wheaton and Pettit were going through the takeoff checklist, Pettit reading the items, Wheaton responding.
"Flight attendant alert?"
"Given."
"Bleeds?"
"They're off."
"Strobes, external lights?"
"On."
"Anti-skid?"
"On."
"Transponder?"
"On."
"[Before] Takeoff [Checklist] complete."
Stanley Gromelski's voice rang through their headsets once again with the anticipated clearance.
"Palm 90 cleared for takeoff."
Pettit responded. "Palm 90 cleared for takeoff."
Gromelski again: "No delay on departure if you will; traffic's two and a half out for the runway."
"Okay."
Wheaton stopped the 737 in position and gave control to Pettit.
"Your throttles."
"Okay."
The first officer's left hand advanced the two throttles to the approximate position of vertical—the point at which the engines would normally achieve stabilized intermediate power. As soon as they had spooled up, he would nudge them into the final setting, a little bit farther forward.
"Holler if you need the wipers," Wheaton said.
The time was 3:59 P.M. and 49 seconds.
The sound of turbines winding up, the engines increasing their power output, filled the aircraft.
"It's spooled." Wheaton again, monitoring the EPR gauges.
The EPR gauge needles shot up as he spoke, advancing toward their maximum thrust target setting of 2.04—and promptly going above that.
"Ho!"

222

"Whoo."

"Really cold here!" Wheaton had grabbed the throttles and pulled them back a few inches, trying to keep the engines from exceeding the maximum allowable thrust setting—apparently assuming the cold weather had somehow caused the familiar vertical throttle position to produce too much power.

Pettit looked over. "Got 'em?"

"Real cold!" Wheaton again, trying to adjust the power setting to 2.04 as Palm 90 began moving forward. "Real cold."

Roger Pettit, both hands now on the yoke, was staring at the engine instrument gauges.

"God, look at that thing!" Pettit seemed startled.

The EPR needles were right where they should be, but the other indications—the two different engine RPM readings for each engine—were all wrong. They seemed low. The fuel flow looked low. It was all very unusual.

Air Florida's Flight 90 was now moving north on runway 36 at just below fifty knots.

The impacts in the cabin from the irregular, ice-covered runway surface were jarring to everyone. Priscilla Tirado patted her young son to counter the startling, jolting influence of the sound and the feeling.

Boom, boom, boom, boom. Slow at first, then faster as the jetliner gained speed. It didn't seem as rapid an acceleration as usual, though. She wasn't being pushed back in her seat.

Kelly Duncan realized instinctively that something unusual was happening. The booming impacts of the rough runway surface were merely annoying, but she knew how this airplane should feel, and it didn't feel right. It wasn't moving very fast. It wasn't gaining speed fast. She didn't feel the usual pressure shoving her back in the seat.

Bert Hamilton put down his book. This was *not* normal. He suddenly realized how slow a takeoff roll it had been—how little acceleration there had been. These jets normally took off very fast. He formed the complete sentence in his mind: We've got a serious problem here.

Joe Stiley knew immediately. He had had a sick feeling as they rolled onto the runway. He was trapped in an airplane flown by a crew he no longer trusted. Now he was facing a dark vindication of his lack of confidence. Now he *knew* something was wrong. These 737s were capable of really shoving you back in your seat—really accelerating. A real scooter. He loved that feeling. He had loved that feeling since he was three years old. Here he was in a Boeing 737 and he didn't have that feeling.

SOMETHING IS WRONG WITH THE ENGINES! The thought invaded his mind like a ten-bell bulletin in a newspaper wire room.

Stiley looked at Nikki Felch again, and she at him. He couldn't see clearly out the window, of course, but he could see the terminal lights passing, and the aircraft was barely moving!

It was exactly 4:00 P.M. and 2 seconds. Roger Pettit had damn good reason to be very concerned that none of this was right. The airspeed had not begun registering yet. They seemed to be barely moving, and the engine instruments were still all wrong. Wheaton, his captain, didn't seem concerned.

"That doesn't seem right, does it?" Pettit asked, apparently expecting a quick response.

There was no response from Wheaton.

Three seconds ticked by like hours. It was now five seconds after 4:00 P.M.

Pettit tried again as the airspeed came off the peg.

"Ah, that's not right!"

Still no response. Maybe he was wrong. Maybe it was his mistake. Maybe Wheaton knew exactly what was going on, and nothing at all out of the ordinary was going on. He'd hate to look like an idiot.

"Well . . ." Pettit trailed off, still watching the engine gauges.

Two more seconds crawled by. The 737 was accelerating slowly, but it was accelerating.

Finally Wheaton broke the silence.

"Yes, it is; there's eighty."

Pettit's mind must have rebelled.

"Naw, I don't think that's right," he said.

There was no additional response from Wheaton. Seventy . . . seventy-five . . . eighty . . . eighty-five. The airspeed was above ninety knots. They had a slick, short runway ahead with the Potomac River beyond and another aircraft bearing down on the runway behind them. Their options were running out.

Roger Pettit had no time to think out clearly a course of action.

He had no time to debate all the fine points of whether or not he should risk being wrong and force Wheaton to abort this obviously fouled-up takeoff. Any pilot in such a position would have experienced the same high-speed feelings of impending trouble at odds with the firm belief that his cockpit—his warm and secure and familiar cockpit in which he had resided for so many hours today—had never killed him before. The airplane always flew. The 737 had plenty of power. Even if there was something wrong, they could probably fly out of it. Something, apparently, was wrong, but to try to stop on such a slick runway would mean risking an overrun accident—a chance of skidding off the end of the concrete at high speed. Wasn't it better to trust the airplane—trust she would fly? Besides, it wasn't his decision. That was the captain's decision. The contradictory thoughts would have darted in and out of his mind at an increasing pace.

Why didn't Wheaton seem concerned? Could he know something he, Pettit, didn't know? Were those confusing engine readings right after all?

Nine seconds had elapsed since his last comment to Wheaton. It was 19 seconds after 4:00 P.M. Palm 90 was rolling at over 110 knots of airspeed. It was obvious now. They were committed. Why fight it. Maybe Wheaton was right.

"Ah, maybe it is [right]."

Wheaton was watching the airspeed.

"One hundred and twenty," he said.

Pettit was still watching the engine gauges.

Goddamn it, it still didn't make sense! Pettit tried again: "I don't know . . ."

The time was 23 seconds after 4:00 P.M.

Captain Jones of Eastern Flight 1451 could see the approach lights of runway 36. They had broken out at 400 feet. He guided the three-engine Boeing over the threshold, flared the aircraft with a gentle but steady pull-back on the control yoke, and chopped the three throttles. The turbojet engines wound down to idle as the main gear touched, and he began lowering his nosewheel to the snow-covered runway surface, pulling out the speed brakes and raising the reverse-thrust levers in rapid succession. Unseen less than 4,000 feet in front of their cockpit, another aircraft, Air Florida's Flight 90, was still on the runway—accelerating slowly toward lift-off speed.

It was 31 seconds after 4:00 P.M. Palm 90 had finally reached an indicated airspeed of 138 knots, although the normal point along the runway for reaching that airspeed was long gone. There was not, in fact, much runway left. The lights of the terminal, the hangar and ramp to the north, and most of the other airport lights had slid past.

Larry Wheaton called the airspeed.

"Vee one."

VR, the rotate speed, was only two more knots. Pettit was too concerned to wait. He had already been holding some back pressure to bring the nosewheel off early; now he couldn't wait for Wheaton to call out VR. Pettit began his rotation immediately.

Pettit put back pressure on the yoke, lifting the nose of the 737 from the runway at a steady rate. The object was to rotate at no more than 3 degrees per second, and stop at about 15 degrees nose up. As Pettit felt the 737 respond, he began to feel something else.

The control yoke felt light in his hand—the forces had changed. She was rotating on her own, and his back pressure had added too much—she was pitching up faster than he wanted.

Wheaton must have noticed immediately. The rate at which the nose came off was too fast.

"Easy," *he said to Pettit, watching the attitude indicator.*

Four seconds passed. "Vee two." *That was safe flying speed. Whatever might have been wrong, they should be okay now. They had safe flying speed, although the nose was too high.*

It was 37 seconds past 4:00 P.M.

At 39 seconds past, the stick-shaker activated. The mechanism that literally shakes the control column to tell the pilot that his aircraft is approaching a stall.

That could not *be right. They had 145 knots of airspeed! They had enough to fly!*

Wheaton saw the nose continue to pitch up. It had kept coming, and it was now way over what it should be. That was causing the stick-shaker. In addition, the left wing had begun to drop. Pettit had rolled the wheel to the right and was pushing forward on the yoke, but the nose was responding slowly—coming down all too slowly.

"Forward, forward," *he told Pettit, who was pushing harder on the column.*

"Easy." *Wheaton was startled—worried. Why wasn't that nose coming down.*

4:00:48 P.M.

"We only want five hundred!" *Wheaton was referring to the rate*

226

of climb. With this pitch-up, it should be excessive, but they only needed 500 feet per minute.

The altitude had reached a couple of hundred feet, but the needles were slowing down. The airplane had stopped climbing. The nose had pitched back down somewhat, but the shaking of the control column had been joined by a bone-chilling shaking of the entire airplane. The stick-shaker was right. They really were stalling!

It had been a very odd sight. Captain Roger Hough, standing by his Learjet on the north general-aviation ramp of the airport, had seen a Boeing 737 break ground near the very end of runway 36, then pitch up to a horrendous deck angle—possibly in excess of 30 degrees! As the 737 disappeared in the murk to the north, he had watched the pitch angle decrease, but Hough was alarmed.

The shaking had begun just after lift-off, just after Joe Stiley had seen the last of the airport lights disappear—just after he had turned to Nikki and said simply, "Nikki, we're in trouble!"

There was a river just ahead and they were going to take a bath. Stiley knew it. This machine was not going to fly! He had not felt it rotate.

Then the shaking. The rapid, rhythmic, bone-shuddering, frightening wham, wham, wham, wham, wham, wham. Less a noise than an awful feeling, accompanied by the most awful rattling of dishes in the rear galley, the lurching of the luggage compartments overhead, things bouncing around the cabin.

As soon as it began Joe Stiley noticed the trees. There were trees visible out the left now. Lady Bird Johnson Park. They were definitely going in.

"Nikki, we're going to crash. We're not going to make it. Do as I do!"

"What?" She seemed puzzled. Time was crawling.

"Just do as I do . . . NOW!"

Joe snapped down into a brace position, and Nikki followed—uncomprehending, but willing to trust.

Joe Stiley's mind was racing, trying to figure out how to survive the impact he knew was coming. The thought flickered across his mind that he'd found a whole new definition for the epithet "Oh, shit!"

Bert Hamilton had no idea what was happening aerodynamically, but he was plenty scared. One of the Fairchild men, in the seat ahead, Joe Caluccio, turned around and looked at Bert, a questioning glance behind an expression of horrified shock and amazement, as if to say "What in the hell is going on here?!"

Bert could only return the questioning look. It sounded as if the aircraft was going to come apart in the air.

On the rear jump seat, Kelly Duncan saw several of the passengers turn and look at her, searching for reassurance. She couldn't respond with a smile. Her face was frozen in apprehension and fear. Kelly had spent the last few thousand feet of the takeoff roll with the sickening realization that it felt all wrong. "Fly, baby, fly. Get this baby up!" She had breathed mostly to herself. Now they were up, and she had no idea what was happening, but it was not good.

Priscilla Tirado squeezed José's hand tightly, as if to say "It's all right, but hold me anyway." She had never experienced anything like this. It was scaring him, and it was scaring her. Maybe it was just bad turbulence. Maybe. Surely those fellows up front wouldn't let them get hurt.

Tom Esdale heard an airplane overhead, much louder than normal. He had just arrived on a Braniff flight and was now in his car heading for the Fourteenth Street bridge. Lady Bird Johnson Park was on his right, and there, just overhead, was a jetliner at treetop level, the nose way up, the engines roaring, and obviously not climbing.

Esdale turned off his car radio and rolled down the window as the blue and green jet disappeared into the snow-filled sky in the direction of the bridge. He was startled. Planes from National Airport usually didn't fly that low.

4:00:50 P.M. The cockpit instruments were shaking so badly it was hard to read them. Both Pettit and Wheaton knew they were in it deep. The airplane had reached almost 300 feet, now it was descending. Pettit had lowered the nose as far as he could, and was still pushing, Wheaton urging him on.

"Come on, forward."

Three seconds, a couple of degrees down, speed still decaying, the bouncing and shaking getting worse, the throttles still where they had been all along—though now neither pilot was thinking of power.

228

Just pitch attitude.

"*Forward!*"

Two more seconds.

"*Just barely climb.*"

The descent began in earnest. The aircraft was still shaking, still descending slowly, the nose coming down, but too little too late. This was a nightmare beyond belief. Surely this was not happening. Surely they would recover, fly out of it, thrust out of it.

4:00:58 P.M.

Thrust! Thrust! That's what they needed.

A hand shot forward and gathered the two throttles, shoving them to the fire wall. It was a gesture that should have been made many seconds before. It was a move that should have been made on the runway.

For the previous one minute and four seconds Palm 90 had been using only 75 percent power! It had never been at full power. Now it was too late!

The noise of the traffic was a dull roar both in front and behind his car, but air-force First Lieutenant Ronald Williams, headed home from his job at the Pentagon, now heard something else. It was the whine of turbine engines, much closer than he was used to hearing over the Fourteenth Street bridge. He looked to the right, and an apparition met his gaze. An airplane, a big one, descending, obviously in trouble, the left wing down slightly, the nose way up, headed almost at him.

Bert Hamilton turned around and looked at Kelly Duncan, his eyes wide, his heartbeat accelerated. There was a woman screaming somewhere in the cabin, and a baby crying, but no other panic, though the horrible shaking was even louder, more seismic than before. They were being beaten to death in the air.

Kelly Duncan's eyes were wide too. She simply stared back.

"*Stalling! We're stalling!*"

Out of ideas, airspeed, and altitude, Roger Pettit hung on to the control yoke that was betraying him and realized they were not going to fly.

"*Larry, we're going down, Larry . . .*"

It was a plea for help, it was a statement of the inevitable, it was an expression of absolute bewilderment. They had always had flying speed, yet they were losing it. There were bridges ahead. There were people ahead; they had two good engines and they couldn't fly. Oh God! Why not?

Larry Wheaton opened his mouth as the Fourteenth Street bridge's south span began disappearing under his cockpit window.

"I know it," he said simply.

Joe Stiley knew what was out there—he knew they were going to hit something—and his mind was flying at lightning speed, wondering how to survive the fire that would result, wondering if he was dead, wondering why the son of a bitch up there had not aborted the takeoff, wondering . . .

Like being rear-ended on a freeway by a speeding Mack truck, the explosive force of the tail hitting the Fourteenth Street bridge was one hell of a shock. As if in slow motion, Stiley was aware of the seat legs failing, the seats collapsing down, shattering his legs and ankles between them. And he was aware of one other thought:

They had not hit all they were going to hit. It wasn't over yet.

And then the river.

There was no way, no language, no words or phrases to describe the monstrous, soul-crushing impact that stopped every cell of his body with incalculable G forces, exceeding by a thousand percent anything he had ever experienced before, as if the entire planet had suddenly exploded with a sound far beyond the capacity to be heard, a force far beyond the capacity to be measured, a psychological reality far beyond the capability to be embraced or understood.

Joseph Stiley III felt himself slipping from consciousness. He knew one thing for certain in that last microsecond. He knew he was dead.

It was 4:01:01 P.M.

Chapter 16

OUT OF REACH

The Government of the United States was effectively shut down at 1:00 P.M.—and that had been a monstrous mistake.

The snowstorm blanketing Washington, D.C., had prompted the decision, but as the tens of thousands of government employees streamed out of buildings all over the District to get into their cars and go home, the masterminds of the unified shutdown belatedly realized that they had tossed a monkey wrench into the works.

A carefully engineered rush hour "plan"—staggering the closing times of various agencies and buildings—normally keeps D.C. traffic in some semblance of continuous motion during the typical afternoon scramble for home. But the plan lay in utter ruins by 2:00 P.M. of January 13, 1982. Almost nothing was moving in the monumental traffic jam.

Those who were left on the eighth floor of the FAA building at 800 Independence Avenue (in the offices of the National Transportation Safety Board)—those who had decided not to be too quick to join their compatriots in the barely animated parking lot outside—were phoning spouses with the word that it would be late afternoon before they could get back to home and family.

Among that number, Ronald Schleede, a forty-two-year-old veteran air-force pilot and veteran NTSB investigator, had left his office just after 3:00 and made his way to Danker's, a favorite watering hole of NTSB, FAA, and Department of Transportation people.

Danker's is located just on the other side of the DOT building down the street. It seemed to Schleede that most of the occupants of those respective agencies were inside the tavern—the noise level was ridiculous and the patrons were five deep at the bar.

The snow outside was relentless—the drifts getting higher by the hour—and the thought of a long battle with an unyielding wall of traffic had made even the camaraderous confusion of Danker's seem like a welcome haven. Ron Schleede finally got his drink at 4:00 P.M.

This was not Ron's month on the Go Team, but standing next to him was another NTSB investigator, Gale Braden, who was on the team, and on call. Braden normally drove a van pool, but he had talked his passengers into waiting out the traffic at Danker's. All of them were there—somewhere—in the loud confusion.

Over in one corner behind the bar a television set was droning away. Only those close to it could hear its various transmissions. One of that number suddenly battled his way over to Ron Schleede.

"Hey, Ron. The TV just said an airplane crashed in the Potomac!"

Schleede looked at the guy. "You kidding?"

"That's what they said."

Why would anyone be flying around in such hideous weather, he wondered. Probably a small private plane, such as a Cessna. Either that or a joke in very bad taste. Schleede turned back to his conversation.

Gromelski had seen one "hit" on the radar scope, then a weak second "hit," then nothing. What the . . . ? Where was Palm 90?

He picked up the tie line and checked with Departure Control. Gromelski had told Palm 90 to contact Departure, but no one had responded. Departure hadn't heard from him either. Where in the hell . . . ?

Gromelski's finger hovered over the crash-alarm button.

A U.S. Air jetliner had landed on runway 36, following Eastern 1451. Now he was rolling out—slowing down. Gromelski picked up his microphone.

"U.S. Air 172, would you do me a favor?"

First Officer Oran Hoover looked at Captain Forrester and keyed his microphone, answering Gromelski in the affirmative.

"U.S. Air 172, can you continue taxiing northbound?"

"Roger."

"Okay, what I'd like you to—taxi—can you go all the way to the end northbound [of runway 36] and make a left turn at the end?"

"Yes, sir." Neither Hoover or Forrester knew what Gromelski really wanted.

"Okay. And give me a check at the end of the runway, U.S. Air 172. I'm looking for a 737."

The fellow was back at Ron Schleede's elbow.

"Hey, Ron—they're saying it was a 737!"

"You must be kidding!?"

"No, I'm not. Right there, on the TV."

"You're *not* kidding?"

"No . . . really!"

Schleede fingered his bellboy pager, which was silent. He had been kidding Braden about being on call earlier. Now he moved over to Gale Braden's side.

"Uh, Gale, I don't want to start a panic or anything . . . but there's a report on the TV that a 737 has gone down in the Potomac."

As Gale Braden looked up at Ron Schleede with disbelief on his face, his pager went off.

The totally unbelievable sight of the blue and green jetliner sweeping with incredible noise and commotion and destructive force across the line of traffic just in front of him had to be a dream sequence—a nightmare hallucination. Things like that do not happen. Jetliners fly over, not across.

There was not a microsecond to think about it. Just accept the fact that he saw it. He would have to work on applying reason later.

Lieutenant Ronald Williams had watched the 737 come out of the snowy, murky skies, engines screaming at what seemed like full power, left wing down, nose way up over 15 degrees, landing gear still extended, and knew instantly it was going to hit. He was on the first few feet of the south span of the Fourteenth Street bridge, a line of traffic in front of him. The Boeing looked at first as if it would descend on his car; then it swept across in front of him like an angel of death, the tail section raking across the line of traffic in little more than a heartbeat, taking part of the right-hand bridge railing, knocking a crane-carrying truck almost over the side, the sound of the impact drowned by the noise of the two jet engines. As it passed, he saw the red car—two ahead of him—spinning in place, its top crushed. He saw and registered—but could not yet think about—the other car tops that emerged from beneath the passing tail, crushed like that of the red car.

The tail of the jet disappeared over the side to his left with a gut-wrenching roar, the nose down now, the tail up, as if it was about to flip over backward. As he watched, the 737 pancaked into the ice of the river with a plowing motion, the cockpit area going under the water.

And then there was nothing. No sound of explosion, no sounds of screaming, no sound at all. Nothing.

Williams jumped out of his car, turning his ignition key to off automatically.

He could almost hear the snowflakes hitting his hair, it was so quiet—so totally without noise. Had he imagined it? Was that . . . that . . . THING a hallucination? His senses were all contradicting what he thought had been reality seconds before. To his left was the river, visible over the edge of the bridge. There was no sound, no explosion, no fire coming from over there.

But there was proof—and destruction—before him. The crushed cars and the broken railing. It was *not* his imagination.

As if compelled, Ron Williams ran forward to the first car he had seen struck by the aircraft, and then to the red car that had been spun around in slow motion as he watched. Other people were getting out of cars behind him now, running forward. He could hear doors slamming and voices.

He stopped on the right side of the roofless red car and looked in. Someone had already been there—someone had placed a cloth on the driver's head. Ron Williams lifted the cloth from the driver, who was clad in an air-force uniform, and realized the fellow was an air-force first lieutenant just like him—except for one significant thing. The lieutenant in the red car was missing the top half of his head.

Roger Olian, westbound, had almost rolled onto the Virginia side and off the north span of the Fourteenth Street bridge when some idiot slammed on his brakes and jumped out of his car. Olian was really ticked. They'd been inching along for over an hour—

234

there was clear roadway ahead of this yo-yo—and now all of a sudden he stops in the middle of the road. Now what?

The man was walking back toward his car. Olian got out too, in time to hear the fellow's excited questions.

"Did you see it? A plane just crashed in the river!"

"See what?"

"A plane . . . a plane just crashed behind us in the river. An airplane. Did you see it?"

"No, I didn't, I . . ."

Roger Olian just wanted to get the guy out of his way. He didn't believe him—not really. He, Olian, had noticed no plane crash. Besides, his mind assured him, that sort of thing just doesn't occur.

But the other traffic had stopped as well. Something, obviously, was wrong.

The man began worrying out loud about an explosion. Were they in danger? Shouldn't they get out of there?

Olian, bearded, six foot one and wearing green work clothes and a green, Eisenhower-style work jacket, took a deep breath and boomed at the fellow, a Brooklyn lilt tingeing his baritone voice.

"If you're so scared it's going to explode, get in your car and get the hell out of here!"

"Right . . . right, I'll do that." The man ran back to his car, jumped in, and took off.

Olian, half disgusted but too curious to leave, drove on a hundred feet or so and pulled into the break-down lane, turning off his ignition and jumping out. Something was very wrong. Cars were halted all over the place.

He began walking back toward the river, passing a truck driver who had stopped behind him and was still sitting in his rig, looking backward out the window.

Olian looked up at the man. "What happened?"

"There was a plane crash . . . back there."

"Do they need any help?"

The trucker looked back at Olian. "Yeah, I'm sure they do. Do you have a CB [citizen's band radio]?"

Olian considered that for a split second. No, he didn't have a CB, but maybe he could help. Undoubtedly the plane would be sitting there in the water. Maybe he could help.

Roger Olian began trotting down the embankment toward the river through the bitter cold and ten-knot winds.

There was nothing visible out there! As he ran, the expected sight of an aircraft sitting in the water refused to materialize.

He was getting closer to the bank, still running. A tail fin could be seen out in the water some distance from shore. Around it to one side there seemed to be about four people.

And he could hear them yelling!

The people already on the bank were hollering back that help was coming, and to hold on. The people in the water were floundering around. Obviously, they wouldn't be able to hang on very long.

Olian was close now, his pace picking up from a trot to a run. In his field of vision was the helpless gathering of individuals on the shore, yelling encouragement just in front of him, a few of them trying to tie what looked like wires and ropes together. On the bridge to his left people were stopping and peering over. Someone was dangling a rope—yards away from the tail fin. Useless.

There were no boats, no helicopters, no help at all. Those people were totally helpless out there.

Roger Olian was overwhelmed with the urgency. Something had to be done for these poor people—right now! No one was out there swimming to them. That's what they needed. That's what he could do.

He knew it would be cold but the thought never matured in his mind. Olian was too busy calculating his trajectory as he ran straight down to the Potomac's western edge and half-plunged, half slid into the coldest water he had ever experienced in his entire life.

Regaining consciousness at all was a shock. Apparently he was alive. Joe Stiley accepted that reality and moved on mentally. He could hear moans and screams from somewhere in front of him. He could feel ice water on his chest. There was pain—searing pain—in his legs and elsewhere, and he was aware that his seat had collapsed into the seat in front of him.

He was also aware that his legs were trapped—he couldn't pull them out.

Thoughts, as ordered as he could make them, filled his growing consciousness. "Gotta have a plan! I know where I am—I'm in the river."

The water was rising.

"Okay, gotta control myself—have a plan—seat belt first, then we go out the back." He was thinking more clearly every split second. Action was critical. "That's the only chance, out the back!

"We? We! Check Nikki!"

Joe looked to his left. She was still alive, too, but water was lapping at her neck.

Words exchanged—feelings exchanged somehow—she's badly hurt too, legs trapped, legs hurt.

Seat belt. Help with her seat belt. Joe reached over and helped claw it open as they went underwater and into pitch blackness.

Somewhere in the next few moments the frantic but ordered process of pulling on his right foot, which wouldn't budge—pulling, yanking on the ankle, trying to get that foot out of the loafer. Break the damn ankle if necessary, thank God I didn't wear my boots. The leg moves both ways, move it backward, free the heel. There, out of the shoe. Left leg, left foot, pull it backward—it's broken anyway.

It wouldn't budge. It wouldn't come out. He couldn't move it!

The thought of the consequences didn't invade his mind. There was no time. He had a problem. His foot was stuck and he had to get it free. It was simple as that. The water was rising, but all he could focus on was the struggle to get free.

As he pulled and yanked, Joe felt the remains of the fuselage along with the seat settle to the bottom with a lurch.

Finally, moving his broken leg and twisting it downward, he began to get it moving. He pulled, it moved more. One final yank and he was out.

Then to Nikki, also struggling with her shoes. Can't see a damn thing. Totally black, blackness, darkness all around in the icy, black silence.

There was no air to breathe—only water around their faces—but in the urgency of the moment and with adrenaline coursing through their bodies, neither Joe nor Nikki was acutely aware of the need to break to the surface and breathe, fast.

Joe was struggling with her leg, thinking to himself the words he could not speak through the black water. Pull the damn thing back, farther, there! It moved. Out, back, up!

Somehow they grabbed for each other and began scrambling for the surface, swimming, clawing, back over what felt like seats, looking for light overhead—thinking all the way about fire.

What if the surface is on fire? Navy training—remember your training, boy! Splash like hell, grab a breath, and submerge. How to tell Nikki. How to . . .

Joe Stiley and Nikki Felch broke through the surface of the Potomac River and into a snowstorm. There was jet fuel everywhere, but no fire.

Joe Stiley looked around at the scene on the surface with a cascade of thoughts. He—they—were alive. At least he and Nikki. He was responsible for her. He had to save her.

There should be a Jolly Green, an air-force rescue helicopter, over the horizon any moment. The airport wasn't far away. They had made it this far; they would make it all the way.

But the shock! He looked around again. The hardest thing of all—far beyond the almost procedural process of getting out, getting to the surface, and even beyond the pain—was the incredible recognition that one moment he had been sitting in the overly warm cabin of a commercial jetliner, and now he was here, swimming in the icy Potomac River in the middle of winter. Hurt. And there was no sign of the airplane! No trace! No . . .

There was debris, and there was Nikki, and as he looked around behind him, he could see part of the tail fin.

And there was someone else, another passenger perhaps, out there on the other side of the tail. As soon as Joe saw him, he lost sight of the man.

"Nikki, my legs are broken."

"Mine are too."

They were both dog-paddling, trying to stay afloat.

"That tail section, let's get over there!"

The two of them started working their way over, using a "snap the whip" technique. Joe would pull Nikki to him, then push her beyond. She would then do the same thing with him. Slowly, painfully, laboriously, they moved to the tail section. Back and forth. They couldn't kick their way over—only their arms were mobile.

Bert Hamilton vaguely remembered the impact—more like an intellectually acknowledged event, not a physically remembered one. The magnitude of the shock was too great to remember. He had never lost consciousness, though.

But where the hell was he? He realized he was still in his seat, but where was the seat?

"We're down. We're down on the runway. We came back down on the runway, and I'm hurt—beat up." The conscious procession of thoughts flashed through his mind one after the other as Bert sat in the rising water.

Water? How could there be water on the runway at National?

A crash made no sense. He didn't think of a crash; that was not even on the periphery of his consciousness. Those fellows had simply landed again—hard. They wouldn't have crashed. A professional crew would not have crashed.

"This is *not* a plane crash!" his mind assured him with finality.

But water? Where was water coming from? That simply would not correlate.

Bert reached down with his left hand and unbuckled the seat belt. He looked in front of him and saw light—a hole—and . . . Water.

He was *floating* in water. He was *in* the water.

But there was no water on National Airport. His mind kept rebelling at the inconsistency, but he'd have to examine it later. Right now he had to work on keeping alive and afloat.

Bert grabbed onto some debris in his vicinity and realized that he was losing one of his shoes. Suddenly that shoe became the most important thing in his life. He began struggling to keep that shoe in place.

"My baby! Help me find my baby, please!"

Priscilla Tirado had surfaced some yards from Joe and Nikki, screaming, thrashing about, frantic for help.

"Oh, please! Where is he! I've got to find my baby!"

She had no clear memory of anything in the darkness below her, just that she had pushed away from someone—maybe José—and risen to the surface. She couldn't understand where her husband was. She wanted to go back down for her baby and her husband. She wanted someone to help her find them before it was too late.

"Down . . . down there . . . got to go back. I can't find him. Where is my baby?" Her voice pierced the frigid air with anguish.

She was, of course, too hysterical to realize it already was too late.

Nine-week-old Jason Tirado had been catapulted from his mother's arms and thrown on impact through the disintegrating cabin like a tiny human cannon ball. He never had a chance.[1]

1. Nor did either of the other two infants on the aircraft have a chance. Nor does *any* infant have a chance when held in a parent's arms rather than restrained by a proper seat belt or infant seat system! No infant who has ever flown anywhere commercially while being held in a lap has ever been safe from instant death in an otherwise survivable air accident. The scandalous failure of the airline industry for decades to properly provide for the safe restraint of infants in commercial airliners is a little-known and little-discussed subject, which finally, in 1984, began to get some attention. The FAA at long last began to consider authorizing the use of automotive infant seats in commercial aircraft—while still permitting the parents to hold them in laps. There is not now, nor has there ever been, any justification or excuse for ignorance of this area of potential infanticide! Nor does the author give an inch in condemning the motives and the laziness of those who have turned away from seeking solutions

Her husband, José Tirado, was gone too. Not being in a brace position, he had not survived the impact with the seat in front of him.[2]

Kelly Duncan, the flight attendant, had also come up somewhat close to the tail section, and she too was screaming for help. "Get us out! Help, somebody, get us out of here." The sight of the activity on the west span of the bridge seemingly overhead was tantalizing. It was to those people that Kelly kept yelling.

Bert Hamilton had seen Priscilla Tirado floundering as he finally got his bearings. He had spotted Joe and Nikki and called to them. Then he yelled at Joe Stiley, hoping Stiley could reach Priscilla. Bert was on the far side of them and out of reach. As he watched, Joe and Nikki Felch pulled her in, Joe swimming out to grab her, almost being dragged under by the frantic flailing of Priscilla's arms and the death grip with which she seized his necktie.

Kelly had been closer, and managed to paddle to the tail section on her own.

It was this group, Joe Stiley, Nikki Felch, Priscilla Tirado, and Kelly Duncan, all yelling for help, whom Roger Olian had spotted near the tail as he dashed toward the river.

---

in decades past because of economics or the difficulty of drawing up uniform rules! The fact that relatively few infants have met death this way is totally immaterial. The real potential, not the statistical reality, should be the governing factor.

2. Also in 1984, a formal test was finally conducted in the opening round of investigation into whether airline seats should face forward or backward. The U.S. Air Force's Military Airlift Command has always installed passenger seats backward to prevent deaths and injuries due to passengers being whipped forward into the seat before them, and impaled on tray tables and seat frames. The entire reason for a "brace" position is to prevent such an injury when an otherwise survivable air accident results in a tremendous forward deceleration—as occurred in the crash of Palm 90. The airline industry has long suppressed these arguments with belittling attitudes based on the assumption that passengers would be aesthetically opposed to flying backward. As one NTSB investigator of many years has said off the record, "Who gives a good goddamn what is or isn't aesthetically pleasing? If you want to save lives, require the things installed backward and the entire industry will have no choice. Therefore, if you want to fly, you fly backward. End of problem!" After all these years (the issue first arose in the thirties) the subject is still "under consideration," and passengers continue to die from injuries sustained in otherwise survivable accidents when they strike the seats ahead of them and break their necks or crack their heads open.

240

Chapter 17

OF ORDINARY HEROES

Ten feet into the icy water Roger Olian had turned around and seen the makeshift "rope" sail past him, within reach. They were yelling to him to tie it around his middle. He grabbed the conglomeration of battery cables, ropes, and whatever, and tied it under his arms—then continued on.

The going was excruciatingly slow. The ice floes were a hindrance, not a help. He couldn't climb on them; he could hardly push them out of the way.

His steel-toed work boots weren't helping, but he hadn't time to take them off. How he wished he had gloves. His fingers were freezing and numb. In fact, every part of him was freezing and numb!

Between strokes he kept shouting to the heads he saw bobbing ahead—yelling at them to hang on. He was coming. Help was coming. People were at least doing something.

They yelled back. They had heard. They were alive.

Park Police Paramedic Bob Galey had heard a strange radio call. There was an unconfirmed report of an aircraft—perhaps a

Cessna—down in the Potomac just north of National Airport. He stepped into the helicopter hangar the Park Police maintain south of the downtown area and told Gene Windsor about it. Windsor and pilot Don Usher, both Park Police officers, started gathering rescue equipment and ground-checking their helicopter, Eagle One, just in case.

Roger Olian had made it some twenty-five feet from shore through sheer muscle and determination. The difficulty of swimming among those ice chunks with his arms and legs freezing was incredible. He stopped occasionally to call out ahead. The voices of Joe Stiley, Priscilla Tirado, Kelly Duncan, and Bert Hamilton were getting clearer and closer. Olian put his face back in the water and started again, still dragging the makeshift, ragtag rope even as the people back on the shore added sections to it.

Bert Hamilton looked up at the bridge, saw a city bus stop, saw the passengers get out and peer over the edge, then get back on the bus.

"My God, we're doomed," he thought. "They can't even see us!" Bert started yelling at them.

Other people began dangling lines over the edge. None of them coordinated their efforts and tied their various lines together. Just a noble but totally ineffective network of short lines hanging over the water from the bridge.

As a fire truck from National Airport screamed up to the Virginia bank, one of the people on the bridge bellowed down to Hamilton.

"Hang on, the fire trucks are just arriving. They'll get you!"

Bert Hamilton looked at the fellow in amazement. What the hell could a fire truck do, he thought. They can't float. Hamilton yelled back at the fellow: "A boat is what we need. Somebody get us a boat out here!"

"I'm not gonna make it."

"Yes, you are, just hang on. Help is coming. They'll be here anytime to get us out."

"No . . . I'm trapped. I can't get out of my seat. I'm trapped in here—something's holding me in. I'm not going to make it."

Joe Stiley was pleading with a man on the other side of the broken rear cabin wall. The fellow was still in his seat, his chest half submerged in the icy water. Joe could just see him by peering over the top of the shattered cabin wall. He and Nikki had grabbed onto the tail section, Joe half balancing on what might be a broken emergency-door section, fighting to keep Priscilla from pulling his head underwater. At first he had tried to push Nikki up the side of the fuselage and out of the water, but the whole section had rolled, as if it might sink. That wasn't such a good idea. They both abandoned the attempt and stayed in the freezing water, watching each other's eyebrows turn to ice and trying to hang on.

The middle-aged man was just inside that part of the shattered cabin, and he was convincing himself of the worst. He had life vests in their plastic bags floating all around him, but he couldn't reach them. Joe tried, but he couldn't quite reach either the trapped man or the life vests. It was horribly frustrating, and all the time the man kept repeating: "We're going to die out here" and "I'm not going to make it."

Kelly Duncan was now next to Nikki. Nikki Felch had reached out and retrieved a floating, uninflated life preserver. But she couldn't tear the plastic bag open. Her hands were too cold, too frozen. The plastic simply wouldn't budge.

Kelly Duncan grabbed the preserver and started gnawing at the plastic bag, finally tearing it open with her teeth, then handing it back to Nikki Felch. Finally, Nikki slipped it over her head.

They were all stabilized for at least a few minutes. Freezing cold, gravely injured, treading water or barely hanging on, but six of them alive. Joe Stiley knew it depended on time in the water. He had already calculated the survival time, measurable in minutes in the near-freezing water. They had only a little while longer before consciousness would slip, the limbs would fail to respond, and one by one they would slip under the surface to join the others who had never even made it this far.

He also realized that the cold was a mixed blessing, blocking the pain of injuries for all of them, giving them abilities of movement that would be fraught with excruciating pain if the temperatures were warmer.

And then there was the painful realization that they were a small group—that there were many more below them who had not bobbed to the surface, and would not.

Of the seventy-nine human beings who had left the surface of runway 36 at Washington National less than fifteen minutes before on Palm 90, only six remained alive as a Park Police helicopter called Eagle One became airborne several miles in the distance, and as television camera crews began a frantic setup of live broadcast feeds from the bridge and the Virginia bank of the Potomac—feeds that would chronicle in living color one of the most dramatic and frustrating rescue efforts ever played out in front of a mass audience.

But in the meantime, despite the beehive of activity on the bridges and the Virginia bank, despite the fire trucks, the sirens, the confusion of the so-called airport-disaster plan, and the unsuccessful search for an immediately available boat, the six people in the freezing Potomac had only one real indication that help was on the way. They could see and hear a single, solitary, determined, and freezing man by the name of Roger Olian resolutely battling his way toward them through the ice chunks, stopping every few feet to call out more encouragement, towing the damnable ragtag rope that held him back as much as it promised salvation. Even the negativism of the man trapped in his seat, moaning that they weren't going to make it, couldn't dampen their determination to hang on until Olian reached them. Even if he couldn't, he was truly their lifeline.

Pilot Don Usher and his partner, Gene Windsor, were astounded at how bad the weather really was as they headed Eagle One toward the Fourteenth Street bridge. Word had finally come that instead of a Cessna, this was an airliner! They were needed instantly. But they were not prepared for the sight that greeted them when Usher brought the Bell JetRanger in low between the two bridge spans.

With only a couple of life rings on board, Eagle One and its crew were the only hope the people below had—that was obvious.

Usher maneuvered the jet-powered helicopter over the tail section of the wreckage, a couple of feet above the water, kicking up a secondary blizzard of blowing spray and snow into the faces of the survivors as Windsor dropped a line to Bert Hamilton.

At first, Bert put it on wrong—draping it around his neck—but with Windsor screaming at him through the open door, Bert finally got it under his arms. Slowly, gingerly, Usher brought in collective (the pitch control) and a touch of throttle, lifting Eagle One and Bert Hamilton from the Potomac, and carefully ferrying him over to the Virginia bank, then returning for the next one.

This time he was right over Joe and Nikki, and dropping a line to Kelly Duncan, who partially looped it around herself and hung on. Once again Usher brought in a touch of power and collective,

244

urging his craft with an unbelievably delicate touch, pulling Duncan straight up from beside Joe and Priscilla, and in the view of live TV cameras feeding the signal nationwide by satellite (though the presence of cameras was totally unknown to the crewmen of Eagle One), ferrying her as well to the Virginia bank.

The remaining group would prove much harder. The third time Usher brought Eagle One right in over the sixth man—the fellow trapped in his seat—after Roger Olian had once again waved him off. Windsor had no way of knowing that Olian was a rescuer, not a survivor. The wave-off had been a surprise. Now, over the main tail wreckage, they had a man and two women, and a second man inside the shell, waving his arm over the top of the fuselage wall.

Joe Stiley knew the fellow was trapped. He knew it would take someone without broken legs and fingers to go in after him—and probably someone with wire cutters or other tools. He tried to yell that information to Windsor, but the noise and the spray were too much.

Windsor dropped two lines to Joe's group. Joe Stiley got them secured as well as possible, hung on to Priscilla, grabbed Nikki around the neck, and gave Windsor the sign to proceed.

Don Usher knew he couldn't pull all three out the same way. For one thing, Gene had told him on their intercom that two of the three were not tied to the rings, but merely hanging on. They would have to be towed, not lifted. Usher began nudging Eagle One toward shore many yards away.

Nikki Felch, holding on to one of the two lines, was pulled away from Joe and Priscilla almost immediately. She tried to hang on to the rope, but slowly—finger by finger—her grip loosened, her frozen fingers refusing to obey, letting her fall away in open water, wearing a life vest, but still far from shore. Joe tried to grab for her, but they were too far apart as Eagle One pulled him and Priscilla past, directly toward a large ice floe.

The impact of Joe Stiley's chest with the ice floe was horrific. It knocked the breath out of him and probably broke a rib—loosening his grip on Priscilla. Joe hung on as Usher continued, unaware of the hazard inherent in striking the ice floes.

A second one crashed into Joe's chest, and his grip on Priscilla's right arm faltered. He struggled to hang on to her—she was a dead weight, not responding, hardly able to move. But despite Joe's efforts, she too fell away, making a feeble attempt to swim, unaware that her weight was suspended on a block of ice.

Finally he felt himself being pulled up onto a hard surface, and began relaxing his grip. But suddenly Joe Stiley was in open water again, unable to see anything clearly, so great was the blast of spray from the chopper and the water flying in his face. Eagle One kept on, however, through the last gap of water to the bank, hauling him up the snowy bank to the hands of several waiting firemen.

Stiley gathered all his strength and yelled in their faces.

"Listen! Listen, I have got to tell the chopper something. Do you have contact with him?"

One of the firemen leaned down. "That's okay, fellow, you're going to be all right. Just take it easy."

The image of the sixth man sinking slowly in the tail section, water up to his chin, his body trapped inside, was burning into Joe's mind.

"Listen to me, damnit, listen to me!"

He was screaming now, or so it seemed. The noise from Eagle One was deafening overhead, the noise of engines and voices were competing with him.

Another rescuer tried to quiet him down with soothing words.

"Hey, LISTEN TO ME! You've got to get a message to that pilot!"

Finally, one of the men bent down. "Okay, okay, what?"

"Tell him they're going to have to go into the water for that last fellow. He's trapped in his seat—he can't come out on a life ring. Understand? They've got to go into the water after him!"

"Okay, we'll tell him." The fellow faded into distance, and Joe Stiley let go, letting his muscles relax as three more people lifted him up and rather unceremoniously dumped him first in the snow, and then on the bare, wet, slushy metal floor of an ambulance.

Nikki Felch at least had the life vest. She kept trying to dog-paddle, but before she could make any headway, Usher and Windsor were back, dropping another line.

Nikki looked up, tried to reach for it—that was literally life dangling there. She knew that, but her fingers weren't listening to her. They weren't responding. She could not hold on. Once, twice, she tried, and fell back.

Windsor reported to Usher what they were going to have to do, and Don Usher dropped a tiny bit of collective, letting the skids of Eagle One actually submerge in the water as Gene Windsor reached down and gathered Nikki Felch in, over the skid, balancing her. In that fashion, Usher pulled the ship up and ferried her to the bank and the waiting ambulance.

And then there were two out there to save.

Again Usher and Windsor wheeled around and went back, this time for Priscilla Tirado, who was losing it fast. Her limits had almost been reached. Windsor dropped the life ring to her and she managed to get hold of it with great effort, her fingers and hands barely moving, but as Usher began towing her off the ice flow and up onto another, she lost the ring.

Priscilla Tirado was dying. She was a few feet from shore, almost frozen to death, and she was dying. Her pitiful attempts to keep dog-paddling were not enough to keep her moving, or enough to keep her head out of the water. She was going down for the last time, and to one man on the bank that was intolerable.

Lenny Skutnik had run down the embankment earlier, after his carpool had stopped to see if they could help. He had been standing on the bank, watching the dramatic events. Now, as he saw Priscilla's last efforts, he whipped off his boots and impulsively plunged into the icy water to grab her. Keeping her head above water, he pushed her back to shore, where several others pulled them both out. It was a heroic deed—but it was by no means the only one.

When Usher saw Lenny Skutnik go in for Priscilla, he headed Eagle One back to the tail section for the last man.

But the tail fin had been almost gone, almost underwater when they had departed before. The wreckage had been sinking slowly since Stiley and Felch had surfaced. Now it was under.

The sixth man, the one trapped in his seat, had been waving each time Usher and Windsor had hovered over the tail to pull out the others. The ropes Windsor had dropped, whipping in the breeze, had been very close to the fellow. It appeared to cameras on the bank that the waving arm might be waving off the crew, letting the others go first. It was hard to tell, even for Windsor.

Of course, Joe Stiley knew the man was trapped. He had tried to get the message to them. But by the time Don Usher and Gene Windsor brought Eagle One back overhead, the man, his seat, and the tail were beneath the waves.

They kept searching, Usher expertly urging Eagle One first to one side, then the other, spotting a body on an ice floe almost under the bridge, but no one alive. They continued orbiting for many minutes, knowing there had been another, and frustrated as hell that they couldn't find him.

One more individual was being pulled from the Potomac as Eagle One kept searching. The would-be rescuer had struggled to within fifteen or twenty feet of Joe Stiley, Nikki Felch, Bert Hamilton, and the others, but Eagle One's arrival had provided a better solution. He had waved off the chopper crew twice as they swooped in low, realizing they would mistake him for a survivor. As the crew of Eagle One concentrated on the desperate, freezing little group floating near the sinking tail of Palm 90, the rescuer began the awesome task of swimming back to the shore. With frozen limbs and lowered body temperature, doubts began to rise up in his mind over his survival. He was figuratively and literally at the end of his rope when those on the bank began pulling on it, hauling him back in. He was shaking uncontrollably when they helped him up onto the Virginia bank, but he waved them off, planning on returning to his car and driving home alone and anonymous. There was great satisfaction in his mind that he had done his best for those poor people.

As Roger Olian tried to get his rubbery legs to obey him and trudge back to the roadside, someone grabbed him and escorted him to an ambulance. Many on the bank had watched his twenty-minute struggle, and they were determined to see him taken care of. (His car, however, would be ticketed and towed away later in the confusion.)

As the survivors were transported from the scene along with Olian, the TV cameras continued to chronicle the rescue efforts. They had not seen Olian enter the water, and they hardly saw him come out. The media had no way of knowing that instead of just one hero—Lenny Skutnik—there were two.

It was Lenny Skutnik who disregarded any consideration of his own health and welfare to save a poor girl who had no time left. And it was Roger Olian's valiant and selfless drive to reach the survivors that had kept them alive for more than twenty minutes— kept them hanging on—long enough for Eagle One to perform its miracle of deliverance. It didn't matter that Olian had never reached them physically. He had reached them on a much deeper level and given them the strength of hope.

And watching the gyrations of Eagle One on TV screens around the nation, helicopter pilots and former helicopter pilots knew that there had been a master at work on those controls. There was true heroism displayed (even though in the line of duty) by Don Usher in the surgically delicate skill with which he flew his machine, providing a rock-steady platform so that his partner, Gene Windsor, could work wonders with inadequate equipment and heroic determination.

As the survivors (and two of their rescuers) were being transported to different hospitals in the D.C. area in various states of shock and injury, the broken bodies of Captain Larry Wheaton and First Officer Roger Pettit remained entangled in the debris of their cockpit beneath the surface of the icy Potomac—temporarily entombed along with seventy of their passengers and two of their stewardesses. To the dead—their remains scattered around the bottom muck of the frigid river—the exact reasons why Flight 90 had failed to carry them safely to Florida hardly mattered. Their lifeless presence simply bore witness—mute witness—to a sad and undeniable fact: These passengers had trusted—and had been betrayed.

They had trusted Larry Wheaton and Roger Pettit to know their jobs and to carry them out properly, carefully, and professionally. They had trusted Air Florida's training and professionalism as an airline, and its presumed dedication to safety before profit. They had trusted Boeing's dedication to safety regardless of cost or impact on sales. They had trusted the Congress to make good on its promise never to let a free-market environment compromise the public's ability to use any certificated airline without taking inordinate chances. They had trusted the system, and most important, they had trusted the FAA to keep the system safe.

Indeed, they had trusted in vain!

As the nation watched TV clips of the dramatic rescue and read the death list in newspapers, and as the press began the inevitable, predictable (and mostly incorrect) public musing over the causes, the government bureau charged with being the watchdog of the system, the NTSB, was shifting into high gear to answer the questions of "Why?" and "How?"[1]

The twists and turns of that investigation during the next six

1. One local TV news report announced with a great air of authority that the crash had occurred because the runway at Washington National Airport was covered with ice and snow—and therefore Palm 90 could not "get enough traction to take off." The newscaster was embarrassed when informed later by a local FAA man that airplane wheels simply rotate freely—the motive force to move the aircraft comes from jet engines (or propellers). Most reports were far more informed but sought to focus on the reported presence of snow on Palm 90 as the cause of the accident.

months—the mistakes and misunderstandings, the posturing and self-protective maneuvers of lawyers and company men, FAA, airline, and even NTSB investigators—would render some aspects of the investigation itself almost as bizarre as the causes of the crash.

And, of course, given the political volatility of deregulation and the vested interests of so many in pretending (at any cost) that the noble experiment with free-market forces could do no wrong, cause no evil, there was another inevitability. The most important point of all would be carefully circumvented: America had experienced its first major deregulation crash—and it would not be the last!

Chapter 18

A HORROR STORY
OF HUMAN FAILURES

"Is this another Grand Canyon?"

The question occurred to more than one veteran air-safety investigator as wreckage recovery attempts began in the Potomac River on January 14, 1982. A major accident at Washington National Airport, right in Congress's backyard, involving what appeared to be a perfectly good airplane flown into the river, was sure to be hammered into a powerful and deadly weapon by the opponents of deregulation—as well as by those who wanted overcrowded Washington National Airport closed or curtailed.

If the freedoms of deregulation had helped set up the Air Florida crash and the public became sufficiently upset over it, Congress might have to change the rules of the system again. Those senators and congressmen who had followed Professor Alfred Kahn's blueprint for unfettered competition might be forced to abandon their blind, bipartisan support for the experiment and face the air-safety issues for the first time. Just as in the 1956 tragedy in the Grand Canyon in Arizona, public outrage might force change. And as many of those in the FAA and NTSB knew (but would confirm only to their families), the airline "system" was becoming far too loose in terms of safety compliance, and was in desperate need of immediate, substantive change that only Congress could provide.

The political implications began to dawn on the NTSB investigators immediately after the crash. These were men and women who knew well the bipartisan political support for deregulation—the intense determination on the part of the administration and Congress to make it work.

These were also investigators who had heard the cockpit voice tape, who knew that the engines had been operating on impact, and who were seeing more and more evidence that whatever had happened to Palm 90, it was not the fault of the airplane.

Or was it?

Several thousand miles across the Atlantic, David Davies and his fellow experts in the British Civil Aviation Authority Airworthiness Division read the early details of the accident and wondered. A Boeing 737, apparently carrying ice on parts of the airframe (according to quoted witnesses), had been unable to sustain flight with both engines operating and had hit a bridge at what some witnesses described as a "nose high" attitude.

When Davies heard the news reports in the days after the crash, he could not help but remember the letter he had written to all United Kingdom operators of Boeing 737s less than two months before. It was a letter notifying the various airlines that Davies's Airworthiness Division was going to take positive action on the Boeing 737 pitch-up/roll-off problem rather than just offer more advice—advice that was not stopping the reported incidents. Davies had become quite weary of waiting for Boeing to do something more substantial than issue advisory bulletins. At the time, he did not realize that the situation in North America had been significantly different.

We believe that tangible actions need to be taken rather than merely offering further advice, if we are not to see a future accident attributable to [pitch-up/roll-off tendencies] of the Boeing 737.[1]

1. Davies's letter of October 28, 1981, had gone on to list the main conclusions of the Airworthiness Division:
"a) The service record suggests that the B737 appears to be unusually sensitive to minor wing leading edge contamination, during takeoff rotation and initial climb. Boeing flight tests with an artificially contaminated wing leading edge confirm that the effect on stall speed is very significant.
b) Present operating procedures place responsibility on the aeroplane commander [captain] to insure that the aeroplane is free of ice at the start of take-

Davies had no desire to be a soothsayer of doom, but was it possible that the regrettable affair in the American capital was the very "future accident attributable to this cause" that his letter had predicted? Quietly, Davies and his people began watching the developing investigation for signs.

But the NTSB was finding a horror story of *human* failures. They didn't need to look for mechanistic causes or peculiar aerodynamic tendencies, especially in an honorable and long-certificated airliner built by the trusted experts at Boeing. The developing list of human foul-ups in the saga of Palm 90 was quite sufficient to explain a half-dozen accidents!

To those in the field of accident investigation who had long supported the lonely battle to investigate the human causes of human failures, Air Florida was unfolding as a classic of major proportions. Certainly it was not the only contemporary accident of a major carrier caused by human mistakes that couldn't be dismissed as simple pilot error, nor (by February) was it the latest.[2] But because of where it crashed and the upstart, expansionistic nature of the airline, it was a potentially powerful accident. The lessons that

---

off and current advice to B737 operators emphasises the importance of thorough and efficient pre-flight de-icing. The incident record suggests that further more positive action needs to be taken.

c) To allow the aeroplane commander discretion in the use of overspeed procedures in circumstances when wing leading edge ice contamination is possible, implies that he will be able to satisfy himself that the wing has not become contaminated during taxying from the ramp to the take off point. This we believe to be unrealistic, particularly at night and at busy airfields where significant pre-take off delays can occur.

d) The use of mandatory overspeed procedure [one recommended Boeing procedure to avoid the effect of leading edge contamination] will have an adverse economic effect for some B737 operators [in reducing the payload they can carry] on certain of the longer routes, through an increased frequency of refueling stops. We believe that this economic penalty is not disproportionate to the significant safety improvement which will result from the use of the proposed overspeed procedures."

2. On January 30, 1982, seventeen days later, a World Airways DC-10 landed too far (and too fast) down a runway at Boston Logan Airport. The crew had been led mistakenly to believe that the runway surface was less slippery than it really was. The FAA controllers in the Boston Logan tower had failed to pass on current information. In fact, the runway was a solid sheet of frictionless ice. The DC-10 skidded off the far end of the runway, the cockpit section breaking off and apparently drowning two passengers (whose bodies have never been found). There was nothing substantively wrong with the DC-10. There was a lot wrong with the human performance of FAA tower personnel and the aircrew's landing and reversing technique.

might be learned from the tragedy of Palm 90 might illuminate the long-ignored reality at last: *Preventable* mistakes *made by people* cause almost all the accidents.

The developing evidence showed that a perfectly good airplane had been flown into the Potomac River solely through a chain of human errors. There was little initial focus on why Palm 90 had (apparently) pitched up smartly on takeoff. If this had happened, and the crew's comments on the cockpit voice tape plus witness statements seemed to confirm it, it seemed merely an outgrowth of the crew's mistake in trying to take off with ice and snow on their airplane. Peculiar pitch-up/roll-off characteristics and the long history of pitch-up/roll-off incidents were never even suspected at first; but then, those characteristics had been kept very quiet and were not widely known.

It would be up to the legions of attorneys (who were already drawing their battle lines by late January) to ferret out all that history with a flurry of costly legal depositions taken in multi-million-dollar wrongful-death and damage suits, principally against Air Florida, Boeing, and the U.S. government. While the defendants battled to minimize their role and their culpability, more facts kept emerging about the Boeing 737, which could not be ignored.

At first, the 737's pitch-up/roll-off tendency was missed by the NTSB (though they were told of the tendency within days by former NTSB investigator Chuck Miller). It was not aware of the long-standing concerns of the British Airworthiness people, nor the 1971 Boeing flight-test discovery, nor Boeing's refusal to do more than issue advisories and bulletins, nor the long European experience with the 737's chilling tendencies to overrotate when the leading-edge flaps were speckled with ice particles.

They were concentrating on the operational mistakes, and there were plenty. The litany of human screwups and bad judgment surrounding the Palm 90 departure was staggering—as was the diversity of those mistakes. Very little if anything had been done right by anyone connected with the flight.

The pilots, of course, had committed an incredible series of errors, displaying sophomoric judgment, careless attitudes, and a monumental ignorance of their responsibilities—not to mention lack of the required technical knowledge. But they were by no means alone. Further links in the chain of destruction were forged by more human foul-ups. The American Airlines people, the Air Florida ground personnel in Washington, the Air Florida executives and training people back in Miami, the air traffic control specialist in Washington National tower, the Boeing men who wanted to minimize the damage of adverse publicity, and many more human beings had contributed to a long causal chain, which ran from Seattle through Miami to the bottom of the Potomac River.

The de-icing sequence alone resembled more of a Keystone Kops operation than a mature, professional airline procedure.

Air Florida had contracted with American Airlines for its ground services in Washington, including de-icing. But American, quite simply, had blown it. Their men had de-iced Palm 90 twice, but as for the anti-icing coat of glycol (which is supposed to temporarily retard the buildup of new snow and ice), the left wing never received a coat. The right one did.

On top of that, it turned out that the wrong nozzle had been placed on the de-icing truck's hose, and each operator (there had been a shift change in the middle of the de-icing sequence) was using a different (and wrong) outside air temperature to compute his ultimately incorrect water-glycol mixture—which couldn't be delivered properly anyway because of the illicit nozzle!

The de-icing sequence violated both Boeing's and Air Florida's procedures. Air Florida had only one maintenance representative at the airport, and with the confusion of being a rapidly expanding carrier, it would not have been surprising to find that one of its people had made a mistake. But Air Florida's man had not run the de-icing truck. No less a huge, established, and professional organization than American Airlines was performing that function— and not even *it* could get it right.

There were rules to be followed, inspections to be made, and guidelines to be met, and someone was supposed to be supervising to make certain it was all accomplished according to plan. But neither Air Florida's maintenance representative nor American's personnel had any idea whose responsibility it was to know which rules applied and who should supervise them. So no rules were applied at all and no one supervised anything. They just more or less played it by ear.

The fronts of the engines of Palm 90 should have been closed off with large, plastic disks during de-icing. They were not.

The sensitive air probes (called Pitot-static tubes), which are vital in providing altitude, airspeed, and EPR settings to the pilots, were supposed to be covered. They were not.

The temperatures used by the de-icing operators should have been accurate. They were not.

But where, the investigators wondered, was the captain all this time—the man legally and morally responsible for such things?

For that matter, where was the first officer/copilot, the second-in-command? In a three-pilot airliner, the flight engineer monitors such things, but in a two-pilot operation like the 737, the copilot would usually keep a close watch on the situation.

As the statements came in and the NTSB investigators kept assembling evidence, it became clear that the two pilots had been sitting passively in the cockpit of Palm 90 the entire time, enjoying the warmth and the reassuring familiarity. Their reality was the cockpit—all that activity and the snowy problems outside could be dealt with at a distance.

Neither Captain Larry Wheaton nor First Officer Roger Pettit had put forth the effort to leave his cockpit and walk around the aircraft since long before the de-icing sequence had begun. (One of them had inspected the front of the aircraft just after arrival, but not in the hours since.) They had no idea whether their airplane was free of ice or snow. They simply trusted the ground crewmen to make sure nothing was wrong.

But the ground crewmen were not aware that such responsibility was resting on *their* shoulders. They did not realize that they were being trusted blindly to make such momentous decisions for the crew. In their minds, such things were the captain's responsibility. Certainly they would have mentioned anything overtly wrong if they had noticed it. And if the captain or the copilot asked for their opinion or asked them to make an inspection of how much snow was on the wings, they would have done so gladly. But in the normal rush of operations, the ground crewmen, struggling with freezing temperatures, blowing snow, pools of glycol, and many other airliners coming and going in the confusion and screaming, cacophonous noise around them, assumed that the pilots knew the condition of their own aircraft.

Even the Air Florida maintenance representative failed to accept the responsibility for checking to see if the aircraft was free of ice or snow.

In fact, no one accepted the responsibility—so no one walked around and looked. What Joe Stiley had noticed—no footprints in the snow around the wings of the aircraft—was in fact quite significant. It graphically demonstrated a looseness to the operation—a lackadaisical attitude of people who were essentially unconcerned. It bespoke a tolerated air of carelessness—and it became worse as the Palm 90 departure progressed.

Larry Wheaton should have known the consequences of using reverse thrust in a 737 on a snowy ramp. It was in his manual. It should have been in his head! Had he been more receptive to instruction, had his training been adequate, it might have been. Had the Training Department at Air Florida been made acutely aware by Boeing of the dangers of using reverse thrust, perhaps it could have gotten through to Wheaton. In fact, Air Florida had never even addressed the issue of using reverse thrust at the gate in blizzard conditions. It simply did not have sufficient experience with such conditions.[3]

American Airlines' procedures *did* prohibit a captain from using reverse thrust at the gate, and the American tractor driver told Wheaton just that as he struggled to push Palm 90 backward. But Larry Wheaton was apparently unimpressed. If the tractor couldn't get enough traction to push the airplane, he would use reverse thrust. Wheaton's maneuver undoubtedly deposited on the leading-edge flaps some of the very ice particles that later caused Palm 90 to pitch up so violently.

And then there was the rather monumental matter of trying to take off in such lousy conditions with only 75 percent of normal thrust.

Even in good weather on dry runways on beautiful days, airliners are not launched into the air with the engines at three-quarter power. On short runways in poor conditions such an act can be fatal.[4]

3. The airline did issue each year a brochure on winter flying, which had been prepared by Captain Jim Marquis in 1981, and in fact, many established airlines do little more. Again the difference is the operating experience of the personnel. A more established carrier—especially one operating in and out of many northern cities—will have well-tempered aircrews who have seen blizzard conditions many times and realize they have an immediate need to know all the procedures and techniques. When the few weak pilots in a less experienced airline do not know enough about cold-weather conditions to even realize that they are inexperienced—when they are not sufficiently knowledgeable to be "scared"—they are vulnerable. Passengers do not pay to be flown around by vulnerable pilots.

4. Most airlines operating large turbojets do have a provision for the use of "reduced thrust" when there is more than enough runway available and the weight of the aircraft is not approaching the limits. The reason is to lengthen the life-between-overhaul of the engines by being a bit more gentle on them. That thrust reduction, however, amounts to no more than 15 percent and is carefully calculated on a chart. Palm 90 was 25 percent under maximum thrust. In short-runway, bad-weather conditions such as those that faced Palm 90, reduced thrust would not be authorized. Maximum takeoff power would be used exclusively.

How on earth did they end up with only 75 percent power? The investigators found the answer before they were fully aware of the question: On the tape from Palm 90's cockpit voice recorder Roger Pettit had read the words "Engine anti-ice?" and Larry Wheaton had responded with a single word: "Off."

The engine anti-ice had never been activated.

That anti-ice system uses hot air to keep ice off the front edge of the engines and keep ice out of the small probes (called PT2 probes) that provide part of the data for the EPR gauges in the cockpit. If ice forms in those tubes, the EPR gauges can read higher than the actual EPR. If the engine anti-ice is on (and the engines are at a sufficiently high thrust setting), the tubes are heated, and ice can't block them. The readings will be correct, and the aircrew will not be lured into setting the engine power at only 75 percent of normal takeoff thrust—as were Wheaton and Pettit.

When Wheaton had advanced the throttles at the end of the runway, he had pushed them forward to a certain familiar position, one that yields (usually) just-under-takeoff power. Both pilots expected the EPR gauges to shoot up and stabilize just under the maximum power setting.

"It's spooled. . . . Ho! . . . Whoo!" Wheaton obviously had been startled to see both needles go far above that setting. It appeared that the engines were overthrusting, and the captain certainly yanked the throttles back to get the needles on the proper setting, thinking that somehow the frigid temperatures were responsible for giving them more power at a lesser throttle position than in Miami.

"Really cold here!"

But the throttles weren't positioned correctly. Both of the EPR gauges were wrong. There was ice blocking the PT2 probes in the front of each engine, and that caused the needles on the EPR gauges to register much higher than the true EPR. So when Wheaton pulled the power back and set the needles on the maximum power reading, the engines were running at far less than maximum power and putting out only 75 percent of normal thrust.

If the engine anti-ice had been turned on at the gate (and the proper engine run-up procedures had been used), the EPR readings would have been correct.

But Wheaton had said "Off" and Pettit did not challenge him, even though there was a crystal-clear mandate in the normal procedures and in the flight manuals they both carried that *required* them to turn the system on just after engine start—at the gate—in such snowy, messy, cold conditions.

In fact, any pilot of a heavy turbojet aircraft, military or civilian, who has had long experience with the characteristics of large turbojet engines, knows that rule: If the temperature is below 8 degrees centigrade (in some manuals 10 degrees) and there is visible moisture in the air, TURN ON THE ENGINE ANTI-ICE JUST AFTER ENGINE START!

Ah, but neither Larry Wheaton nor Roger Pettit had a history of long experience or training in such aircraft.

Certainly Wheaton had not. He had never flown a turbojet before checking out in Air Florida's jetliners (the DC-9 and 737).

And Roger Pettit had been a fighter pilot, flying the latest state-of-the-art engines, which handle and operate differently from airliners.

Neither man had received enough emphasis on the subject in his Air Florida training, and that is a self-proving failure: If their training had been sufficient, the engine anti-ice would have been turned on.

Because Air Florida had expanded so rapidly, both pilots had progressed far faster than in most large, stable Part 121 airlines. If they had joined Delta or Eastern, for instance, they would have started as flight engineer/second officers, sitting in the backseat in the cockpit of Boeing 727s and learning their profession by actually observing what very experienced pilots do when faced with snowstorms at National Airport. They would have put in years as engineers before having the opportunity (and the tough training) to qualify them as copilots or first officers. Even then they would have had to undergo a well-polished and rigorous series of simulator and flight check rides designed to see if they really were good enough. Only after ten to fifteen years of such tempering would either man have had the chance to upgrade to captain.

But at Air Florida, Larry Wheaton had become a turbojet captain less than two years after his very first flight in a turbojet airliner! He had compiled only 1,100 hours as a Boeing 737 captain since that upgrade, and twice had failed check rides—once being removed as captain until he could requalify.[5]

5. Failed ("busted" in the language of pilots) check rides are not unknown among airline pilots; otherwise the system of flight checks would hold no deterrent value to those who might be tempted to backslide on their proficiency and continuous study obligations. However, for a professional airline pilot and a captain, such failures are unusual—and deeply embarrassing. When a captain has failed two flight checks at different times over a few years, something is wrong! Either the pilot is not getting sufficient training, is not paying attention and working at his qualification, or is having trouble doing the job. Larry

259

In addition, Wheaton and Pettit were flying in one of the most labor-intensive environments in commercial aviation—a two-man cockpit. The latest versions of the two-pilot DC-9s (called the Dash-80s) and the newer Boeing 757s and 767s use automation to control systems that, on earlier airliners, the flight engineer controls. The Boeing 737s flown by Air Florida have partially automatic pressurization systems, electrical systems, hydraulic systems, and many others that the copilot usually has to set and monitor. Although 737s and DC-9s were never designed for more than two pilots, the cockpit work load (including company communications, paperwork, and systems operation) is often equivalent to that of a three-man 727. When two pilots do the work of three, things can get very busy in the cockpit, and in flight the addition of a weather problem, a fuel problem, or any other problem can overload even the most seasoned professionals.

However, paying two pilots is cheaper than paying three. For airlines like Air Florida, which must maintain low fares and a low overhead, two-pilot airplanes mean less expense. Therefore the older DC-9s and Boeing 737s have rapidly become the preferred airliners of many low-cost carriers.

There is, however, a hidden flaw in such economic logic. The youngest, most junior, and many times most inexperienced pilots working for the most rapidly expanding carriers often end up in the cockpits of such airliners. In far too many instances, it doesn't take much to overwhelm their level of experience with problems they may never have seen before. Sometimes, as with Palm 90, the overload leads to fatal mistakes, and fatal mistakes can destroy the economic viability of a low-cost carrier. If people believe they can't trust it, they won't fly on it.

Larry Wheaton had flown into ground-icing conditions only eight times in his service with Air Florida. Neither his experience in the Florida Keys with Air Sunshine nor his time in the south with Air Florida had adequately prepared him. With inadequate experience and insufficient training, Larry Wheaton was abysmally unprepared for operations in ice and snow. It didn't matter that the information was contained in his flight manual—his passengers had trusted that the information was contained in his head—and it wasn't.

Wheaton had "busted" two flight checks at two different times while with Air Florida. Yet he was allowed to continue carrying passengers in all types of weather conditions because he was technically legal.

The management of Washington National Airport (owned by the federal government and operated by the FAA) should have been no stranger to ice and snow and wintry conditions. It should have had a smooth, practiced, and inherently safe method of dealing with snowstorms and departure delays in freezing weather—but it did not. The airport's failure forged still another link in the causal chain.

There was too little gate space to let the outbound flights hold at the gate. Logic would seem to dictate that the airliners should be kept at the gate until the last minute, then de-iced and headed out for takeoff. But even in the worst snowstorm of the season National's Ground Control let the airliners leave their gates and line up like a procession of lumbering elephants—the snow relentlessly destroying any beneficial effect of whatever anti-icing they might have received back at the terminal. That was, quite simply, standard procedure at National.

Palm 90 had been de-iced at the gate thirty minutes before taxiing out to join the lineup. Twenty minutes more went by before it was number one for takeoff—a full fifty minutes since the last de-icing, which itself was badly flawed.

At some airports in the world (such as Montreal, Canada), one method for dealing with snowfall on taxi-out closely parallels the idea discussed by Wheaton and Pettit in the cockpit prior to the departure of Palm 90: Departing jetliners taxi between a pair of de-icing trucks seconds before entering the runway for takeoff. But at National, that had never been tried, even though it was obvious that the airliners would be accumulating quite a bit of snow before beginning their takeoff roll.

"Boy, this is a . . . losing battle here, trying to de-ice those things. It [gives] you a false feeling of security, that's all it does," Pettit had commented as he sat in Palm 90's cockpit and apparently realized how much snow was building up on everyone. The reply from Wheaton was telling:

"That, ah, satisfies the feds."

Wheaton and Pettit knew they were accumulating snow, but there was considerable operational pressure and peer pressure on them to continue taxiing out for takeoff. They had a place in line. No one else was giving up and taxiing back to be de-iced again. Not one of the airline captains in that lineup broke formation and headed back to the terminal, though they all knew the conditions. (A few of the aircraft had been de-iced less than twenty minutes before, were still in reasonably "clean" condition, and would not have needed to go back.) If just one had done so, Larry Wheaton might have decided to follow.

Instead, he may have tried to de-ice Palm 90 by taxiing close behind a New York Air DC-9—even though such a procedure was dangerous and contrary to Air Florida procedures. It should have been obvious that with the temperature outside well below freezing, snow melted on Palm 90 by the exhaust from the DC-9's engines just ahead could run down the wings and refreeze—possibly on the control surfaces. But Wheaton and Pettit merely joked about it.

Pettit by that time was already seeing fluctuations in the EPR gauge—but Captain Wheaton was unconcerned. He kept whistling while his copilot wondered.

And therein lies another predictable mystery—another human propensity seen time and again in airline accidents. Why did Roger Pettit, the copilot, let Larry Wheaton, the captain, continue the takeoff when he knew—KNEW—something was seriously wrong? Why didn't he grab the throttles, yank them to idle and then to reverse, and commit Wheaton to stopping a rapidly developing problem?

Principally because Wheaton was the captain, and the captain has the sole authority to decide whether to reject a takeoff. Even though Pettit had a reputation for being knowledgeable and assertive (and he was certainly not an intimidated new employee, as had been Twin Otter copilot Richard Roberti of Air New England), even though he mentioned his worries no fewer than three times as they were bumping down runway 36, he could not bring himself to demand that Wheaton abort—or to do it for him.

Pettit was not sure, and there wasn't enough time to evaluate the problem. He could see that the gauges didn't look right, the thrust felt weak, the throttle position was all wrong, but was it him? If he had only had more experience in the 737, he might have known.

Suppose Larry Wheaton had it under control after all? Suppose the captain knew nothing was wrong? Roger Pettit couldn't run the risk of committing them to the possibility of skidding off the end of the runway just because he *thought* things weren't quite right. Pettit was not completely sure that the engine-thrust setting was dangerous; therefore he couldn't risk doing anything about it. All he *could* do was point it out—over and over. He couldn't arm-wrestle Wheaton for the throttle controls; therefore he had to let it go. He had to take a chance.

Copilots and flight engineers are supposed to bear equal responsibility for the safe conduct of a commercial flight. In theory, if the captain is about to do something dangerous and will not heed a fellow crew member's warnings to stop, the other pilot can take over.

That, however, is a very rare event. Very few copilots ever physically seize control from a captain. The captain "owns" the airplane. He makes the final decision. Copilots and second officers defer to *him*. That has always been the unspoken reality—and of course it clashes head on with the responsibility of *each* pilot to keep the flight safe.

For decades there has been little if any support for copilots and second officer/flight engineers who questioned a captain's judgment, and the problems cut across the industry. On small carriers the size of Downeast Airlines to large carriers the size of giant United, there has been a strong, traditional reluctance of subordinate pilots—copilots and flight engineers alike—to be assertive. They may know that they have a concurrent legal and moral duty to keep the flight safe, but it is too easy to wait too long to say something. After all, if you offend the captain you're paired with for the next four weeks, it's going to be a difficult month. Being diplomatic gets in the way of being responsible—especially when your airline refuses to encourage assertiveness.

And then there are the autocrats—the "iron pants" captains who never relinquish control for a second. There were (and still are) far too many such men who simply cannot be told anything by their subordinate pilots, and who consider themselves unquestionable lord and master. It takes a strong copilot indeed to change such a captain's course of action—unless the airline has made it clear that the captain *must* listen and yield to the advice of his fellow pilots. United Airlines has instituted such a program with moderate success. Air Florida had never had time; they were too busy expanding.[6]

So Roger Pettit essentially let Larry Wheaton kill him—along with seventy-two others in the cabin. Of course, Stanley Gromelski's takeoff clearance to Palm 90 with an Eastern flight less than two miles behind it could not have helped. Larry Wheaton

6. Cockpit resource management training, like its predecessor assertiveness training, was pioneered by United Air Lines and developed in conjunction with the human-factor studies of NASA Ames Research Center at Moffett Field in California. The program was undertaken as a result of a recommendation written in part by NTSB investigator Alan Diehl following the United DC-8 crash in Portland, Oregon.

may have been unmindful of the potential problem, but perhaps Roger Pettit realized what would happen if he seized control and forced an aborted takeoff. Even if they managed to stop the aircraft before skidding off the far end of runway 36, they might be hit from the rear by the Eastern jet. What role such a realization might have played in staying his hand—limiting his recognition that something was wrong to words instead of action—no one will ever know. But Gromelski's illegal sequencing simply added another partially contributing cause, which could have evolved into a completely different disaster.

Behind that question lies a lesson most of those associated with Palm 90's demise would miss entirely. The "probable cause" of the accident—the central determination that the NTSB is required by law to discover and report—is inconsequential and misleading if it obscures the other contributing factors. All the failures and problems that *contributed* to the accident, or that *could* have contributed to *any* accident, must be spotlighted and their causes thoroughly investigated. Not just the "star" failure. Not just the probable cause.

But when companies are trying to avoid the lion's share of the legal liability, and when an aircraft manufacturer is desperate to avoid bad publicity for one of its products, this truth can be eclipsed. Such would be the case with Air Florida.

Chapter 19

STACKING THE DECK

Captain James Marquis of Air Florida, for one, was hopping mad. The investigation was going in the wrong direction as far as he could see.

Marquis had been included as part of the NTSB accident-investigation team (on several of the fact-finding groups) as a result of his position as a captain with Air Florida as well as his extensive background as an aeronautical engineer and former test pilot.[1] He could tell in the first few days that it was unfolding as a human-failure accident. But soon after the investigation had begun, he

1. It is standard practice for the NTSB, as it forms its fact-finding groups (such as the Witness Group, the Power-Plant Group, the Weather Group) to include experts from the air carrier involved, the aircraft manufacturer, the FAA, and other qualified people as well as the NTSB staff members. Such a mix brings to bear a far greater range of experiences and viewpoints. Seldom, however, do such NTSB groups include an active airline pilot with as extensive a technical background as Jim Marquis. In previous years, as a former test pilot and aeronautical engineer for Chance Vought Corporation, Marquis had participated extensively in flight testing the A-7D/E, a military jet, which is one of the few aircraft that has a sealed, drooping leading-edge flap like that of the Boeing 737.

had started to hear about the previously obscure history of pitch-up and roll-off incidents in the Boeing 737.

It was not terribly hard for Jim Marquis to believe that the pilots on Palm 90 had made mistakes that caused a fatal crash. He was a pilot too. He was well aware that pilots, no matter how professional, are not perfect. But Jim was also very familiar with Larry Wheaton, having taught ground-school classes Wheaton had attended (before Marquis had quit the extra-duty teaching assignments out of disgust over the complaint by the vice-president of operations that he was "expecting too much" of his students). He was well aware that Wheaton considered himself so well versed as a pilot that he did not need to study or prepare, or to steep himself in the unfamiliar techniques of handling a large turbojet airliner. Wheaton had maintained the attitude that it would simply come to him. He would learn on the job. After all, he had been a DC-3 captain; he could handle the 737.

Jim Marquis also knew Roger Pettit. He knew him to be a fine pilot, a sharp student, and an engaging fellow who would not remain silent when something was wrong. But he also knew Pettit was serving as copilot to a captain who was essentially weak and would not admit it.

Marquis had heard the voice tape of Wheaton and Pettit's final minutes. He had reviewed the litany of failures. He had heard Pettit speak up again and again, and he had heard Wheaton ignore the inputs again and again. He also knew his company's training had been too lax—that was why he had quit as an instructor.

But something in the thrust of the investigation was not right. Palm 90 should have been able to fly out of the stalled condition with Roger Pettit pushing hard on the control yoke—despite all the other failures. The 737, after all, is a powerful aircraft. The fact that the aircraft actually left the ground—achieved flight—was of great significance to Marquis. The fact that the V2 flying speed of 142 knots had been reached spoke volumes to a man well versed in the behavior of high-speed swept wings and jet engines. The discovery that many other 737 crews overseas had experienced similar pitch-ups just after lift-off (and the natural desire to prove his airline somewhat *less* guilty) launched Jim Marquis on the trail of the aerodynamic realities of Palm 90's condition—a lonely quest, which was to receive little support from even his own company.

Early on, Marquis came to a radical, tentative conclusion: The probable cause of the crash might in fact have nothing to do with the crew's using only 75 percent power! If Palm 90 had pitched up too steeply—say above 30 to 35 degrees—the tremendous amount of aerodynamic drag that angle would have created on the airplane might have been far too great for the engines of Palm 90 to equal or overcome—even if the throttles had been pushed to the fire wall![2] In other words, it was possible that the Boeing 737 jetliner could do far more than just experience a pitch-up or roll-off on takeoff. It could pitch up so far as to become uncontrollable, unflyable, and unrecoverable—an inherent aerodynamic flaw that, if it existed, would mean that in icing conditions, the 737 was a potentially dangerous airplane. Of course, the mere mention of such a possibility in public would send cold chills through the heart of

2. When an airplane wing moves through the air, it creates lift, but it also creates drag. You have probably experienced drag personally when, as a child, you held your hand flat against a forty-mile-per-hour (or faster) windstream out of a car window. The pressure felt by your hand is the "drag"—its degree of resistance to moving through the air. The drag experienced by an aircraft moving through the air is the same thing—the degree of resistance.

When a Boeing 737 (or any jetliner) rotates—lifts its nose—on takeoff, the wings increase their pitch in relation to the wind. That "pitch" is called the "angle of attack" of the wing, or the angle at which the oncoming wind strikes it. As that angle of attack is increased, the lift produced by the wing increases, sending the aircraft higher. But the drag is increased as well. Where there is lift produced, there is drag produced. Now, if the angle of attack is that of a normal takeoff, the power of the engines—measured in pounds of thrust—is substantially greater than the amount of drag, also measured in pounds. Therefore the aircraft will accelerate, or climb, or both. If, however, the pitch-up is too great and the wing's angle of attack is increased too far, the drag produced by the wing can increase beyond the capacity of the engines to equal it or overcome it. For instance, if the engines are capable of producing 30,000 pounds of thrust, but the drag has increased to 40,000 pounds, the aircraft is not going to increase its speed—it's going to slow down. If that nose-high, high-angle-of-attack condition is maintained for too many seconds, the deteriorating airspeed will cause the wing to stall (the airflow will separate from the top of the wing and the lift produced will drop off dramatically), and the aircraft will begin to descend. At an extremely high angle of attack, which could result from a severe pitch-up, the 737 could act like a giant version of your hand held out of a car window—simply plowing through the air rather than producing lift. The burbling effect of the separated airflow from the stalled wings will flow back over the tail like rhythmic turbulence, causing the aircraft to shake violently. If the pilots are unable to push the nose down, or do not act fast enough, or fail to use the proper control inputs to bring the nose down, the aircraft will lose several hundred feet in a matter of seconds as its descent steepens. If the jet is only a few hundred feet off the ground to begin with, it will crash—even with the engines screaming at full power and the nose up as if trying to climb. The condition of having more drag than the maximum thrust from the engines can overcome is known as being "behind the power curve."

the Boeing Commercial Airplane Company back in Seattle. Yet if he could prove it was so, Air Florida might get off the hook.

But all the other reported incidents of Boeing 737 pitch-ups and roll-offs in Europe had resulted in successful recoveries. In each of those cases (unlike that of Palm 90) the pilots had already had full engine thrust when the incident began, and most had slammed the throttles as far forward as they would go as soon as they knew they were having trouble.

Someone in the cockpit of Palm 90 HAD slammed the throttles forward mere seconds before impact.[3] But why wasn't that done before, when the stick-shaker began its death rattle? Who were they saving the engines for, the NTSB?

Marquis was painfully aware that pilots with limited experience on large turbojets often don't realize that EPR limits are only operational limits. The engines will not necessarily be damaged if the throttle is jammed full forward with no regard for the EPR limits. Damage to the turbine blades or the "hot section" may occur, but the engine will (in most cases) keep on running. The life of the engines may be shortened, but the life of the passengers may be extended by the extra thrust.

Airline ground schools and instructors and even the air force teach pilots strict compliance with EPR limits, however, in order to preserve the operational life of the expensive jet engines. It takes at least a few years of maturation and observation for a pilot to make the personal decision that if the airplane is threatening to crash anyway, he or she will disregard those operational limits and take every ounce of thrust available—by slamming the throttles to the fire wall.

Neither Wheaton nor Pettit, though, had reached that stage of experience. That instant reaction was not yet ingrained in their minds. The rattle of a stick-shaker (meaning the approach of a stall) should automatically cause a pilot to go to full power and lower the nose. With all the problems Roger Pettit and Larry

3. The increase in engine speed—hence the revelation that the throttles had been pushed up—was determined by NTSB evaluation of a blade-passing frequency-analysis test recording done by Boeing and the NTSB with a single 737 in a hangar in Renton, Washington. The sound characteristics of the known power settings from that test were compared with the sound patterns on the cockpit voice-recorder tape of Palm 90. Although the results seemed to clearly confirm that the power increased dramatically only seconds before impact, there remains some disagreement over when the power change occurred, and whether the pilots pushed the throttles up just after the stall warning began.

Wheaton were experiencing the second they broke ground, even the mistaken belief that they were already *at* full power—maximum EPR—should not have stayed their hands. But on Palm 90, when the stick-shaker had begun, the throttles had been left untouched—until it was too late.

Jim Marquis was determined to find out if it had been "too late" all along, and an upcoming series of 737 simulator flights at Boeing's factory in Renton, Washington (a suburb of Seattle), would provide the opportunity. The NTSB had scheduled the test "flights" to re-create Palm 90's flight path and find out what Wheaton and Pettit had really faced. The results of those tests might solve the mystery.

It was June when Jim Marquis arrived in the Seattle area for the simulator tests and other performance research. The horrible publicity over the crash of Palm 90 and the passenger reaction had taken a toll on everyone at Air Florida. The airline had already been in a struggle due to the loss of former Chairman Ed Acker, who had defected early in 1981 to Pan American—taking his knowledge of the confidential plans of Air Florida with him. Pan Am had almost immediately stepped up its competition on Air Florida's most highly traveled routes just as the recession began. That had put the airline in a constant financial bind. Now the load factors were dropping off further in reaction to the crash, and the upbeat spirit around Air Florida had withered.

The flurry of lawsuits against Air Florida was an added agony, along with an irritating NTSB inquiry, which had begun in mid-spring. Pursuing the new NTSB interest in researching the human-performance background and causal connections of airline accidents—an interest the Downeast Airlines investigation had helped spark—a staff psychologist from NTSB headquarters began poking around in Miami. The psychologist was a human-factors specialist who wanted to dig deeply into Larry Wheaton's and Roger Pettit's backgrounds. She began demanding records and interviews and making the already defensive management very jumpy. The feeling among the Air Florida people became a bit paranoid: The NTSB, they were convinced, was out to tar them with blame for the entire tragedy.

In reaction, Air Florida and its attorneys in Washington began toying with the defense that its pilots had done nothing critically wrong! They claimed that the problem was with the airplane—the Boeing 737—in that it (allegedly) had a potentially fatal aerodynamic characteristic that Boeing had not warned them about with sufficient emphasis. Although Jim Marquis was in Seattle as a representative of the Air Florida Pilots Association and as a member of the NTSB Performance Group, his airline back in Miami was hoping he could validate that defense.

But with a flight simulator?

Marquis knew the nature of Boeing's 737 engineering simulator. It was a computer. True, it consisted of a working replica of the 737 cockpit with visual displays of runways and countryside and whatever else the computer was asked to generate on those screens. And the "feel" of the "box" was so realistic that a pilot could almost convince himself he was actually in the airplane. It was an incredible array of electronic wizardry.

But it was not an airplane. It was essentially a computer-driven box on hydraulic arms. It would do only what it was programmed to do.

Normally the simulator—any simulator—is programmed with extremely complicated formulas that match the way the real airplane performs in any normal regime of flight. The data from which those formulas are calculated come from information gained in the aircraft itself during actual flight tests. Therefore, the performance characteristics of the simulator become practically indistinguishable from those of the real airplane—in normal flight regimes.

But the "flight regime" of Palm 90 had been anything but normal. Mathematical calculations of the energy involved (done for the NTSB Performance Group by NASA's Ames Research Center in California) had shown that *something* was causing the 737 to perform very differently from a standard 737. Something had changed its aerodynamics, and until they knew exactly what, and to what extent, there was no way to program any simulator to duplicate precisely the tragically abbreviated flight.

Yet Boeing was getting ready to conduct simulator flights to duplicate Palm 90's handling characteristics in those brief twenty-eight seconds in the air.

How?

Jim Marquis began making quite a point of looking for that answer. The Boeing men told him that since they had done some flight tests in late 1980 (in response to the European pitch-up incidents), they *did* have some flight data showing how the Boeing 737 would fly and handle with ice-particle-contaminated leading-edge flaps.[4]

4. As reported in the Boeing customer magazine *Airline* (October–December 1980 issue), Boeing tests in the fall of 1980 had involved a Boeing 737–200

The Boeing test pilots had been startled by the results of those tests. The effect, it turned out, of very small accumulations of ice or frost on those critical leading-edge flaps was very significant—and scary. The 737 would pitch up smartly, and if the contamination was made greater on one wing than the other, it would roll off dramatically on that side. The test pilots had to act rapidly to oppose the pitch-up and maintain control.

The startling aspect was that with a contaminated leading-edge flap, the stall speed was significantly increased as much as fourteen knots! In other words, a flight crew could be all the way to what it thought was a safe flying speed and still be stalling.[5]

That 1980 test series was the origin of the data the Boeing men had plugged into the lift formulas for the simulator's computer. But Palm 90 had reportedly pitched way up, possibly above 30 to 35 degrees of deck angle.

Did the flight tests, asked Marquis, go that far? Did the pilots

---

advanced model flown to Moses Lake, Washington, for a series of stall tests to determine what they called the aerodynamics "of contaminated leading-edge slats [flaps] typical of: Hail . . . erosion [such as through exposure to Middle Eastern operations on sandy, gritty runways] . . . and frost and ice accumulations that can occur during ground operations." The tests were in response to both pressure from the British CAA and the increasing number of reported European incidents. The aircraft's leading-edge [flaps] were given simulated frost configurations "by coating the surface with an epoxy potting compound and then roughening the surface with a paint roller"—corn ice was simulated with a heavier version of the same coating. (This was a much more subtle contamination than the plastic ice forms that had been tested several years before by Boeing.) Among the published conclusions, Boeing's men stated: "Stall characteristics with both symmetric and asymmetric simulated frost were characterized by a very apparent pitch-up, yaw rate, and roll-off. . . . Generally, increased roughness resulted in increased pitch, yaw, and roll transients."

It is noteworthy that the suspected cause of the majority of the pitch-up/roll-off incidents over the previous ten years had been snow, ice, and frost accumulations, yet even in the Boeing *Airliner* article, mention of "frost and ice" was buried deep in the initial narrative. The genuine fear that the 737 might be saddled with a bad reputation for dangerous behavior in ice or snow conditions was on everyone's mind. Even though it had been sufficiently concerned to conduct the flight tests, Boeing wanted the results kept very low-key and technical.

5. Boeing's first indication of this tendency had come in 1971 when famed Boeing test pilot Lew Wallick and his crew experienced a pitch-up and roll-off tendency in icy winter tests of the 737-200. The problem became so bad on succeeding takeoffs that the pilots stopped and inspected the leading edge of the wing—and discovered a thin coating of roughness, little more than a frosting of ice on the leading-edge flaps. Once it was removed with de-icing fluid, the airplane returned to normal behavior.

investigate what the aircraft would do in a deep, extreme stall with an extreme deck angle such as Palm 90 might have had?

"No, we didn't," one of the test pilots told Marquis, because, he said, with the simulated ice on the leading-edge flaps, the test aircraft became "simply uncontrollable" when they let the speed diminish too far. It was too dangerous to slow down further or investigate a wild, 30 to 35 degree pitch-up. They might have crashed!

Yet the test crew did contend that they had recorded enough data. And from that they had concluded that whatever pitch-up might occur with leading-edge ice, and at whatever airspeed, the airplane could still be recovered. The heartbeats of the pilots might be racing as they flew out of it, but if they jammed the control column forward fast enough (and perhaps used nose-down trim), the airplane would keep flying. Of course, that assumed that the aircraft was using maximum thrust.

But there was one more fact. They had not used a "spin chute," and as Marquis knew well, there was no way the test pilots were going to be able to investigate the full range of the 737's wild performance with contaminated leading-edge flaps unless they were prepared to "lose" the aircraft, then recover it by popping a stabilizing parachute from the tail.

Without a spin chute, they could not have gathered enough flight-test data to duplicate the area of wild aerodynamics Palm 90 had invaded.

But Boeing tried to reassure Marquis that it had used data from scale-model wind-tunnel tests of the 737 wing in that extreme, deeply stalled, high-deck-angle regime Palm 90 had experienced. Then, it told him, it had extrapolated the data. Therefore, although not exact, the formulas would probably be very accurate. The simulator could be made to reproduce the flight of Air Florida's jet into the Fourteenth Street bridge.

Jim Marquis knew that wind-tunnel data often bore little resemblance to actual flight-test data. It was merely a starting point. He asked to see the formulas. Boeing refused, saying he couldn't understand them.

"Try me—I've got a degree and experience in these areas."

"No way."

Marquis was growing very suspicious.

When the NTSB team arrived, they had first set to the task of determining the power settings of Palm 90's engines by making a tape recording of different power settings on a 737 run in the hangar, and then comparing the "passing blade frequencies" to those on the voice recorder recovered from the river. That series of tests had proved that Palm 90 was at 75 percent power on take-off and had stayed at 75 percent until just about two seconds before hitting the bridge.

Then they began the simulator flights, and as Marquis expected, a problem occurred.

They couldn't get the simulator to hit the bridge!

The test pilots had to hold the control column back in their laps aggressively to approximate the flight path of the Air Florida 737. But if they simply relaxed the back pressure, the simulator would "fly out" of the stall and miss the bridge, or miss the bridge and crash on the far side.

Over and over again they tried, but not once would the simulator duplicate Palm 90 without the artificial back pressure on the control yoke.

The reason seemed obvious to Marquis—the flight formulas were totally wrong for such a deeply stalled condition, and the tests were effectively worthless. Yet some of his fellow Performance Group members were becoming convinced that Palm 90 could have flown out of its death stall if the pilots had just added full power the second the stick-shaker began vibrating—while others were trying to convince themselves that Roger Pettit could have frozen on the controls, holding positive back pressure all the way to the bridge despite his captain's continuous exhortations to move the yoke "forward, forward . . . come on, forward."

While Marquis was becoming increasingly upset with the tests and the results, the Boeing men were becoming increasingly upset over him—especially over his questions and his suspicions. It was obvious to them that this captain from Air Florida was trying to find a way to pin the entire blame for the accident on the pitch-up characteristics of the 737. What was most frightening to them was his background. Marquis had enough training and knowledge to pull it off. If he had enough information, he just might convince the NTSB, and he might take it to the press.

In addition, Jim Marquis had found out not only about the first flight-test pitch-up incident in 1971, but about the entire history of leading-edge-flap contamination problems. If he blew the cork on the anthology to the press, the damage to Boeing 737 sales could be enormous. The public might decide the craft was unsafe. The FAA might ground it worldwide. Indeed, Boeing's worry about such a potential had been the governing factor in just how much noise it had made about the repeated incidents over the years. That had been one of the principal reasons why the European 737 operators (who were the only ones reporting pitch-up/roll-off incidents) had received more attention from Boeing than domestic United States 737 operators. The other reason was the interesting fact that airlines in North America (including Alaska) had never reported such an incident.

As they all knew, there might be a very good reason for that dearth of reports.

North American airlines flying 737s flew them into and out of severe ice and snow conditions every winter, and had for the same ten-year period that European airlines had been having problems. The U.S. and Canadian aircrews were just as subject to human error as their counterparts in British Airlines, so what was the difference?

In a word, immunity.

In the United States, especially, if a pilot reports to the FAA that he almost lost control of his 737 on takeoff because (he suspects) he took off with frost, snow, or ice adhering to the leading-edge flaps, he is admitting that he violated the Federal Air Regulations. The act of trying to take off in such a state is a violation. His company will not want the FAA or the public thinking that such an act is standard practice, or that any other pilot at that company has ever done anything like that before, so its reaction will also be protective and disciplinary. He contravened an FAR and he did it in violation of company procedures and policies. Suddenly he's not a hero, he's an outcast. The possible contribution to safety of his report—his admission—is on no one's mind. His threat to the reputation of the airline and the FAA inspectors responsible for seeing that "their" airline does not do such things is on a lot of minds. The man must be disciplined, and in fact, he could lose his license and his career. Such "rewards" do not do much to elicit reports of pitch-up/roll-off excursions in any airplane.

In Britain, however, such a pilot is protected. He is in the act of contributing knowledge to the safety system, and his individual "transgression" must go unaddressed in the interest of assuring that others will have the opportunity to learn by his mistake.

In the United States, the FAA began an incident-reporting system, but since few aircrew members trusted them with self-incriminating information on aerial mistakes, NASA took over. NASA's immunity provisions, however, later became substantially

274

restricted in the late seventies through the actions of former Administrator Langhorne Bond, who, while engaging in what seemed to be a personal vendetta against airline pilots, felt that they were getting away with misuse of the immunity system.[6]

As a result of Bond's crusade of accountability, any U.S. pilot who alighted at his destination (adrenalized and breathing heavily after a near-disaster with a severe pitch-up in his 737) and picked up the phone to report the incident in detail would be taking an enormous risk. Even in the major, established airlines, his career could be imperiled. In the new-entry carriers or rapidly expanding ones with no time or money for FAA battles, and no union protection, such a phone call could be professional suicide.

In such an atmosphere, how could anyone know how many pitch-up/roll-off incidents might have occurred in years past, or how many times unknowing passengers might have had a brush with death? There could have been scores of close calls, all unreported.

The fact that U.S. pilots cannot make such reports openly and safely to an agency with the legal teeth to take immediate action to prevent further problems (and not simply to stomp on the pilot) is a major, festering flaw in our system. Information and knowledge

6. NASA and the FAA signed an agreement in 1975 by which NASA would establish a Safety Reporting System based at NASA's research center for human-factor studies at Moffett Field in California. The object was to let pilots and controllers (or anyone else) report a problem, even one resulting from their own negligence, without giving the FAA the right to attack and prosecute them. To protect them and make the system operable, the FAA agreed to give the reporting crew member immunity, and NASA agreed to give him anonymity. Of course, human nature dictated that certain less scrupulous pilots would take advantage of the system by hurrying to report an otherwise actionable "mistake" before the FAA found out, just to obtain the immunity. In some cases the report contributed to greater safety awareness anyway; in other cases the only beneficiary was the guilty pilot. When Langhorne Bond (a Carter appointee) took control in 1978, he declared war on such people and set about to destroy the immunity structure of the NASA reporting system. Despite vociferous warnings from all over the nation that his scorched-earth policy would muzzle the vital exchange of safety information and destroy the embryonic reporting system, Bond persisted—and managed to win a reduction in the scope of the immunity grant. By doing so, he effectively succeeded in emasculating (but not destroying) the system. The NASA Safety Reporting System today (because of the many safeguards of the reporter's anonymity) is far too general and uncategorized to be of immediate use in system reform. Unfortunately, the FAA pays hardly any attention to NASA's system.

are strength, and in the high-speed field of commercial aviation (especially under the confusion and internecine wars of deregulation) anything that retards the exchange of information vital to safety is lethal. Who knows better what is really happening on the front lines than the troops? But if the troops can't talk, the generals can't listen.

Of course, at Boeing in the early summer of 1982, the concern was that at least one "troop" *would* talk. Jim Marquis was being perceived at higher and higher levels within Boeing as a major threat to the reputation of the 737. He might influence the NTSB to shift the emphasis from the human failures associated with Palm 90 to the aerodynamics of the aircraft. He might convince someone that Boeing had stacked the deck with the simulator tests and its previous performance data, and had covered up a family aerodynamic secret. He might make his point that the flight tests had been inadequate, or that the dissemination of information on the results had been suppressed. After all, the Boeing officials reasoned, hadn't Marquis already told the attorneys in a deposition that all those human failures had little to do with the cause of the crash? He was telling even Boeing's technical people that Palm 90 had pitched up so far that no amount of power could recover it, and that it appeared they were trying to cover it up.[7]

During the simulator flights, one of the Boeing engineers had indicated to Marquis that he was not surprised that the simulator refused to pitch up uncontrollably as Marquis was certain Palm 90 had done. It did not amaze him that the test pilots were having to hold back pressure.

"We've seen it before," he told Jim Marquis.

"What do you mean, exactly?"

"Well, when we were adjusting the simulator before you people got here, we found we had to hold back pressure."

Marquis was astounded. What sessions of "adjusting the simulator" was this fellow talking about?

---

7. For that matter, the FAA's Northwest Region, in charge of certification of the 737, was also on tenterhooks. If Boeing was perceived as being negligent in suppressing information or not conducting the proper tests, the FAA (who had responsibility for following the aerodynamic capabilities of the 737 every inch of the way) would be at fault as well. All the men in the Northwest Region knew what had happened to their counterparts in Los Angeles in the wake of the DC-10 scandals in the mid-seventies, and the intense embarrassment that had followed widespread publicity. Though Langhorne Bond was no longer in office to do to the 737 what had been done to the DC-10 (a nationwide grounding), the apocalyptic possibility loomed. Jim Marquis was a threat to the FAA too.

"You flew these tests before the team came out here?"

"Sure, to adjust the parameters. With the interpolated data we needed to get it as close as possible, so we flew it and kept adjusting the formulas until it would perform properly."

Jim Marquis had returned to his Renton hotel room almost in a daze. He was thoroughly shocked and disgusted. Here was the NTSB, flying what it thought were untested simulations, and it had all been set up and adjusted ahead of time. If the original formulas had caused a true, uncontrolled pitch-up in the simulator, Boeing would have had the opportunity to eliminate it by changing the formulas! The Boeing 737 would not be allowed to look bad. At least, that's how it seemed.

Marquis had not hidden his feelings at the news. If this wasn't a cover-up, it sure smelled like one.

And Boeing's men hadn't missed the point. Indeed, Jim Marquis was becoming a major threat to the financial security and future of the Boeing Commercial Airplane Company, and by definition, the city of Seattle and the state of Washington. That threat would not remain unaddressed.

When the phone by Jim Marquis's bed rang in his hotel room that night, he was deep in thought. He was feeling very ineffective. All his concerns were being ignored, as if he were yelling into a stiff wind. He was not having much impact—or so he thought.

His answer into the telephone receiver was routine, and his recognition of the voice on the other end was slow in coming.

"Hello. Jim Marquis?"

"Yes, this is he."

The voice on the other end had a slightly familiar drawl to it—a voice he vaguely remembered hearing somewhere before. Not an accent, just a slow, measured, method of speaking.

"This is Senator Henry Jackson, Captain Marquis. Do you know who I am?"

Sure, he knew. Jackson was a United States senator. A very senior one. It was recognition of a simple fact. He would evaluate it later. Why would such a man be calling here?

"Yes, Senator, I know who you are."

"Well, I think we've got something in common. We both have an interest in people involved in this accident you're working on."

"Yes?" This didn't make much sense. It was not a secretary or an aide on the other end asking to hold for Senator Jackson. This WAS Senator Jackson. Why, wondered Marquis, would he be bothering with a small player like me?

"The reason I'm calling is that I'm concerned about my people in the Seattle area. I'm concerned what might happen to them if you get everyone worked up over the Boeing 737."

Jim Marquis thought for a fleeting moment of the recent discoveries from Boeing's files. The interoffice memo that directed other Boeing people to "suppress this information" regarding the pitch-up/roll-off characteristics; the information given to European operators, which was not given to U.S. airlines; the Boeing memo that cautioned against recommending the use of 10-degree flap settings in icing conditions (which would effectively eliminate the problem) because the aircraft could not carry as much weight and would be less competitive with the DC-9 and other airliners. In those few seconds, Marquis knew why the senator was calling, but he still couldn't believe it. And he certainly wasn't prepared for the next question.

"What are you trying to do, Captain Marquis, put Boeing out of business?"

The senator let that sink in for a second before continuing.

"Are you on some kind of vendetta against Boeing?"

"No, sir, I'm sure not. I've just . . ."

"Well, do you want to shut down all the 737 flights in the U.S., is that what you're after?"

"No . . . no, I'm not. I feel a moral obligation to the rest of the traveling public to have this issue surface. That's the whole concept of accident investigation, y'know, and I think it's being suppressed."

Jim Marquis couldn't believe that Jackson thought he could have that much impact on the situation. He couldn't even get his fellow NTSB Performance Group members to listen. How could he damage Boeing?

"Let's talk about moral issues a minute, Captain." The voice remained slow and calm and measured.

"Well, let's talk about the effect of shutting down 737s all over the nation or the world. Think of the people you're going to inconvenience. Think of the people you're going to put into despair if you put 'em out of work at Boeing by shutting down the 737 line and disrupting sales. What happens to them if Boeing collapses? Have you thought of that moral issue?"

"No, sir, I . . ."

"Well, you need to think of that. That's a moral issue too, isn't it? You know, out of all those people, two or three [of them] will probably be so despondent they may go over the brink. They may

278

be in such a state from losing their jobs and having no income that they may even commit suicide. What would you think about that as a moral issue? Would you like to be responsible for that?"

"Senator, I appreciate what you're saying, but there are plenty of passengers who may lose their lives if this problem is not solved."

Jackson was obviously a very smart fellow. He wasn't demanding anything, and he wasn't trying to intimidate. He was counseling—like a father to a grown and intelligent son. He was trying to change Marquis's attitude with logic. Jim Marquis was impressed, but he wasn't persuaded.

The two dissimilar men, the senator and the airline captain, talked a few more minutes before Jackson ended the conversation.

"Well, Captain, I wish you success in accomplishing whatever goals you set for yourself on this issue. I hope someday we have a chance to meet."

"We may, sir. I undoubtedly will be in Washington a lot."

And just as quickly the phone was back on the hook.

Marquis sat there for the longest time with a wildly fluctuating set of emotions. Jackson had asked who he thought he was to make such decisions—decisions on the fate of the Boeing people and their company. And Marquis wondered just that. He hadn't thought he was having an impact. Good Lord, apparently he had scared them all the way to the chairman of the board, or someone high and mighty enough to pick up the phone and ask Scoop Jackson to intervene. Of course, there was a question of propriety, as well, over Jackson's making the call. Should a U.S. senator do such a thing? But Marquis kept coming back to Jackson's question. Did Jim Marquis indeed have the right to imperil such a company just because he thought there was a safety point being swept aside? Was Jim Marquis that powerful? Was he right?

And, of course, the question the call was meant to raise: Should he give up?

Chapter 20

SIDESTEPPING THE REAL ISSUE

The words of the NTSB Blue Cover report issued on August 10, 1982, were an honest and thorough attempt to address the myriad of reasons why Palm 90 ended up a broken mass of metal and people in the frozen Potomac—but like so many involved in the investigation, the NTSB missed the most important point.

The National Transportation Safety Board determines that the probable cause of this accident was the flightcrew's failure to use engine anti-ice during ground operation and takeoff, their decision to take off with snow/ice on the airfoil surfaces of the aircraft, and the captain's failure to reject the takeoff during the early stage when his attention was called to anomalous engine instrument readings.

Contributing to the accident were the prolonged ground delay between deicing and the receipt of ATC takeoff clearance during which the airplane was exposed to continual precipitation, the known inherent pitchup characteristics of the B-737 aircraft when the leading edge is contaminated with even small amounts of snow or ice, and the limited experience of the flightcrew in jet transport winter operations.

Boeing had escaped from its nightmare of a potential public-relations disaster. It had worked hard to drive home the argument that pilot error, ground-crew error, air traffic control, and airport procedure problems had caused the crash—not an unrecoverable aerodynamic quirk of the 737. It was illegal to have attempted a takeoff with ice on the airplane, they said, and nothing else mattered.

But to Jim Marquis that was a smokescreen. Since Boeing had not used a figurative two-by-four to slap Air Florida into awareness of the pitch-up/roll-off problem and its seriousness, Boeing was responsible for the crash. In his view, the pitch-up of Palm 90 had been too severe to be recoverable by any flight crew with any power setting, and therefore the plane was at fault.

Back and forth, charge and countercharge, the exchanges continued, Boeing becoming frightened enough to enlist a U.S. senator and a lengthy round of arm twisting in Washington and playing fast and loose with simulator test parameters and programming out of economic fear, and Air Florida taking the essentially silly position that the acts of Wheaton and Pettit were fully professional and without fault.[1]

And in the end they were both right and they were both wrong.

They were right in that the charges each was leveling at the other identified important failures and problems that needed immediate attention. Human-performance problems, management problems, training problems, communications problems, testing and aerodynamics problems, moral, ethical, and fiduciary problems involving the public trust, and more. All were serious failures that needed full and impartial probing to prevent another disaster from occurring from related causes.

And they were wrong in that they were too busy trying to pin the blame on each other to focus on the same truth Marquis had touched on in his chat with the senator: The purpose of an accident

1. Although Jim Marquis helped draft the Air Florida recommendation of findings and conclusions to the NTSB, which sought to exonerate its pilots, the Air Florida attorneys in Washington, D.C., wrote the final version, and their defense of Wheaton's and Pettit's actions was couched on the premise that harsh criticism of their performance was unjustified because the pitch-up alone had caused the crash. Marquis had never maintained such a stance, though the management of Air Florida had. Marquis maintained that the only significant cause of the accident was the pitch-up, and that the aircrew's many mistakes were immaterial to a finding of probable cause. Jim Marquis had said that if the pilots had survived, and the choice had been his, they would have been, at the very least, "busted" (if not fired) for the stupidity of their actions.

investigation is not blame; it is exposure and understanding of the problems, *all* the problems (especially the human problems), so that passengers will be safer in the future.

The struggle for the NTSB's attention had itself been an instructive exercise in human dynamics, filled with politics, self-protection, intrigue, and maneuvering—all designed to put the blame where each party had convinced itself it belonged: on somebody else.

But to its credit, the NTSB managed to address each of the links in the causal chain, though the exact relationship or severity of many of them was unclear or hotly disputed.[2]

No one knew, or would ever know, exactly how high Palm 90 had pitched up. Without that knowledge, no one could ever know whether Wheaton and Pettit could have flown out of it—like so many others before them.

Jim Marquis had not given up, despite the feeling that no one was listening to him (and despite three separate death threats—two by anonymous phone call, one by anonymous note). What he lacked was the one final piece to the puzzle—irrefutable proof that the pitch-up was unrecoverable. Without that last piece (and in the face of the NTSB's partial reliance on the simulator flights), he could not prevail.

Under the heading "Other Factors Relevant to the Accident," the NTSB talked about the pitch-up characteristics in detail:

> The high pitch attitude occurred because the flightcrew failed to, or was unable to, react quickly enough to counter the aircraft's longitudinal trim change. . . . The Safety Board concludes that the pitch-

2. The list of findings in the Blue Cover accident report pinpointed American Airlines' failures in de-icing Palm 90 and the failure of both American's and Air Florida's personnel to follow the proper procedures; it chronicled the voluminous and continuous mistakes of Palm 90's pilots, the troubles with Washington National's procedures and the FAA controllers' mistakes, and spotlighted the pitch-up propensity of the Boeing 737 while concluding that Palm 90 would have been able to fly out of its stalled condition with three-quarter power and no ice, or full power and leading-edge ice, but not both. The conclusions (and those in the text of the report) left few stones unturned and even mentioned the dubious validity of the Boeing engineering simulator tests. Nevertheless, in the eyes of Boeing and Air Florida, the report was badly flawed because the "blame" was spread around and not placed on one particular cause or one particular party. Please read the NTSB's list of findings in the Acknowledgments. (These conclusions are from NTSB-AAR-82-8, pages 79–82.)

up tendency of the aircraft because of leading-edge contamination contributed to the accident. However, to place this contributing factor in perspective, the Board notes that no aircraft design requirements include the ability to perform with snow or ice contamination and that any known contamination, regardless of the amount or depth, must be viewed as potentially critical to a successful takeoff. For this reason flightcrews are not only dissuaded, but are prohibited, from attempting a takeoff with such contamination.

The Safety Board, however, agrees with the United Kingdom Civil Aviation Authority that there are times, such as night time operations, when a small amount of contaminant may not be detectable by the flightcrew and that precautionary procedures should be developed and implemented to reduce the potential of control problems if a takeoff is conducted under those circumstances. The occurrences of pitch-up or roll-off were first reported over ten years ago and although they prompted The Boeing Company to examine the B-737's flight characteristics during flight tests, preventive actions taken by both the manufacturer and the Federal Aviation Administration have been limited solely to the dissemination of advisory information. Even this information is couched in a manner which may fail to impart the hazard potential to the reader. . . . The Safety Board concludes that Boeing should have placed greater emphasis on the prohibition of takeoff if leading-edge contamination is observed or even suspected.

Moreover the Safety Board is aware that the Boeing Company has been considering and evaluating modifications to the B-737 wing thermal anti-ice system which would permit that system to be used during ground operations to prevent the formation of ice on the wing leading-edge devices. In view of the span of time over which the pitch-up/roll-off incidents were reported, the Board believes that the Boeing Company should have developed this modification and promulgated corresponding operational procedures more expeditiously. . . . The manufacturer and the FAA should move rapidly and before the next winter season to assure that wing thermal anti-ice system modifications and related operational procedures are implemented or takeoff speed margins are added to prevent further pitch-up/roll-off occurrences of B-737 aircraft during cold weather operations.

But the Boeing people got off lightly. None of their frantic "damage control" efforts, none of the interoffice memos, none of the shenanigans involving the engineering simulator or the ten years of nervousness over the skeleton in the 737's aeronautical closet would capture the imagination or undivided attention of the press.

There would be no public recoil from the 737 (and indeed none was deserved), and there would be no public accountability for the acts of good company men protecting their economic backsides. Unlike McDonnell Douglas and the DC-10, Boeing didn't get caught in the cross hairs of public scandal.

Boeing had not escaped completely, however. The Safety Board's admonition that Boeing had not done enough to make airlines aware of the pitch-up problem, and had not done enough to adopt a "fix" in procedures to prevent the problem, stung badly back in Seattle.

Air Florida had a point when it said Boeing should have warned it. Boeing should have realized that in dealing with upstart or newly expanded airlines exercising their freedom in the new unfettered airline system, they were not dealing with the same quality of organizational excellence that had always been expected of a United or a Delta. Boeing should have known that in the confusion of expansion and cost controls, it could not assume that anyone at Air Florida would do more than post bulletins, warnings, and articles.

While a major carrier might be expected to understand and immediately incorporate the bulletins into its training and procedures (and that may be stretching the imagination), Boeing should have known that Air Florida had reached no such level of organizational maturity.

Boeing, above all, had good reason to know that the 737 was becoming a favorite of upstart and newly expanded airlines with the greatest propensity for fielding the least experienced crews on the thinnest of training budgets in the worst possible flying conditions.

But if it knew it, it ignored it. Deregulation had changed the formula, and even Boeing had been unprepared to deal with the new order.

And therein lay the major—though honest—failure of the NTSB's otherwise extensive and comprehensive report on the crash of Flight 90. The element it failed to spotlight was the primary force that permitted all other factors to come together to cause the accident: Congress's 1978 passage of deregulation.

The true significance of the crash was that a newly expanded airline taking full advantage of deregulatory freedoms had grown

284

too fast to maintain the proper safeguards and standards the flying public has a right to expect of Part 121 carriers. The selection of aircrews, the training function, the standards set by management, and the operational control of maintenance and ground-support functions were all culprits, and they all stemmed from the ability to expand without restriction under deregulation—and the inability of the FAA to keep up with its duties of surveillance and control. Larry Wheaton would not have been in the left seat of Palm 90 if the Deregulation Act of 1978 had not passed. In fact, in all likelihood there would have been no Palm 90 at Washington National Airport in the absence of that legislative experiment. Therefore, the crash of Air Florida's Palm 90 was without question a deregulation accident.

That is not to say that deregulation is bad, unworkable, or by definition, responsible for crashes. Certainly, history has proved that any major carrier could have had a similar accident—a weak pilot slipping through the net and making a host of bad decisions at exactly the wrong time, running afoul of an aircraft-design problem that had been minimized, for economic reasons, by decent people working for an honorable manufacturer.

But the fact that such a tragedy could have occurred before deregulation is essentially immaterial. What matters is that a darling of deregulation failed to rise to the near-perfect standards expected of it, and in so failing, set up the deaths of trusting passengers. That unavoidably and legitimately calls into question Congress's promise in 1978 that safety would never be compromised.

Did this accident indicate that the safety level is deteriorating, at least in this type of carrier at this stage in its development under deregulation? The question has yet to be answered officially because, despite its extensive treatment of the human-performance background of Wheaton and Pettit and the expansionistic pressures on the carrier, the NTSB never faced the question head on.

Perhaps this was a political failure, since the board was certainly sensitive to the embarrassment that a causal connection between Palm 90's demise and deregulation could bring (indeed the board member assigned to the accident was sweating out reappointment by the White House). Yet, despite the sensitivities of the NTSB staff, they seemed simply overwhelmed by the complexity and the multiple causes of Palm 90's demise. They may never have focused on the prime issue *as* the prime issue.[3]

3. The work and passion of the NTSB staffers that went into the Air Florida investigation was monumental. Tempers flared on occasion, and in one instance two investigators nearly came to a fistfight in a hallway over the issue of deregulation's involvement.

What they *did* have to say about it, however, was helpful:

Flightcrew Experience and Training

Both the industry and the traveling public have come to expect the highest degree of performance and professionalism from flightcrews of scheduled air carrier operations and particularly from airline captains. It would further be expected that the basics of turbo jet operations would be clearly ingrained in the mind of an experienced, well trained airline captain, and that under the weather conditions existing at Washington National Airport on January 13, 1982, these basics would have dictated checking the wings for snow and ice, using engine anti-ice, and rejecting of the takeoff when the engine instruments appeared anomalous. An airline captain should have assimilated or gained thorough knowledge of these procedures and of the conditions which warrant their use. He should have done so both through actual experience and through formal training as he progressed through the various stages of his career. By the time a pilot qualifies as an airline captain, he should be capable of detecting and coping with not only the situations demonstrated in this accident, but with every phase of reasonably anticipated transport aircraft operations.

The Safety Board concludes that the flightcrew's limited training and low experience in jet transport winter operations in snow and ice conditions were contributing factors in this accident. The Board believes that the captain of Flight 90 missed the seasoning experience normally gained as a first officer as a result of the rapid expansion of Air Florida, Inc., from 1977 through 1981, wherein pilots were upgrading faster than the industry norm to meet the increasing demands of growing schedules. The Safety Board's informal survey of major trunk carriers showed that pilots upgrading to captain had served an average of 14 years as a first or second officer with the carrier.

The Department of Transportation periodically makes quite a point of how safe the air-transportation system has become under deregulation—by citing accident statistics. But accident statistics are historical measures. When the rules have changed, they have little bearing on what will happen tomorrow. The statistics quoted on January 10, 1982, did nothing to prevent the loss of Priscilla Tirado's husband and baby. They had no effect whatsoever in preventing the World Airways DC-10 from skidding off the end of

Boston Logan Airport seventeen days later. In fact, no amount of accident statistics can properly answer the question that means the most to the individual who gets ready to trust his or her life (or that of a family member) to an airline: How well or how poorly is deregulation working in relation to the standards of safety that the Congress promised in 1978?

No amount of smokescreening can obscure the relevance of this question. Congress changed the system. Congress promised the change would be safe. Therefore, the entire system that has resulted (and especially those carriers that were born of deregulation in whole or in part) must be examined to see how well Congress's promise is being kept. Glib assurances of safety are essentially meaningless—and rather insulting to what we assume is an informed electorate.

Is the system as safe as the American public has "come to expect"—as the public *demands* that it be? Can it be measured by yesterday's accident and death statistics, or does it make more sense to examine in precise detail what is actually happening out there? Just as "probable cause" is essentially immaterial if it obscures any of the other factors that could have contributed to an accident, any examination of the current air-transportation system is immaterial if it fails to probe in great detail the actual operational problems that exist this very day and hour in carriers large and small, new and old, throughout the land.

Waiting for the human deficiencies in a chaotic human system to metastasize into a devastating accident before acknowledging that a problem exists is just as stupid as placing a label of "pilot error" on an accident like Downeast's and closing the books without further probing.

Nothing will be learned, and nothing will be improved, if deregulation is to be accorded the protection of political interest and blind trust.

If the emperor is nude, it is up to the members of a free society to point out the indecent exposure!

Chapter 21

TO ERR IS HUMAN

"I don't know what the hell's holding that . . . in! Always something—we coulda [made] schedule."

Al Stockstill was not getting anywhere trying to force the small light bulb into the socket over the landing-gear handle. The green light that normally indicated that the nose gear was safely down and locked had refused to illuminate as they put the landing gear down on approach to the airport. Now they were flying around out to the west under radar control while trying to fix the problem. Stockstill assumed it was a burned-out bulb, but the replacement bulb wouldn't fit in the socket.

Captain Bob Loft was leaning over the center pedestal, watching intently—unaware of a small "click" as he nudged the control yoke.

Stockstill kept fumbling as Don Repo, the second officer, stood in the darkness of the electronics bay, one floor below the cockpit of the Lockheed 1011, straining to see through a tiny window. Repo was looking for a small, painted line on the nose gear in the wheel well behind the bay—a sight that would confirm it was down and safe—but the floodlight in the wheel well was out, too, and he couldn't see a thing.

The melodious sound of an electronic, musical chord in the key of C sounded in the cockpit. The chord came from the altitude-alert device. It signaled the fact that their aircraft had descended more than 250 feet below their assigned altitude of 2,000 feet.

Loft and Stockstill were devoting all their attention to the light bulb. They didn't notice the C chord—nor were they aware of the needles on their altimeters, which were starting to unwind—slowly at first, then increasing in rate of descent.

In the darkened interior of the radar-approach control room at Miami International Airport, the FAA controller handling Eastern 401 realized the flight was approaching the boundary line of his area. As the phosphorescent blip representing 401 continued its progress, he also noticed something else. The small, alphanumeric "data block" printed by the computer on the screen next to the blip was showing EAL 401 at 900 feet instead of his assigned altitude of 2,000 feet. Nevertheless, it wasn't unusual to have the computer "drop" the altitude information for a few hits of the radar beam. The 2,000-foot reading should return in a few seconds.

The controller keyed his microphone.

"Eastern, ah, 401, how are things coming along out there?"

Captain Loft picked up his microphone.

"Okay, we'd like to turn around and come . . . come back in."

"Eastern 401, turn left heading one eight zero."

That sounded okay to the controller. The altitude display was obviously wrong—they must be at 2,000 feet as assigned.

But Eastern 401 was not at 2,000 feet. The 900-foot reading had been accurate (but was no longer). In the time it had taken for that radio exchange, the Lockheed 1011 had descended to a mere 400 feet above the pitch darkness of the nighttime Everglades—and was still going down.

And at last someone in the cockpit realized something was wrong.

First Officer Stockstill suddenly focused on the altimeter in front of him.

"We did something to the altitude!"

They had been fooling with an electrical system in this new aircraft; perhaps they had fouled up the altitude display.

Captain Loft seemed uncomprehending.

"What?"

"We're still at two thousand, right?"

The altimeters on both sides of the cockpit were unwinding

through 250 feet, still moving, the giant airliner in a slight left bank, the autopilot disengaged, and nothing but black on the other side of the cockpit windscreens.

Captain Loft stared at the altimeter. It didn't make sense. It didn't compute. It could not be right.

He expected to see 2,000, but what he saw was nearly zero feet.

"Hey, what's happening here?"

Just as the captain spoke the word "hey," a beeping noise filled the cockpit, generated by the radar altimeter, which had routinely sensed the fact that the ground was very close—a fact neither pilot had yet figured out.

Captain Loft was probably in the process of closing his hand around the control yoke at the very instant the left wing of the Lockheed 1011 dug into the swamplands west of Miami at just below 200 miles per hour, tearing the fuselage to shreds and throwing the passengers and crew into a black-hole nightmare of death and confusion. It was December 29, 1972.

The 163 passengers had boarded the red-eye special in New York expecting to arrive at Miami International just before midnight. Now they were scattered in the blackness of the swamp, 94 of them dead, 69 of them alive—some just barely. Of the 13 crew members, 10 lived through the crash, including Captain Loft and Second Officer Repo. (Captain Loft would die within minutes—Second Officer Repo would live for some thirty hours.)

The wasteful crash of Eastern 401 was a classic human-factor/human-performance accident. The crew had permitted their perfectly flyable aircraft to fly itself into the ground because they had been distracted with a minor problem. They didn't realize they had disconnected the autopilot with just a gentle pressure on the control column. That was a human failure, first and foremost; but the NTSB accident report would (as usual) focus only on mechanistic problems and mechanistic solutions: The autopilot had disconnected with too little pressure because the wrong switch had been installed; the light in the wheel well was not operating; the C chord was insufficient to alert the crew to the altitude deviation; and the approach controller was not required to be concerned about their altitude readout.

There were members of the NTSB staff who knew that sort of report would do little or nothing to prevent another crew at a later date from doing a similar thing. Deeper research was needed into the causes of pilot distraction. The NTSB report needed to address the question of how airlines can guard against such mistakes.

But all the board would do was cloak the point in the all-too-typical accusatory form. The captain had been (in effect) told to go fly his trip and refrain from crashing his airplane, and he had failed. Therefore it was the captain's fault. The finding read simply: "The captain failed to assure that a pilot was monitoring the progress of the aircraft at all times."[1]

It did not happen in 1972, but had Eastern 401 hit the swamps ten years later, the tragedy would have received substantial attention from NTSB human-factors investigators looking into the behavioral causes behind the human failures.

After all, there were some legitimate and critical questions there. Why, for instance, had the crew behaved as they did—becoming myopically preoccupied with the landing-gear light? Could their lack of awareness of what was happening have stemmed from upset and stress (as in Downeast), lack of sufficient training and aircrew screening (as in Air Florida), or some other failure not yet studied in depth? As yet, that sort of in-depth inquiry into the real causes of the human failure that led to the accident was just not done—just "not considered suitable matters for investigation," as Al Diehl was told constantly.

The most vital and important "message" from the death of EAL 401 had an impact on the airline industry despite the shortcomings of the NTSB's report: Regardless of what goes wrong in the air, maintaining aircraft control is the prime task of the pilots, and it's up to the captain to specifically assign himself or his copilot to that task alone. In human-factor terms, such considerations are known as "cockpit resource management."

The U.S. Air Force and Navy had been working on such problems for at least a decade. There was already a large body of information about the interactions in the cockpit of a large aircraft, and from that research it was very clear that someone had to be in control to properly manage the talents and the attention of the others (though the NTSB knew little of such research). It was also clear that such resource-management talents could be acquired only through specific training.

1. In the body of the report (NTSB-AAR-73-14) there were a few other brief mentions of human considerations. "Subtle incapacitation" was discussed because the autopsy on Captain Loft disclosed a slow-growing brain tumor (which was determined to have played no role in the accident). The excessive reliance of the crew on the autopilot and other automated features of the new aircraft was also mentioned, as were flight-crew distractions. For 1972 it was a slight advance from straight findings of "pilot error," but it was still nowhere near enough.

But the airlines knew little and generally cared less for such research. To them, such pronouncements bordered on psychological mutterings restating the obvious, and were no more acceptable in 1972 among serious airline pilots than they had been in the fifties. The entire subject could be summed up by saying that when you were in command, you were responsible for any mistakes, so everyone in the cockpit did things your way (or else).

That macho attitude had not helped Captain Bob Loft, of course. He and his copilot were doing things "his" way in trying to fix the gear-light problem. Yet they crashed and died. Perhaps he could have benefited from some formal instruction about the methods of directing his crew and exercising his right to command. Perhaps some formally taught methodologies would have helped him stop and say to Copilot Stockstill: "You struggle with the bulb while I concentrate on flying the plane. Tell me when you're ready for the approach."

In 1973 even people like Chuck Miller (chief of the Bureau of Aviation Safety of NTSB and an expert in human-factors research, who had formed the NTSB Human Factors Branch in 1969), couldn't devote more than passing comment to the most vital human-factor issues and human-performance flaws without upsetting the precise engineering mentality of the board.

The recommendations to the FAA from the crash of Eastern 401 were mechanistic: Add a light switch for the nose gear, install a placard, and modify the altitude-alert system. The idea of recommending to the FAA that they require airlines to hold classes in cockpit resource management, and specifically that airlines be forced to teach their pilots to assign someone to fly the aircraft first, *then* work on in-flight problems, was too far afield.

Such changes might not have prevented preoccupied pilots from crashing perfectly flyable airplanes, but the attempt couldn't have hurt.

Accident after accident in the seventies begged for full human-performance investigations, and the humans whose performance needed investigating were quite often the airline managers. Inadequate flight training, retention of incompetent pilots, bad management of maintenance, poor dissemination of information to the flight crews, poor air-traffic-control services (by men and women who were expected to deliver 100 percent effort with 100 percent perfection 100 percent of the time), and the perennial problem of crew fatigue and time-zone (circadian) disruption were essentially management failures. But when these things caused crashes, too often the probable cause was "the failure of the flight crew to . . ."

There were problems as well with getting the Safety Board to understand certain human limitations that might actually cause or contribute to accidents—accidents that otherwise looked like pure pilot negligence. Visual illusions on landing, especially in heavy rain, the difficulty of seeing other aircraft from the restricted windows of a modern jetliner (the "see and be seen" concept all over again), and the existence of something called wind shear and microbursts (at first ridiculed by investigators, later cited as a killer of airliners), were just a handful of the concerns.

The need for research into human factors and human-factor-performance research, which had always been regarded by tough, objective NTSB and FAA men as nothing more than whimpering excuses, would finally have to be recognized. There would be several tragedies in the late seventies that would be inexplicable otherwise—and could not be ignored because of the magnitude of the loss of life.

The troubles that humans are heir to in operating aerospace machinery are vast: information-transfer problems, including breakdowns in communication, misunderstanding radio calls, language barriers in international aviation, and pilots misunderstanding each other; cockpit resource management, the breakdown of which in one case would spark a pace-setting training program by United Airlines; pilot judgment, the heretical concept that good judgment needs to be taught and reinforced by the airline's training function, not just left to chance and native talent; and, one of the major killers, task saturation, the ability of the human mind to tune out vital information when overloaded with problems or other stimuli. All these elements, coupled with fatigue, subtle incapacitation in the cockpit (recall Air New England and Captain Parmenter), psychological problems, upsets, and distractions, all needed to be given constant attention by the FAA, the NTSB, and the airline industry. Instead, only a handful of physiological problems had been widely accepted as valid areas of concern by the time Eastern 401 hit the Everglades. Human factors up to the mid-seventies meant crash survivability to far too many people, and occasionally, physiological problems such as the tumor in Captain Loft's head. Yet in disaster after disaster (not to mention near-disasters somehow averted), the true causes radiated from a network of subtle human vulnerabilities affecting far more people than just those in the cockpit.

In most accidents, for instance, somewhere in the background could be found a contributing history of managerial negligence. The company men and women who control airlines, and more specifically the executives and managers who head up flight departments, maintenance departments, and servicing or support departments, are forever setting the stage for disaster through ignorance, misunderstanding, or neglect. The pilots may pull the trigger, but all too often management has loaded and cocked the gun.

Texas International Airlines Flight 655 was having one hell of a time getting from El Dorado, Arkansas, to Texarkana, Texas, on the night of September 27, 1973. There was a classic southwestern cold front complete with a line of lightning-spitting thunderstorms sprawled right across the flight path that Texas 655 was supposed to use, but the crew of Texas 655 was planning to get through one way or another.

The Convair 600, a twin-engine, piston-powered airliner built in 1948 and converted to turboprop engines in 1967, was the workhorse of these short routes for Texas International, which was known as a regional airline, and a Part 121 carrier.[2]

Forty-one-year old Captain Ralph Crosman and thirty-seven-year old Copilot Fred Tumlinson had tried to fly north around the thunderstorms. There was a somewhat primitive radar set aboard the Convair, and they had been fiddling with it, trying to find the holes in the weather ahead. The direct-line heading from El Dorado to Texarkana would have been about west-northwest, but Tumlinson (whose leg it was) had been aiming the airliner nearly due north for over twenty minutes, trying to find a way to get farther west. They were in instrument conditions, but they were trying to stay visual, which was a violation of regulations to begin with. They simply could not stay out of the clouds.

Tumlinson had the Convair down to 3,000 feet now, and he was

2. Texas International, which previously had been named Trans Texas Airlines, had begun with DC-3s and a real assortment of pilots, many with World War II flying backgrounds. In the fifties and sixties it was known by various colorful titles such as "Tinker Toy Airline," "Tree Top Airways," and "Try To Arrive." Despite the not-so-good-natured ribbing, TTA was a relatively solid carrier, which provided reasonably reliable service to many smaller communities throughout the Southwest. TTA, however, had never enjoyed a reputation for strict professionalism in its flight operations.

obviously worried. He apparently didn't know this part of Arkansas, and he was flying in instrument conditions on a visual flight plan. In such conditions, it was difficult to stay low enough to keep the ground in view, so as the base of the cloud cover kept getting lower to the terrain, so did Texas 655.

"We don't want to get too far up the creek here; it gets hilly," Tumlinson told the captain.

"Yeah . . ." Crosman replied, ". . . stars are shining. . . . Why don't you try two thousand?"

They had gone much farther north than he had expected, and it was becoming increasingly obvious the two pilots weren't really sure where they were.

"If we get up here anywhere near Hot Springs, we get in the . . . mountains," Crosman added.

Tumlinson's voice came back. "Uh, you reckon there's a ridge line around here somewhere?"

"Fred, you can quit worrying about the mountains 'cause [two thousand feet will] clear everything over there."

"That's why I wanted to go to twenty-five hundred feet. . . . That's the Hot Springs highway right here, I think."

Crosman agreed. "You 'bout right."

The need to identify where they were was growing. The two pilots needed to find a highway or a town they were sure of.

Tumlinson began again, apparently looking to his left. "Texarkan . . . naw, it ain't either . . . fuckin' Texarkana's back here."

"Texarkana's back over here, somewhere," Crosman agreed.

"Yeah . . . this ain't no Hot Springs highway."

"Well, thirty degrees . . . thirty degrees takes you right to Texarkana, doesn't it?" the captain asked. "Hot Springs . . . here we are sittin' on fifty."

"Yeah," Tumlinson agreed.

"If we keep [going northwest] indefinitely, we'll be in Tulsa."

The two kept jockeying the Convair to the north-northwest, in and out of clouds, weaving their way through the clear areas on the radarscope. For twelve minutes they continued, descending to 2,000 feet to stay visual at one point, and talking back and forth.

Crosman, it seemed, had aimed them right into a cloud with his last reading of the radarscope.

"First time I've ever made a mistake in my life."

Tumlinson didn't respond at first. The demands of trying to see through their rain-smeared windscreen would have distracted any pilot. "I'll be damned. Man, I wish I knew where we were so we'd have some idea of the general goddamn terrain around this fucking place!"

"I know what it is," The captain replied.

"What?"

"That highest point out here is about 1,200 feet."

Tumlinson did not seem convinced. "That right?"

"The whole general area, and then we're not even where that is, I don't believe."

"Two hundred and fifty, we're about to pass over Page VOR . . . You know where that is?"

"Yeah," assured Crosman.

"All right."

Texas 655 had been airborne thirty-three minutes and thirty-four seconds, and was now due north of Texarkana nearly a hundred miles, and flying at only 2,000 feet.

Copilot Tumlinson's voice was already betraying concern. He had more than ample reason to conclude that the whole flight had been a mistake—that they should have been on instruments—that they were in strange territory flying around in clouds pretending to be visual. Yet Captain Crosman kept handing out glib assurances.

"About a hundred and eighty degrees to Texarkana."

"About a hundred and fifty-two."

Tumlinson was apparently trying to read his navigational chart, finding the general area through which they were groping along, and looking for the minimum en route altitude (MEA), the figure that indicated the lowest altitude an aircraft could safely use to fly through that part of Arkansas without hitting something.

Copilot Tumlinson found the figure, and it was shocking.

"Minimum en route altitude here is forty-four hund . . ."

Tumlinson's words were interrupted by the high-speed impact of the Convair with the rocky north slope of Black Fork Mountain in the Ouachita Range near Mena, Arkansas. The Convair, Crosman, Tumlinson, one stewardess, and eight paying passengers disintegrated in less than a fifth of a second. The point of impact was 2,025 feet in altitude. The minimum safe altitude had, indeed, been 4,400 feet.

The incredible irresponsibility of Crosman and Tumlinson in their cowboyish conduct of Texas 655 surprised even the most hardened mavericks in the industry. The crash was far more than pilot error; it was pilot negligence of the worst sort—and it begged for a full investigation of the backgrounds of both the pilots and of their company's management of training, safety compliance, attitude, and testing of their aircrews. The report, however, bypassed all such questions except for the technical noncompliance of Texas International's dispatcher, who had failed to follow the flight properly. The probable cause was cited as "the captain's attempt to operate the flight under visual flight rules in night instrument meterological conditions without using all the navigational aids and information available to him; and his deviation from the preplanned route, without adequate position information. The carrier did not monitor and control adequately the actions of the flightcrew or the progress of the flight."

With such objective conclusions, the board could have just as effectively concluded that the crash occurred because the aircraft hit an immovable object. In terms of true prevention, the report was all but worthless. It told *what* the crew did, but not *why* they did it. In fact, at one point in the report, the NTSB wrote simply: "The Board could not determine the reasons for the captain's departure from established procedures." Without taking apart all elements of the human organization for which Crosman had worked, there was no way to discover such reasons.

The management of Texas International had employed two pilots who had been allowed to develop habit patterns not fit for private pilots—let alone commercial-airline pilots. Yet the NTSB sidestepped the issue. To the pragmatic, it was an overwhelmingly simple case of pilot error. To the human-factors experts, it was a prime example of what can happen when airline managements abdicate their responsibility to maintain control.

The idea that pilot error can be laid at the doorstep of management is still somewhat heretical. But an airline flight department is a human system, and all the normal influences of any social group operate within such a system. If the management doesn't set the tone and maintain it—show by strict imposition of training and discipline that those who violate the spirit or intent of the rules will not be tolerated—a laxity will eventually eat away at the carrier like a cancer. (Indeed, at the moment that Flight 655 thundered into the mountain, the same forces were at work in tiny Downeast Airlines up in Maine, and in many other airlines around the nation.) The same thing can happen in maintenance, in training, in ground-service areas, or in virtually any other subpart of a human organization. The safest way of doing things is seldom the easiest way, and therefore human nature dictates that if not artificially enforced, the "safest way" will be replaced with the "easiest way."

In an airline operation, that translates to a high propensity for disaster.

The last ten years hold many tragic examples of accidents traceable in whole or in part to a breakdown in management's efforts and understanding of its role in the safety equation, and one of them surrounded the landing accident of Pan American Flight 806 on the rainy tropical night of January 30, 1974.

The Pan Am 707 was under the command of fifty-two-year old Captain Leroy Petersen, a seventeen-thousand-hour veteran, and First Officer Richard Gaines, a thirty-seven-year-old pilot with five thousand hours. Just before midnight on approach to Pago Pago in American Samoa, the Pan Am Clipper dropped under the glide path to the runway and crashed nearly 4,000 feet short. Of the 101 passengers aboard, only 5 survived. The awful truth was that almost everyone survived the impact, but those who died either burned to death or were killed by smoke before they could get out. There was massive confusion in the cabin, evidence that passengers piled against doors, which could not then be opened, and evidence that some evacuation slides may have failed.

It had been Captain Petersen's first trip after four months of grounding for a health problem. It was only his second approach into Pago Pago in his entire career. The cockpit coordination broke down completely, none of the required altitude call-outs were made by the first officer, and the possibility that massive visual illusions misled the captain into letting the jet sink too far too fast was considered only briefly by the Safety Board (although that one human factor alone justified a major study).[3]

But there was no discussion of crew fatigue, no information on when the crew had gone on duty, and no mention of the work-rest schedule over the previous few days, even though such aspects could be critical in determining whether the crew might have been far less sharp and awake than usual. There was no discussion of Pan Am's scheduling policies on such grueling international trips, and no information on what training Pan Am had been providing on how to combat fatigue (even though in the forties and fifties

3. In fact, wind shear had been a factor, but the NTSB would refuse to consider that for years until the phenomenon of "microbursts" was accepted as a dangerous fact of meteorological life. The accident was later reopened by the NTSB and the probable cause revised.

298

Pan Am was one of the world leaders in fatigue consciousness and research).

And most important of all, despite some tips and rumors that the area needed investigation, there was no in-depth probe into Pan Am's flight-management department to determine whether the standards for training the crews of the "World's Most Experienced Airline" might have slipped—or become subject to cronyism.

There was an entire anthology of subjective questions involving human behavior and human factors in a human system invested with a solemn public trust, and they were never asked, let alone answered. The crew was "legal and qualified," the company had fulfilled its legal minimums of training and qualification of its crew, and therefore it was up to the pilots not to crash. Since they crashed their perfectly flyable airplane anyway, it was their fault, and thus pilot error.[4]

In Charlotte, North Carolina, on the eleventh of September in the same year, 1974, still another flight crew flew its perfectly operable aircraft into the ground—and still another finding of pilot error was announced by the NTSB.

Eastern Airlines Flight 212, a twin-engine DC-9 jetliner flown by Captain Jim Reeves and First Officer Jim Daniels, Jr., was approaching the airport at Charlotte as the two pilots chatted about a host of subjects unrelated to their duties. In doing so, they either lost track of their altitude or misread their drum-pointer altimeter as they descended over an area of patchy, low, ground fog. With First Officer Daniels flying the DC-9 and the captain monitoring, neither pilot apparently realized how close they were to the ground until it was too late. They allowed the aircraft to descend far too fast, plowing into trees miles short of the runway and killing seventy-one of the eighty-two persons on board. The captain died in the wreckage—the first officer survived.

This time the NTSB inched closer to the underlying causes, discussing (and condemning by reference) the casual conversation and

4. The phrase "pilot error" seldom, if ever, occurs in an NTSB report, but has come to be used by the media and various pilots associations (ALPA chief among them) as a catchphrase to reduce somewhat precise legal and technical language to its base meaning. Often-used NTSB phrases such as "probable cause of the accident was the failure of the pilot to . . ." are, quite simply, findings of pilot error. There were no common, corresponding findings of management error, if for no other reason than that pressing an investigation that deep would create a fire storm of protest and political pressure from the airline being examined.

the apparently casual attitude of the two Eastern pilots, treating the possibility that the illusions produced by the low fogbanks had lulled the pilots into believing they were higher, and touching on some of the other human frailties and susceptibilities that could have been contributory.

But the thought of investigating Eastern's system—investigating the way it trained and monitored the professionalism of its flight crews—was anathema to the board. The habit of never considering management's possible role in perpetuating, tolerating, or contributing to pilot attitudes and pilot compliance levels remained unbroken with the Charlotte accident.

Though it touched on such subjects in a lengthy discussion in the report, the board appeared scared to death that some in the industry might accuse it of (horror of horrors) dabbling in subjective matters not subject to ironclad objective proof. After that section it threw in an obviously worried disclaimer:

It should be emphasized that the possible explanation discussed [previously in the report] is based not only on evidence that is tenuous, at best, but also on the inferences to be drawn from such evidence as to what thought processes were evolving in the minds of the flightcrew. Obviously, such an explanation is, to a considerable degree, speculative in nature. It is nevertheless the intent of the Board that, by including this discussion in the report, pilots will be alerted against the possibility of lapsing into such a pattern [of laxity and casualness in procedures]. We also hasten to add that, even if it is assumed that the sequence of events described in the [previous] discussion in fact occurred, this should be taken to reflect adversely *not* [emphasis added] on Eastern's system, but rather on the flightcrew's implementation of that system in this instance. . . . There is no causal factor beyond the flightcrew itself which would account for their failure [to utilize their unquestionably adequate training and approach aids].

Once again the carrier was exonerated without adequate investigation into anything more than the legality of its training program. In fact, Eastern's excellent training department and flight-management department probably were innocent, but without all such areas laid bare in a full human-performance probe, there was no way to know. The question of whether the breakdown in professionalism and procedures on the part of the pilots was a typical occurrence tolerated by Eastern would never be answered.[5]

The recommendations to the FAA did, however, contain the seeds of future emphasis on the proper areas of research. They asked the FAA to initiate "ways and means to improve professional standards among pilots," citing five previous air-carrier approach accidents as "examples of a casual acceptance of the flight environment." The FAA agreed, and, to its credit, began setting up liaison groups with pilot organizations and airline managements to discuss the subject. In 1975 (the date of the report), there was still adequate cooperation between the FAA and the industry for such discussions, even if the FAA tended to drag its feet on substantive reform in other areas.[6]

Though it was painfully obvious to those on the NTSB staff who

5. Airline cockpits are not the sterile, tense environments filled with serious faces and barked orders that the Hollywood of the past would have us believe. A cockpit is an office—an extremely small, technologically complex and incredibly demanding office—but an office, nonetheless. Since pilots work in that office, there needs to be a rather constant give-and-take atmosphere of conversation, friendliness, a sense of humor, and a feeling of being at home with the controls and the demands of the job. Tense cockpits breed mistakes—but there are limits, and the pilots involved in the Charlotte crash (as the NTSB properly pointed out) exceeded them clearly. On takeoff or approach, the high-speed demands of airline flying dictate a steady professionalism, and long discussions of unrelated issues while flying a difficult approach is neither professional nor conducive to careful flying (nor legal below 10,000 feet). But to take every word from a cockpit voice recorder and interpret casual conversation at *appropriate* times as scandalous would be wrong. Knowing what is appropriate and when is the mark of a professional airline pilot—enforcing and policing that professionalism is the mark of a professional management. Both are necessary for a safe operation.

6. With the arrival of the Carter administration and the appointment of Administrator Langhorne Bond, the FAA would embark on an era of punitive (and sometimes vindictive) enforcement activities across the board, which would all but destroy its ability to create cooperative approaches to problems in the industry. The extremely important surveillance and enforcement functions of the FAA, which were already demonstrably less than adequate in many areas (such as Part 135 carriers), were further compromised by the industry's perception that the FAA could no longer be trusted with information lest it turn and attack. There is a delicate balance between cooperative attitudes and tough enforcement and surveillance, and whatever else Bond may have accomplished in specific enforcement actions (and he did clean up certain problems in the industry), his undermining of the cooperative nature of the FAA's oversight had one very detrimental effect: Those carriers prone to lie and cheat in any degree became more clever at hiding their propensity. In no way were they changed by Bond's methodologies.

had been trained in any aspect of human factors, the majority of the staff, and the board itself, wanted no part of what they considered unsupported speculation in any accident report. To examine the effects management might have had on its pilots through its training program would of necessity involve "speculation," and that was unacceptable.

As always, the events of the following few years would force a change, gradual and glacial though it might be. The NTSB would end up faced with accidents that simply could not be explained away in objective terms.

Chapter 22

A FAILURE
TO COMMUNICATE

"Say again?"

The phrase "information transfer" meant little in the mid-seventies to accident investigators not schooled in human factors, but failing to understand, or misunderstanding, information provided by another—especially through the spoken word—is as endemic in the airline industry as it is in everyday life. It is as ludicrous to tell pilots to refrain from misunderstanding as it is to place on the instrument panel a placard reading: CRASHING THIS AIRPLANE IS PROHIBITED.

The reasons for information-transfer problems are many and vast, and can be understood only through research. Though the NTSB was undergoing a painfully slow metamorphosis and was gradually recognizing the need to probe previously untouched areas such as the adequacy of management, it was still (as was the FAA) turning a blind eye to information-transfer problems in airline cockpits—though there were plenty of deadly examples.

Near Berryville, Virginia, for instance, on the afternoon of December 1, 1974, the flight crew of TWA 514 tried to fly an approach that had already drawn complaints from another aircrew weeks earlier. The problem was a potential for misinterpreting what altitude the aircrew could descend to on approach to Washington Dulles once the approach controller at Dulles cleared them for the approach. There was high terrain in the way if the aircraft was too far out, yet the approach plate seemed to indicate that 1,800 feet was safe. It was a textbook example of failure to communicate (by paper and by radio) exactly who was responsible for maintaining clearance above the ground. The captain of TWA 514 studied and studied the approach, and decided that the controller must be the responsible party. If he said "cleared for approach," then they could go down. The approach plate and the controller's instructions assumed it was the crew's responsibility to stay at the appropriate altitude until they were close enough to the airport to start obeying the altitudes shown on the approach plate. The captain of TWA 514 descended to 1,800 feet in solid clouds—and killed himself and his ninety-one passengers and crew members in an awful explosion of people and machinery against Mount Weather, Virginia, at about 250 miles per hour.

This time, the NTSB waded in and started asking the right questions. The issue was unavoidable: A perfectly good aircraft had been flown into the side of a mountain by a perfectly qualified aircrew doing a professional job in everything except interpreting an FAA-approved approach plate, which had been complained about two months before. The NTSB's report said: "The basic questions requiring resolution are (1) why did the crew knowingly descend to 1,800 feet in an area where the terrain obstacles extended almost up to that altitude; and (2) why did the approach clearance not include an altitude restriction under the circumstances of this case."

The NTSB investigators dug further into the background of the carrier's training than on previous accidents, because the crash could not be explained with total objectivity. Here was an area of misinterpretation not just of the approach to Dulles but of the worldwide use of the phrase "cleared for the approach." What, exactly, did that mean in terms of altitude? The NTSB staff were startled to find that pilots differed throughout the country in their interpretation, and that many, many pilots would have arrived at the same fatal conclusion as the captain of TWA 514.

They were also startled to discover that the FAA had no formal method of receiving complaints on approach procedures and doing something substantive and immediate about them—and too many pilots were afraid to report such situations for fear of FAA retaliation! The report included an interesting and revealing look at the FAA's practiced "ass covering" as well as the need for a comprehensive reporting system capable of granting immunity:

At the Safety Board's public hearing, FAA witnesses testified that they were not aware that there was any potential misunderstanding on the part of pilots as to the meaning of the term "cleared for the approach," in a case where a nonprecision approach is made, particularly when the clearance is issued a long distance from the airport. The evidence, however, does not support this conclusion, since, for several years prior to this accident, various organizations had perceived a problem in the use of the term "cleared for the approach."

Ironically, approximately 6 weeks before the TWA accident, an air carrier flight, after being "cleared for the approach," descended to 1,800 feet while [at approximately the same location as TWA 514] on approach . . . at Dulles. The carrier involved had implemented an anonymous safety awareness program, was in fact made aware of the occurrence, and subsequently issued a notice to its flightcrews to preclude the recurrence of a near-fatal misinterpretation of an approach clearance. The Board is encouraged that such safety awareness programs have been initiated. It is through such conscientious safety management that the expected high level of safety in air carrier operations can be obtained.[1]

The text went on to lament that such a reporting program did not exist everywhere in the system, and that the specific report that could have saved the lives of ninety-two people could not prompt the FAA to admit that it had a terminology or communications problem.[2]

1. The airline relayed the report to Dulles FAA personnel, who failed to pass it on to Washington FAA headquarters because there was no procedure for formally reporting such troubles to higher headquarters. Consequently nothing was done to clarify either the terminology or the approach plate.
2. At times, the FAA's air traffic controllers have been blamed for causing accidents that were essentially "set up" by other circumstances. Four months after TWA 514's demise, an air-force C-141 returning from Japan was erroneously given descent clearance to 5,000 feet while flying inbound over the Olympic Mountains of Washington State. The controller thought he was enunciating the call sign of a navy airplane that he wanted to descend, but the Military Airlift Command (MAC) jet, a huge four-engine transport as large as a Boeing 707, called in at precisely the wrong second, and the controller mixed up the call signs, issuing the clearance in MAC's name. The problem was that

The crash of TWA 514 occurred on December 1, 1974, some four years before the passage of the deregulation bill. The NTSB in its report called for the establishment of a no-fault reporting system that could do more than NASA's subsequent contribution (which combines anonymous reports into somewhat generic areas). A system was needed that people would not be afraid to use and that would result in instant action when necessary.[3] If the government wouldn't do it, then industry should do it, implied the NTSB. But when the great free-market wars would get under way in 1979 and 1980, few if any carriers would have the time or money for establishing and administering such a system. And in too many cases, the new carriers born of deregulation would neither understand the need nor have the sense of responsibility to trouble themselves with such affairs as airline safety offices and anonymous reporting programs.

And then came the granddaddy of them all—an accident so horrendous in terms of totally avoidable loss of life that it galvanized the civilized world. Even the crash of a Japan Airlines Cargo DC-8 in Anchorage, Alaska, on January 13, 1977, which was caused by a drunken American contract captain losing control of his aircraft, was invisible to the public compared with the events of March 27, 1977.

In short, a very senior and very experienced Boeing 747 captain under substantial stress decided that what he had heard from the control tower was what he had been expecting—takeoff clearance.

---

the MAC crew was exhausted after nearly a twenty-four-hour crew-duty day, they had received too little rest in several days of flying, the two pilots were very inexperienced, and there were several other factors relating to operational control, training, and management of the MAC system. The tired pilots accepted the clearance blindly, never thinking to check whether they had terrain clearance. The controller had pulled the trigger of a gun loaded and cocked by negligence in air-force management of human fatigue and other considerations. The jet obediently descended toward 5,000 feet, and at 7,000 slammed into the side of the last snow-covered ridgeline between its position and the safe harbor of McChord Air Force Base in Tacoma, Washington. Ten crew members and seven military passengers died instantly. When the air-force accident board issued its report, the accident was blamed solely on the controller for issuing the clearance and on the crew for accepting it. The controller had merely "pulled the trigger" through a human error, but the command management policies that had loaded and cocked the gun were left unscathed.

3. The idea actually originated with NTSB investigators Jack Carroll and Chuck Miller some four years earlier in 1970.

Seconds later, Captain Jacob Louis Veldhuyzen van Zanten, fifty years of age, a senior pilot with eleven thousand hours of flight time, and head of the KLM training department, saw a sight ahead of him on the runway at Tenerife in the Canary Islands that exceeded his wildest nightmares. As KLM 4805 passed 115 knots of airspeed and began emerging from a fogbank that had rolled across the runway, another Boeing 747, Pan Am Clipper 1736, sat right in front of it.

Captain van Zanten had no choice. He could not stop. He could not swerve at that speed. His only chance was to leap over the Pan Am Clipper. The KLM captain yanked the control column as far back as it would go, the KLM 747 reared up in the air, the tail striking the runway and embedding metal in sixty feet of concrete, and finally became airborne. For a second or two as its nose gear and front section passed safely over the Pan Am Clipper, it seemed that they might make it. But the rotation had been too late—too slow. They were too low.

The main landing gear of KLM 4805 smashed into the other jumbo, destroying the top of the Pan Am fuselage and ripping away KLM's main gear.

Mortally wounded, robbed of airspeed by the energy transferred to the impact with Pan Am, KLM settled back to the runway at less than 100 knots, skidded, and burst into intense flames. Not one human being escaped from the inferno, which consumed the 234 passengers and 14 crew members.

On the destroyed and collapsing Pan Am Clipper, some were escaping the twisting, burning airframe as the intense flames and smoke enveloped it, but in the end 335 would be burned to death—the passengers all vacationers on a charter trip.

The disastrously mistaken assumption of Captain van Zanten that the runway was clear and they were cleared to take off, was the saddest—and most dramatic—example of information-transfer failure in commercial aviation history. That one decision killed 583 human beings and maimed many more. It was the accident the entire industry had dreaded—the collision of two jumbo jets. No one, however, had predicted that such would happen on the ground.

The two 747s had diverted into Tenerife for fuel because of a terrorist bombing that had shut down Las Palmas airport in the Canary Islands, which had been their destination. The controller's native language was Spanish. The Pan Am Clipper crew's native language was English, and the KLM crew's native tongue was

Dutch. All three, however, in their radio communications, were speaking the international language of aviation: English. Unfortunately, accents and phraseology in different parts of the world make such communications in a common language difficult at times. It is very, very easy to misunderstand—which is exactly what happened at Tenerife.

Captain van Zanten also had a crew-duty time problem. If his crew did not get back to Amsterdam before the end of their maximum authorized crew-duty day, they would be in trouble. Overflying one's crew-duty day was a serious violation for the Dutch, but shutting the flight down somewhere else with a charter group would mean incredible expense and confusion involving housing and feeding of passengers. Yet the unexpected diversion had put them in a bind. The captain was very anxious to get out of Tenerife and on his way to Las Palmas, where another charter group awaited pickup. Yet they had had to refuel and wrestle with a minor hydraulic problem at Tenerife first, and during that time (due to congested parking) their huge 747 had blocked the Pan Am Clipper from leaving. With all those frustrations, Captain van Zanten was upset.

The runway at Tenerife was being obscured by clouds, which were periodically moving across the runway, providing a localized fog condition. KLM taxied out first, reaching the end of the runway as Pan Am was taxiing down the runway behind it. Since the taxiways near the terminal were too narrow for a 747, the jumbos had to use the runway until about midfield, then taxi off onto a parallel taxiway, and use that to the end of the runway. The Pan Am Clipper had missed its first turn and was trying to find the second one when Captain van Zanten's right hand began nudging the four KLM 747 throttles forward.

His copilot, thirty-two-year-old Klass Meurs, was instantly alarmed.

"Wait a minute, we do not have an [Air Traffic Control] clearance."

The captain brought the throttles back to idle. It was irritating for one of KLM's most senior captains to have to be reminded by his copilot of something so very basic.

"No, I know that; go ahead [and] ask."

Meurs picked up his microphone.

"Eh, the KLM 4805 is now ready for takeoff and, eh, we are waiting for our ATC clearance."

The heavily accented voice of the Tenerife tower controller came back, reading the clearance in English, and Meurs repeated it after the captain had nodded and said, "Yes."

As Meurs was finishing his read-back of the clearance, the captain's hand once again began nudging the throttles forward.

"We go—check thrust."

Meurs was startled. His finger was still holding down the microphone button, he could see what the captain was doing, and he was confused. That was an ATC clearance, not a takeoff clearance. But this man was his boss. He had already corrected him once. Maybe the captain had heard something in that clearance that he, Meurs, had missed.

Without letting go of the transmit button on his microphone, Meurs stated the obvious—in the hope it was already approved.

"We are now—eh—taking off."

The sound of the 747's huge, high-bypass engines winding up filled the cockpit as Meurs waited for word from the tower.

Finally the tower controller keyed his microphone, and Meurs heard him speak the word "Okay . . ." before a loud squeal blocked the rest of the transmission.

In the Pan Am cockpit, and in the tower, Meurs's words had been heard as "We are now AT takeoff." Mentally, the Pan Am pilots assumed the word "position" should have followed. Certainly he was reporting in takeoff position. The phrase "at takeoff" was bad English, but "at takeoff position" would make sense.

Nevertheless, Pan Am Copilot Bob Bragg keyed his microphone to make sure they knew Pan Am was still on the runway.

"And . . . we're still taxiing down the runway, the Clipper 1736."

At that very same instant, the tower controller was telling KLM: "Okay, stand by for takeoff—I will call you—"

On the flight deck of KLM, the two messages canceled each other out in a loud squeal. All the KLM crew heard was the word "Okay . . ." followed by the squeal, and the words ". . . Clipper 1736."

The tower controller answered the Pan Am Clipper.

"Papa Alpha 1736, report runway clear."

"Okay. We'll report when [we are] clear."

"Thank you."

First Officer Meurs was concentrating on the takeoff. The "Okay" from the tower had come over the KLM crew's radio with clarity, and with that it was reasonable to conclude that they were cleared to take off. Besides, the captain acted sure of that fact.

But the second officer/flight engineer, Willem Schreuder, seemed concerned. The KLM was gaining speed and entering another of the clouds that had been blowing across the runway. He could see nothing ahead. What did Pan Am mean: "We'll report when [we are] clear"? They were already clear—weren't they? How could Captain van Zanten have started the takeoff if not?

"Is he not clear then?" Schreuder asked.

The captain asked in a clipped manner, "What did you say?"

"Is he not clear that Pan American?"

The KLM 747 was now at 80 knots and accelerating.

Both the captain and First Officer Meurs responded to Schreuder's tentative inquiry.

"Oh, yes!"

Schreuder shut up. The captain must be right.

Twelve seconds later the Pan Am Clipper loomed up in the windscreen.

The last recorded words from the crew of KLM was a familiar epithet to those who regularly hear cockpit voice-recorder tapes. More than one pilot had uttered the same phrase when confronted with the possibly irreversible reality of impending disaster.

Captain van Zanten said simply: "Oh, shit!"

Tenerife is in Spanish jurisdiction, so the Spanish authorities conducted the investigation while the NTSB watched rather ineffectually from the sidelines. The result, while an honest treatment, which did give some attention to the problems the KLM captain was having with operational pressures, was not a very useful document.[4]

The tragedy at Tenerife held many lessons for commercial pilots and managers alike. Indeed, Captain van Zanten *was* a manager—the head of KLM's training department. (His picture, in fact, had just appeared in worldwide color advertisments for KLM when the crash occurred.) But the fact that he represented an extreme au-

4. The NTSB does not issue a report of such overseas accidents even though they may involve U.S. airlines. The Air Line Pilots Association decided to tackle the accident and issue a U.S.-style report. With the collaboration of ALPA human-factors specialist Bill Edmunds, Pan Am Captain P. A. Roitsch, and United Captain G. L. Babcock, it produced an extensive document, which examined in great detail the human-performance aspects of the tragedy. In terms of sensitivity to human factors and human-performance considerations, the report was an excellent work. Its bibliography also provides an interesting reference to current works and authorities in the field of human factors.

thority figure to his subordinates intimidated them, whether he had tried to achieve that result or not. The accident stands as a singular and irrefutable justification for intense training at all airlines in aircrew assertiveness in the cockpit—the willingness to make certain that the captain understands what the other crew members know, even if he doesn't want to listen. If such assertiveness had occurred the night of December 28, 1978, over Portland, Oregon, a United Airlines DC-8-61 would not have ended up in the middle of a suburban Portland neighborhood.

Captain Malburn McBroom of United Flight 173 had been flying for over an hour around Portland while he and his crew grappled with a landing-gear malfunction. McBroom had received numerous warnings from his flight engineer about the status of his fuel, but he was apparently too busy concentrating on the gear problem and elaborate precautionary landing preparations in the cabin to pay attention. The fuel system on the model of DC-8 he was flying is notorious for giving erroneous fuel-quantity indications, which can vary as much as four thousand pounds. Most pilots would never operate a DC-8 on less than five thousand pounds, and even that figure is enough to make most DC-8 crews very nervous.

Yet McBroom permitted his aircraft to burn down to four thousand pounds indicated.

In fact, the gauges were incorrect. They did not have four thousand pounds left. They had nothing left. Suddenly the gauge indications began winding down toward zero and all four engines flamed out.

The first officer's last radio transmission to Portland Tower was chilling: "Portland Tower, United 173 heavy Mayday! We're . . . the engines are flaming out, we're going down, we're not going to be able to make the airport."

First Officer Roderick Beebe would be gravely injured within two minutes. The flight engineer, Forrest Mendenhall, would be dead. Captain McBroom would survive—and would later be fired by United. The big DC-8 came down six miles southeast of Portland International Airport, destroyed a vacant house, and smashed into a wooded area, destroying the cockpit and the first-class section of the airplane. Somehow, the majority survived. Pictures of the tail of the DC-8 showed that as it crossed into the woods it had snagged a group of high-tension power lines, which acted like the arresting gear of an aircraft carrier, and the fuselage had been brought to a halt—minus wings.

Of the 181 passengers and 8 crew members aboard, 8 passengers and a flight attendant were killed, and 21 passengers were seriously injured.

The accident investigators from the NTSB rapidly confirmed what had happened. There was no fuel in the tanks. The crew of United 173 had committed the ultimate sin in aviation: They had run out of gas!

The crash of United 173 led directly to the establishment by United of an assertiveness-training program, which now leads the industry in human-factors/human-performance practical application. The point of assertiveness training is to ingrain in every pilot that it is his responsibility to speak up and refuse to be squelched if a fellow pilot is doing something questionable, wrong, or dangerous.

Unfortunately, few of the other major carriers have instituted such a program, and none of the new-entry or cut-rate carriers has begun such instruction. The FAA has yet to seriously consider making such training a requirement. Part of the reason revolves around the original congressional charge to the FAA to "promote" aviation. To require economically pressed airlines to add more courses in the middle of the deregulatory wars would add to their financial strain—and too often these are carriers who have already cut their training costs to the legal minimum to stay afloat.

The glacially paced development of assertiveness-training courses in the industry is very representative of deregulation's impact. In 1978 when deregulation was passed, assertiveness training was a radical, experimental, and costly idea. Even if deregulation had not passed, such a course would have taken perhaps a decade to become standard among the air carriers of America—if then. But with the severe economic pressures imposed on existing (and often struggling) carriers by deregulation, and with the growth of new-entry and rapidly expanded airlines, which simply do not possess (in many cases) the time, the sophistication, the experience, or the money for developing such a course, the lessons learned by United from the failure of Captain McBroom and crew may never become standard—may never be applied to the rest of the system.

The chances of United's pilots ever duplicating the crash of United 173 in Portland have been substantially diminished because the airline spent the time and money and overruled the objections of some of its senior pilots to make sure the lessons are understood by its people—but the same cannot be said of the majority of airlines who do no such training. They have not instilled those lessons

312

in their pilots. Their propensity for repeating the mistakes of Captain McBroom and crew is just as high as before the crash, and that borders on the ludicrous. Such, however, are the risks of cut-rate fares and free-market competition imposed on a subjective, human system that was not perfect to begin with.[5]

Deregulation forced the industry to worship the god of strict cost controls, and in that religion there is little or no room for the expenditure of money and no time for anything not required by regulations. Therefore, in the eighties, if the FAA doesn't understand it and require it, the majority of carriers will not do it.

There was a time when no respectable major airline would have been satisfied with just being "legal," because the "legal" standards were far below the industry standards that had built airline safety in the first place. Those legal standards have changed little over the decades, but far too many carriers now embrace them as "their" standards. And the airline of the eighties that meets the "legal" minimums and nothing more is far, far below the safety level that built the U.S. (and international) aviation system into the safest form of travel. "Legal" and "safe" have two entirely different meanings!

5. The Air Line Pilots Association in December 1983 held a well-attended conference of industry people, titled "Beyond Pilot Error," in which the recommendation was made that all airline training programs incorporate cockpit resource-management training as a requirement. With the deregulatory battle for survival, few if any carriers have shown any serious intent to start such courses. In fact, the widespread reluctance paralleled the historic refusal of most carriers to form safety departments. As one major airline president put it, "Why do I need a safety department? That's what we have insurance for!"

Chapter 23

MANAGEMENT IS THE KEY

Human failures have been the prime cause of airline accidents since long before senators Ted Kennedy and Howard Cannon began pushing airline deregulation toward passage. That congressional act, however excellent and progressive it may be in terms of economic efficiency and cheap ticket prices, has exposed and exacerbated the weaknesses in the safety of airline flying for everyone.

Deregulation has retarded dramatically and dangerously the spread of a very basic understanding: People will make mistakes; those mistakes can be anticipated through human-factors and human-performance research and investigation; and what the industry can learn from such research can be applied in direct practical ways to prevent those predictable mistakes from causing crashes and killing passengers.

The management failures that breed crashes like those of Downeast in Maine and Texas International in Arkansas, or that possibly contribute to disasters like those of Eastern 401 in the Everglades, Pan Am in Pago Pago, or any number of other examples, can be prevented—but only if studied and understood.

The information-transfer problems, which led TWA 514 into a

mountain in Virginia, which became the final straw in an air-force C-141 crash in Washington, and which caused the worst commercial-aviation carnage in history at Tenerife, can be short-circuited and avoided through knowledge and understanding.

The timidity and lack of assertiveness by subordinate pilots, which permitted United's DC-8 to run out of fuel in Portland and was underlined so graphically by the failure of First Officer Roberti of Air New England in 1979, can be addressed—and are being addressed—at least by a few carriers.

As 1979 dawned and pilots like Jim Merryman of Downeast were caught in the conflicting tides of opportunities born of an expanding system and the agonies of constant change and disruption, the staccato drumbeat of airline tragedies resulting from human error continued—in some years more than in others. The tragedy—indeed the scandal—was that the causes were always variations on mistakes that had been seen (and too often ignored) in years and decades past.

The NTSB has grown in sophistication and understanding with the investigation of Downeast and later accidents—incorporating human-performance background data and legitimate speculation that would have been hooted out of the hearing room ten years earlier. But the airline industry (for the most part) has been too involved with simple survival to absorb the lessons learned—unless they are procedural lessons (such as the clear terminology problem that led to the Berryville crash of TWA 514).

Just before the Portland crash of United 173, a horrible midair collision between a PSA Boeing 727 and a small Cessna sent 137 people to their deaths in a San Diego suburb—9 others on the ground died in the wreckage of 22 houses. The basic cause was an old, familiar malady: the same bankrupt concept of "see and be seen" (now called "see and avoid") that had led to the collision of United and TWA airliners over the Grand Canyon in 1956. The PSA pilots had been engaged in some possibly distracting cockpit conversation on a routine flight in a familiar area during a tiring work schedule before beginning a visual search for "traffic" that San Diego tower had called out to them. The copilot thought he saw the other aircraft and figured they were going to be clear of it—but he was apparently mistaken. They were *not* clear. The converging single-engine airplane was below and in front of the 727, and headed in the same direction. The two pilots in the Cessna couldn't see the approaching airliner, and PSA's crew did not spot them until too late. The big Boeing overran the small Cessna, turning it into a twisted ball of metal, which, in the process, ripped away too much of the 727's right wing for it to sustain flight. One of the last comments on the cockpit voice recorder was the simple and emotionally powerful phrase from one of the pilots who realized, as the ground rushed up at him, that his life was about to end.

"Ma . . ." he said, " . . . I love yah."

Months before the PSA disaster in San Diego, the fate of a National Airlines flight headed for Pensacola provided more graphic proof of how long and ludicrous (and predictable) the chain of human errors can be in causing aerial mishaps. Pilot, management, maintenance, controller, and procedural errors came together with a massive information-transfer problem on the foggy evening of May 8—and three passengers died as a result.

The flight crew was conducting a too hurried nonprecision approach to Pensacola in a Boeing 727, using sloppy cockpit procedures and an alarming descent rate on final approach (aided by controller mistakes that led to the accelerated pace), when the incredibly loud noise of the ground proximity warning system (GPWS) filled the cockpit. The captain had apparently misread the altimeter, as had the copilot. The GPWS was therefore indicating that they were too low for that descent rate. The flight engineer tried to yell over the noise to ask the captain if he wanted it turned off. The first officer yelled at the captain that the descent rate was setting it off. The captain responded by beginning to shallow out his descent rate. With the noise destroying all crew coordination, the flight engineer thought he heard a "yes" and activated the "GPWS inhibit switch" on his panel. The captain, now thinking that he had corrected the problem by slowing his descent, satisfied himself that the warning had nothing to do with altitude, and that they were not too low.

But they were.

Seconds later, National 193 plowed into Escambia Bay off Pensacola. Only the presence and fast reaction of a tugboat crew who pushed their empty barge over to the airplane to serve as a massive life raft (and the fact that the aircraft settled onto a sandbar and did not sink) saved the occupants. The passenger life vests were contained in plastic so tough that most survivors could not even gnaw them open. National 193 held no life rafts for the passengers. They were not required to have them, since the Mobile to Pensacola flight was not considered an overwater flight—even though it was indeed over water.

*　　*　　*

After the bloody year of 1979 (American Airlines' DC-10 disaster in Chicago, Downeast in Rockland, Air New England in Hyannis, and a long list of other commuter tragedies), more than two years would pass with no major Part 121 carrier accidents in the United States before Air Florida Flight 90 struck the frozen Potomac. There were, however, numerous close calls—one of them involving Air Florida and one of its leased DC-10s just four months before the Potomac tragedy.

The DC-10 was on its takeoff roll from Miami in September 1981 when one of its wing-mounted engines blew apart. A metal tool left deep inside one of the engines—apparently during a previous overhaul by United Airlines in San Francisco—had finally disintegrated the engine. During the process, hydraulic lines were ruptured, and the leading-edge devices on the right wing began retracting without the knowledge of the pilots.

It was the same basic problem (asymmetric lift) that had killed men, women, and children in the crash of American's Flight 191 in Chicago more than two years before (though the Air Florida pilots had been trained as a result of that accident to use higher climb-out airspeeds than the speed schedule that led to the American disaster). The engine failure occurred early in the takeoff roll, and Air Florida's pilots—unaware their leading-edge devices were affected—properly decided to abort the takeoff. There were no injuries. Had they continued the takeoff, the result could have been different.

Even in the eighties, the old human mistake of leaving tools in an engine had occurred once again. The frightening aspect was the discovery that the disintegrating tool had been in the engine (sealed in an inner section) for months, and had progressively torn away at the retaining bolts holding one of the turbine wheels. The failure of the engines could just as easily have come at a time when the pilots could not abort—and there was always the possibility of another Chicago.

When Boeing designed its advanced 757 and 767 airliners with sophisticated instruments and automated systems in the two-man cockpit, part of the reason was to short-circuit the opportunity for human error by taking human interaction out of as many systems as possible. As one Boeing engineer put it privately, "If we can someday reduce it to a lever which is marked simply: park, taxi, takeoff, fly, and land, we'll get rid of pilot error altogether."

Unfortunately, Murphy's Law and its endless corollaries have not (and will not) be retired by advances in the machinery. If it can fail, it still will—and computerized cockpits are definitely no exception.

In fact, human-factors experts are becoming quite concerned that increased cockpit automation will create a generation of pilots whose basic flying skills will have deteriorated from lack of practice. When manual control becomes necessary, they may not be up to the challenge—and as Air Canada discovered in 1983, there will always be times when the fate of even the most sophisticated jetliner depends on the basic skills of its pilots.

One of Air Canada's newest 767s was cruising smoothly westbound at 41,000 feet on the afternoon of July 23, 1983, with a computerized cockpit replete with state-of-the-art instruments and controls designed to make the pilot's life very simple. It was a study in space-age automation—until a lengthy chain of human mistakes caught up with the machine: It ran out of gas!

One of the sophisticated fuel readouts had malfunctioned the day before, and a second fuel-gauge system died that morning. But the maintenance supervisor at Ottawa had talked Captain Robert Pearson into taking the airplane anyway—even though he had no operable fuel gauges! The maintenance crew would "dip" the tanks (using measuring sticks installed beneath each fuel tank) and arrive at an exact figure, which would assure them of plenty of fuel with which to fly to Winnipeg.

But the mechanics fell victim to the constant battle between the metric system and English measurements. The dipstick system was calibrated in centimeters. The reading should have been converted to liters, then to kilograms and finally to pounds. The 767 was Air Canada's first airliner to use the metric system. The mechanics, however, converted it into pounds through the wrong formula. The captain left the gate in violation of the minimum equipment list (MEL, the approved list of items without which the aircraft may not legally be flown in revenue service), thinking he had more than enough fuel. In fact, he had less than half the required amount. Over Red Lake, Ontario, that supply of fuel ran out, and the 767 became a state-of-the-art glider.

Suddenly Captain Pearson and his copilot, Maurice Quintal, were searching for an emergency landing field—and they began to realize they were not close enough to glide to Winnipeg. Thanks to Quintal's previous experience as a sport glider pilot and his knowledge of the area, they were able to bring their powerless 767 down on an abandoned runway of an old Royal Canadian Air Force base in Gimli. Though the runway had a fence down the middle and people on the far end (it was being used as an auto racetrack), the wide-body jet managed to straddle the fence and stop without hurting anyone.

Earlier, on April 2, 1983, a Republic Airlines DC-9-80, another advanced-cockpit-design twin-engine, two-pilot aircraft, had flamed out both engines at night over the wild southern Utah terrain. The pilots had missed a checklist item during a busy departure and had forgotten to turn on the fuel boost pumps in the center tanks in accordance with the checklist. At 35,000 feet over Hanksville, Utah, the fuel in the wing tanks ran out and both engines flamed out (stopped running). With the engines went all electrical power except the battery. The center tank had plenty of fuel, but the boost pumps couldn't be operated to get the fuel out of that tank and back to the engines without the electrical power from the engines, which couldn't restart without the fuel. All they had was a battery-powered boost pump in the wing tanks—and those tanks were empty.

Talking to a startled Salt Lake Center air traffic controller, the two pilots asked for radar headings to an emergency airfield forty-eight miles ahead (they picked Page, Arizona) as they lost altitude. Somehow, at 13,000 feet, the frantic efforts of the copilot to restart an engine succeeded. Apparently there was enough residual fuel in the sump box of one wing tank for the battery-operated fuel pump (called a "start pump") to feed it to one engine, which allowed a momentary restart. That few seconds of engine power gave them electrical (alternating current) power, which caused the center-tank boost pumps to start working, which in turn fed more fuel to the engine, which then became self-sustaining and kept running. With electrical power and fuel pumps, restarting the second engine was no problem, but the shaken crew made an emergency landing in Las Vegas, Nevada, as quickly as they could. There were 140 very worried passengers sitting in the darkened cabin of Republic Flight 303 as it glided from 35,000 feet down to 12,000 and an uncertain fate. The readouts from the aircraft's flight-data recorder later indicated that the crew probably would not have been able to reach the airport at Page. They would have been forced to crash-land in the dark of night on unknown terrain at over 150 miles per hour. Survival of anyone on board would have been questionable.

Less than two months later on May 28, 1983, *another* Republic DC-9 (this one an earlier Model 30) almost ran out of fuel headed from Fresno, California, to Phoenix.

319

The two pilots had taken the DC-9 from Phoenix to Fresno the night before. When they left Phoenix, the fuel gauges showed 15,400 pounds. When they did their preflight inspection of the aircraft the next morning, the fuel gauges still showed 15,000 pounds. Though they knew that no fuel had been added in Fresno, the contradiction did not register. There was no way they could have flown from Phoenix to Fresno on 400 pounds of fuel. Something was wrong.

Approaching Phoenix on the return leg that morning, the two pilots suddenly noticed caution lights indicating low fuel pressure. The copilot looked at the fuel gauges—the fuel gauges showed 15,000 pounds still! He looked at the circuit-breaker panel behind the captain's seat and found that the breaker controlling the fuel gauges had popped. When he pushed it in, the fuel gauges went to zero.

Republic Flight 366 started down immediately and, with the help of air traffic control, made an emergency landing at Luke Air Force Base on the western edge of Phoenix. One engine had flamed out and had been restarted before landing.

When representatives of the NTSB tried to drain the fuel tanks, they found 4.7 gallons remaining! There were eighty-one passengers aboard. Fortunately, no one was hurt.

The crew of Republic 366 had flown nearly two legs with the fuel gauges frozen and never noticed it until almost too late. True, a circuit breaker had tripped, but that should have been discovered. If not before, the copilot should have discovered it on his preflight inspection at Fresno.[1]

1. The pilots insisted that they had pushed the test button on each fuel gauge before departure, and that the needles had obediently moved to zero and back, indicating the instruments were in good condition. It is true that on some Republic DC-9s such a test can be done even when the fuel circuit breaker is pulled, but an examination of the circuit-breaker panel is also a preflight requirement.

The Republic incidents were all the more surprising because of the quality of the airline. (Republic has become a major carrier and is an excellent operation with highly experienced people.) One of the problems behind the first incident, however, was Republic's recent history of expansion. Republic Airlines is actually the end result of a late-seventies merger between North Central Airlines and Southern Airlines, and the later purchase of Hughes Airwest—all of which were high-quality regional Part 121 airlines before deregulation. The merger of Southern and North Central, their name-change to Republic, and their purchase of Hughes Airwest were defensive maneuvers partially in response to the threats of deregulation. The cockpit checklist procedures from each of the

There is hardly any imaginable excuse for airline pilots' running out of fuel—except for the inescapable fact that airline pilots are humans, and humans can be expected to make mistakes. When airlines cut fuel loads to the bone to save money (a common practice in the energy-conscious late seventies and eighties), the potential for trouble is increased. If the weather is worse than expected, or too much holding time is consumed, or an airport closes ahead of a crew already flying on fumes, or a miscalculation has been made, an in-flight flameout could be the result. The thinner the margin of reserve fuel, the thinner the margin of safety.[2]

Between the two Republic fuel incidents, Eastern Airlines came

---

three carriers were somewhat different, but the philosophical approaches of the Southern–North Central combination and the West Coast–based Hughes Airwest were in sharp contrast. At Airwest, there were strict procedures governing such items as boost pumps and when to turn them on (in fact, the center pump was always turned on before takeoff). At Republic (Southern–North Central), the captain had great latitude for personal discretion in procedures. In fact the acronym "SCD" was used in the operations manuals, meaning "subject to captain's discretion." When Airwest was merged with Republic, the Boeing 727 crews had their procedures standardized over a period of time with many hundreds of hours of coordinating work by committees from both airlines. As a result, there were few problems. The DC-9 pilots, however, were handed a different situation. On a single day, all Airwest pilots were required to change to Republic procedures. The result was often confusion, especially when a former Airwest copilot flew with a former North Central captain.

Even a superlative carrier such as Republic, with the best crews and the most severe dedication to safety far above the legal minimum, can suffer human-performance problems in the midst of an expansion.

2. In fact, in 1981 a Pan American Boeing 747, which had taken off with a fuel load calculated to New York with a five-minute flight alternate of Newark, New Jersey, was vectored for a lengthy period north of the New York City area in bad weather, when the pilot could not get into John F. Kennedy Airport in New York. As he rolled off the runway at Newark, two of the four engines flamed out from fuel starvation. On the taxiway, a third died. Two more minutes in the air and the jumbo would have crashed short of the runway with more than two hundred passengers aboard! In the age of deregulation, it is not uncommon for some of the new-entry airlines to arrive at their destination gates with merely minutes of fuel remaining and virtually no margin for error left, because airline managers are constantly pressing for the crew to carry smaller loads of reserve fuel. To carry unused reserve fuel around costs more money because the aircraft has to burn more fuel to carry the reserve fuel. Therefore, an economist's point of view would be to cut the reserves to the absolute minimum. But when unexpected delays occur, that reserve may become a very precious commodity (civilian airliners cannot be refueled in mid-air). With this attitude controlling the industry, the potential for a fuel-starvation accident with a tragic ending builds daily.

very, very close to putting a jumbo Lockheed 1011 into the Atlantic Ocean off Miami (or into downtown Miami)—and the principal cause was management error.

The night before, two Eastern mechanics had changed a small pluglike part called a master chip detector, which monitors the flow of hot oil in the turbojet engines and catches metal chips that are later analyzed to determine the "health" of the engine. To keep the oil from leaking out around the detectors, O rings are required. The two mechanics had a precise procedure to follow in changing the master chip detectors on the three engines of a particular Lockheed 1011, and that procedure required them to install the new O rings on the new master chip detectors, which they were to get from their foreman's office.

But there were none in the foreman's office, so the two mechanics went to the airline's stock room and found three of the detectors. Always before, the detectors they had received from the foreman had the O rings installed. Both men assumed these were identical, and inserted them in each of the three Rolls-Royce engines.

However, not one of the three new chip detectors had the O rings installed. The mechanics had not followed the proper procedure.

They reported on the maintenance log that they *had* followed the procedures, and they signed the log, but in fact the two men never did the required test of motoring the engine and searching for leaks. Nor had they taken a close look at the detectors before putting them in the engines. This was the very failure that had become a close-watch item for Eastern's maintenance supervisors who had posted bulletins—and who assumed their personnel would read them. In fact, the two mechanics had unintentionally set up the aircrew that would pilot ship N334EA to Nassau in the Bahamas within hours.

At 9:15 A.M. the following morning, May 5, 1983, while descending through 15,000 feet for a landing at Nassau, the pilots of Eastern Flight 855 got an oil warning light on the center engine located in the tail of the huge 1011. They shut down the engine, and decided to return to Miami. While en route back to Miami, however, the oil pressure for the wing-mounted engines fell to zero and both failed in sequence.

Suddenly Flight 855 was a heavyweight glider at 13,000 feet and too far from shore to avoid ditching. One hundred sixty-two passengers and ten crew members prepared for the worst while the pilots and flight engineer tried to restart number-two engine (which they had shut down before it had failed). At 4,000 feet, moments before they would have had to put the big ship in the Atlantic, number-two engine restarted. They were twenty-two miles out.

The captain made the decision to try for Miami International, and climbed back up to 3,900 feet (they had sunk to 3,000 before leveling off). With smoke pouring from the overheated bearings of number-two engine, the crippled 1011 made it "over the fence" and landed safely at Miami International. All the oil in all three engines had leaked out around the chip detectors because the O rings were not installed.

Startled NTSB investigators soon found that the problem was not new—Eastern had suffered nine previous incidents involving improper installation (or lack of installation) of chip detectors—five of them involving missing O rings. While the mechanics involved in each incident had been disciplined for their mistake, management at several levels had failed to make sure the problem was solved. Changes were made in the procedures and bulletins were posted, but no one, apparently, made certain that each and every mechanic involved in changing chip detectors on Rolls-Royce engines knew what to do and what not to do.

And as a result, 162 passengers came very close to tragedy. The error was purely human—supervisory, managerial, technical—but human. And the error could have been prevented.

Management is the key. The way the company is run—the way the people are trained and motivated, controlled and supervised at all levels—has a direct and vital impact on whether or not it can operate airliners safely. Cost accountability is necessary for survival in the free-market age, but when cost accountability dictates the degree to which an airline will comply with the regulations—when it motivates management to accept "legal" rather than "safe" as its overall guideline in deciding how much to spend on maintenance or training—it has compromised the safety of the flying public.

In large, established airlines the symptoms of too much cost accountability can be subtle, ranging from increased "pull" times for various items in maintenance and additional requests to the FAA for exemptions from various requirements to reduced and consolidated training. In a carrier under intense financial pressure, reduced outlooks, labor cutbacks, and wrenching uncertainty, it can breed a disgruntled, careless malaise born of frustration and anxiety.

In smaller carriers—especially those whose rapid expansion into the opportunities of deregulation has been built on the weak and shifting sands of inadequate maintenance, training, and control to begin with—the results of financial pressure can be a willingness to cut corners and take chances to help the company. In the fall of 1983, just such a lethal combination was about to give the nation its second bona fide deregulation accident.

Chapter 24

A HERO NO MORE

The cabin lights went out.

Twenty-nine-year-old Barbara Huffman was the only crew member in the passenger cabin of Air Illinois Flight 710. Barbara was the well-liked senior stewardess and director of inflight services for Air Illinois. She was well known for taking good care of her passengers and well attuned to anything that might upset them—such as the cabin lights going out without explanation in bad weather.

Flight 710 was a heading-for-the-barn flight for Air Illinois. The forty-two-passenger Hawker-Siddeley 748 airliner (a twin-engine turboprop) had picked up some of its passengers at downtown Meigs Field on Chicago's waterfront, and after a stop in Springfield, Illinois, was headed for the company's home base in Carbondale, to the south.

But something was wrong. The flight was already running nearly forty-five minutes behind schedule when it lifted off at 8:15 P.M. from Springfield. It was obvious that it had not climbed as high as usual—and now the lights were out.

The seven passengers this evening of October 11, 1983, were an interesting and varied little group—representative of how routine air transportation has become to so many.

Sitting together in the cabin, Judy Chantos and her two-year-old son, Jonathan, looked out the window by their seat row and wondered what was going on. She and her husband, Robert, had tried to have a child for the first eight years of their marriage and finally, with medical help, had succeeded. Little Jonathan was the light of their lives, as well as a source of pride for her parents—whom they were going to see. Robert Chantos was a chef in a Springfield, Illinois, restaurant and had to work this evening, but Judy and Jonathan were going to take the forty-five-minute flight to be with her folks for a few days in Jackson, Missouri, just over the Mississippi river from the Carbondale area.

Dalbir Singh, an Indonesian businessman and a computer-software specialist, was to be married in less than two weeks.

Fifty-two-year-old Dick Baker, coordinator of the Rehabilitation Institute at Carbondale's Southern Illinois University, was flying back to the campus with the director of the institute, Jerome Lorenz, a thirty-nine-year-old native of Virginia.

On the other side of the cabin sat Regina Polk, a thirty-three-year-old business agent for the Teamsters Union in Chicago, and a fast climber in that organization. As she was aware, there were many who expected her to rise to great heights in national union affairs in the next few years.

A senior official of the Illinois Department of Labor, fifty-two-year-old Jerome Brown, also occupied a seat in the darkness of Flight 710's cabin, in which the lone stewardess was trying to decide how to handle a worrisome situation.

She knew the captain. Captain Lester Smith was familiar to most Air Illinois crews as an extremely moody individual. Tonight, however, he seemed in a fairly decent frame of mind—though he and his copilot, Frank Tudor, had been deadheading (traveling as passengers to position for a flight) all day. Smith and Tudor had started at 10:00 A.M. and would be getting tired.

But in the cockpit of the Hawker, Smith and Tudor were facing a crisis. Just after takeoff, the left electrical generator had failed. The right one was putting out power, but Tudor couldn't get it to engage.

Without either generator on the line, they had only batteries left to power the flight instruments, the cockpit lights, the de-icing equipment, the radios and navigational instruments—and the cabin lights.

And that battery was supposed to be good only for thirty minutes—fifteen minutes less than the length of time Flight 710 was scheduled to be in the air on its way to Carbondale.

Springfield, Illinois, had been in visual flying weather—and still was. It was only eight minutes behind them. As Frank Tudor well knew, there was bad weather ahead between their position and Carbondale, and there was only one right—and legal—thing to do now: turn back and land at Springfield.

But Frank Tudor also knew, as did so many of the Air Illinois copilots, that Les Smith ran a one-man show in the cockpit. When it was his leg, he did things his way. To suggest to the thirty-two-year-old captain that maybe he should do something else often met with a churlish—or childish—response.

Tudor was no shrinking violet, but he had to fly with the guy. If he told Les they ought to go back to Springfield, Smith would probably go on to Carbondale—through bad weather in deteriorating visual conditions with no power and batteries that might fail, leaving them with absolutely no way to fly the airplane other than visual reference. Surely Les wouldn't try that.

But Captain Les Smith had made the decision already. They would stay low and press on to Carbondale. They might have to land in a dark cockpit by reference only to the outside airport lights, but they could do it. He didn't want to ground the Hawker at Springfield where Air Illinois had no maintenance facilities. He knew they had been having chronic problems with the right electrical generator for a month or so, and the airplane would be best off back at Carbondale.

Perhaps Les Smith had decided to be a hero for his company again, and bring the Hawker home.

Air Illinois had come into existence in 1970 as a commuter carrier—founded and run with an iron hand by a local businessman named E. R. Dzendolet, and starting with one de Havilland Twin Otter. The atmosphere from the first was autocratic, and persistent stories of captains being fired for refusing to take flights only added to the problem. Captains well knew they would be fired if they didn't take flights, regardless of what problems or legalities might confront them.

The company had expanded into Chicago by 1973, and Air Illinois bought the Hawker 748 in October of that year—in their estimation the best aircraft to get paying passenger loads out of the convenient (but tight) Meigs Airport in downtown Chicago.

In 1979, after Dzendolet's resignation under financial pressure (and the purchase of the airline by an international financier), Air Illinois began jumping into the big time on the springboard of deregulation, picking up routes abandoned by Ozark and Delta airlines, and entering (and subsequently abandoning) a large number of markets through 1982. In 1980, it acquired a second Hawker 748 (which was never matched with a profitable route), and by 1982 it had decided to go for the mainstream and enter the Chicago O'Hare Airport market, which led to the subsequent purchase of a slightly aging BAC (British Aerospace Corporation) 1-11, a twin turbojet airliner (a type used by Braniff in the mid- to late sixties).

In fact, the president of Air Illinois as of mid-1979 had been recruited from Braniff in Dallas. Roger Street, a marketing man, came on board with Jim Benge, another Braniff veteran, and immediately discovered he had joined a carrier in need of great reform.

Benge in particular was amazed at the way Air Illinois had traditionally operated. The pressure on the pilots, the accepted method of letting captains operate with almost monarchical authority over station agents as well as their own flights, the financial problems, and the lack of a dispatch system amazed him. As Benge knew well (coming from a mainstream Part 121 carrier like Braniff), a large and growing airline flying under Part 121 would have to have a well-developed dispatch system.[1]

The Hawker, under a special section of Part 121, could be operated with a nonbinding "flight following" office back at Carbondale. The employee in flight following did not have joint authority, though. He or she could not overrule a captain. But the pure-jet BAC-1-11 *had* to have a full dispatch system—and the FAA General Aviation District Office that supervised Air Illinois finally decided it had become a big-enough operation to require a dispatch function even with the Hawker.

Nevertheless, change had come slowly to the pilots of Les Smith's vintage (Smith had joined the company in December 1978 and was number five on the seniority list by 1983), and one of the reasons, according to many of the younger pilots, was the tough management control of former marine Major Ray Schrammel, who had been head of the operations function (and finally vice-president-operations) since the company was formed. Schrammel was not known ever to direct people to break rules, but he was known for being intolerant of pilots who shut down flights for what he considered "no good reason."

The efforts of Jim Benge, senior vice-president of transportation

1. In such a system, the dispatcher (who is FAA licensed) is equally responsible with the captain for all aspects of a flight, and must keep abreast of the flight's progress, the weight and balance, the weather, and all other facets. Shared authority was hard enough to impose on Braniff captains—Air Illinois captains wanted no part of it, and resisted it every inch of the way.

services, and President Roger Street to whip Air Illinois into a thoroughly professional organization had run into financial trouble as well as morale trouble despite their best efforts. And whether or not they knew it, there were major problems in the maintenance area.

Some of the maintenance men who worked on the aircraft overnight—specifically the Hawker 748—had a habit of dealing with maintenance write-ups on scrap paper, or verbally. Too many times, important problems with the airplane, which legally must be put into the logbook of the airplane, were not. If scraps of paper were used, or just verbal briefings, the FAA would not be able to second-guess them and give them a hard time for missing certain required inspections or failing to solve a particular electrical problem—and also failing to ground the plane!

Over a period of time, some of the maintenance people had developed the habit of maintaining a paper airplane for the FAA, and the metal airplane for the good of Air Illinois.

Such had been the treatment of a particularly vexatious problem with the Hawker's right generator, which they were still "working on," exchanging telexes daily with the manufacturer to try to find the problem—while keeping the airplane in revenue service.

Les Smith had received more than the required training at Air Illinois. The company's training was reasonably good for a small carrier—especially on the Twin Otters.

But Les Smith had received no instruction on crew coordination, no directives to let his copilots participate in the decision process, and no discipline concerning when to abandon the idea of pressing on for the good of the company.

In fact, Lester Smith (and others in the senior-pilot ranks) had developed a relish for becoming "barroom heroes"—for bringing back company flights a "lesser" pilot would have grounded. It was a form of macho, a form of bravado, and a form of potential homicide that neither Ray Schrammel nor the chief pilot below him had corrected.

And there was something else about Lester Smith that had seriously upset at least one of his copilots, Jeanene Urban. If they tried to influence him not to do something, he would be likely to do it anyway to prove his command status. Urban, for one, had gone to great pains to avoid flying with him—especially after he deliberately ran their flight (in the Hawker) through a thunderstorm cell when she had made the mistake of pointing it out to him. She was convinced he was dangerous, and she wanted no more of it.

She had also gone to the management with her concerns, but other than a mild conference with Smith, nothing more had been done.

Lester Smith was still in command—and by 8:30 P.M. of October 11, it was becoming increasingly apparent that he was also in trouble with Flight 710.

Barbara Huffman was on the interphone, Les Smith on the other end. She had told him that her passengers wanted to know what was going on.

"They want to? . . . We have a little bit of an electrical problem here but we're going to take it to Carbondale. . . . We had to shut off all the excess lights," Smith told her.

"I've only got the reading lights, the front lights by the bathroom, and the baggage light, and the, ah, entrance light," Huffman confirmed. "What time do we get in there?"

The sound of . . . something . . . interrupted her thoughts.

"Is that rain?" she asked Smith.

It was 8:33 P.M. and Flight 710 had indeed entered an area of showers as Smith headed southward toward Carbondale. The radar was off now—it could not be used with only battery power.

Smith couldn't remember what time they had left.

"What time did we lift off, Frank?"

No response. Apparently Tudor was busy.

"Frank?"

Finally he responded. "There about on the hour."

Tudor could hear Barbara's question from the back. The worrisome blackness out in front of them took precedence, however. Without radar they were blundering blindly into this stuff.

Barbara Huffman replaced the intercom handset. Lightning was visible in the distance off to the right of the aircraft, and the passengers had seen it too. On the hour, Tudor had said, and the time now was . . . 8:35.

Flight 710 had been on nothing but battery power for fifteen minutes. Barbara Huffman and her passengers did not know it, but their batteries were supposed to be good for only fifteen minutes more—and Carbondale was twenty-five minutes ahead.

At 8:50 P.M. it began to come apart.

"I don't know if we got enough juice to get out of this."

330

It was Tudor's voice in Les Smith's ear. But they were in and out of weather now, trying to stay visual in clouds at 3,000 feet. They had asked for a lower altitude, but the Air Route Traffic controller couldn't give it to them without losing their radar return.

"How come?" Smith asked.

No reply from Tudor.

"Squawk your, ah, radio failure [code]," Smith instructed. The code was a series of numbers a pilot could set in his transponder to tell the controller their radio was out.

The controller at Kansas City Center was calling them. He had lost radar contact because the same transponder Smith was referring to (an electronic box that generates blips on the controller's radarscope) had finally dropped off the line. There wasn't enough power left.

The Hawker's batteries were dying—on schedule.

". . . know your radio failure code?" Smith tried again.

Still no answer.

The controller's voice came through their headsets:

"Illinois 710, Kansas City."

Tudor was trying to answer, but now the radio was gone. Not enough power to transmit.

"Frank, do you know your radio failure . . . [code]?"

Finally, "Yeah."

"Squawk . . ." Smith began.

"Yeah," Tudor replied again.

"Watch my altitude, we're going to go down to twenty-four hundred [feet]." Smith was in great need of seeing the ground. If the batteries died, he would have nothing but a compass and airspeed. Even the bank-and-turn indicator would be gone.

"Okay," Tudor replied simply.

"You got a flashlight?" Smith again.

"Yeah."

"Here we go . . . go, you want to shine it up here?"

Tudor heard Smith's words. His reply was rapid. "What do you need?"

"Be ready for [the flashlight]."

"What do you need?" Tudor repeated.

"Just have it in your hand, if you will."

Thirty-one very long seconds ticked by as the truth bore in on Frank Tudor. They were out of radio contact, out of good contact with the ground, and almost out of electrical power. They might not make it!

"Well . . ." Tudor began, ". . . we're losing everything . . . down to about thirteen volts."

"Okay . . . watch my altitude, Frank."

"Okay . . . twenty-four hundred."

"Do you have any instruments?"

"Say again?"

"Do you have any instruments? Do you have a [artificial] horizon?"

The time was 8:54. The cockpit voice recorder had finally stopped recording—the power too low to continue.

Barbara Huffman and her increasingly concerned passengers noticed the last of the lights dim and die. Now only the small batteries in the emergency exit lights over the doors could be seen in the dark interior—except for flashes of lightning filtering through the clouds outside the windows.

Everything else was gone—except for the engines, which kept up their steady hum.

Arilla Fisher and her seventy-eight-year-old husband, John, had been spending a quiet evening on their small farm nestled next to a pond in the rolling countryside of southern Illinois. It was raining steadily, and they had both been aware of lightning flashes and the bass drum roll of thunder.

Their farm was compact but beautiful. Their home, small but comfortable. Two other houses were just behind theirs to the north—good neighbors, also at retirement age. To the south was a cornfield, sitting in a protected draw behind an eighty-foot ridgeline studded with sturdy oak and sycamore trees. To the south of that were fields, and another neighbor.

The sound of the rain on their roof began to give way to something else, the rising sound of aircraft engines obviously getting closer, then passing by overhead.

Arilla was somewhat puzzled. The plane seemed awfully low.

The Fishers' neighbors to the north had heard it too.

And then it returned.

The airplane was lower now, apparently circling overhead.

But why in this weather, they wondered?

The engine sounds died down again and disappeared into the white noise of the raindrops.

Flight 710 was in a right bank and descending. There was no instrumentation left to help Smith and Tudor know whether their wings were level or not. They were down to a flashlight, an airspeed indicator, and an altimeter, in total instrument conditions.

The lights of farmhouses northeast of Pinckneyville, Illinois, were visible through the steady rain to anyone standing near the ridgeline south of John Fisher's place as the crippled Hawker began its third, circling pass over the same area. Even if the lights had filled Les Smith's eyes as his airliner broke out of the clouds and turned northbound, there would have been no time to react to the sight of the small, grassy hill as it loomed up in the windscreen at over 150 miles per hour. There would have been no time to say anything as the right wing tip slammed into the ground, the right wing shattering, the nose striking the soft earth, and the kinetic energy of the Hawker compressing and fragmenting the structure, flinging props and engines, tail members and metallic members up the small incline in a deadly spray of aircraft parts and human bodies, smashing through the trees on the ridgeline overlooking John Fisher's cornfield, disintegrating the fragments as the already lifeless bodies of Judy Chantos and her little son, Jonathan, thudded into the bank of the creek bordering the corn rows, the assaulted body of Dalbir Singh catapulted nearly a thousand feet at over a hundred miles an hour into the edge of a pickup-truck camper, the right engine of the Hawker sailing beyond the Fishers' house, and the broken remains of Richard Baker, Jerome Brown, Jerome Lorenz, Regina Polk spilling in a grisly, deadly rain over the darkened pastoral landscape mixed with the ruined bodies of their crew, Tudor, Smith, and Huffman. It was a singular impact, a loud noise, like a clap of thunder at very close range. It was the end of Flight 710.

Les Smith would be a hero no more.

A month later the hearings began in Carbondale, with NTSB Chairman Jim Burnett presiding. The interviews with various Air Illinois personnel had begun turning up the problems with maintenance, the problems with morale, the problems with Les Smith, and a long litany of pressures and problems that could only be laid at the doorstep of the freedom for expansion granted by deregulation.

At first, the local FAA men who had been responsible for supervising Air Illinois (GADO inspectors) reinspected and quickly gave the carrier a clean bill of health. They even flew to Washington, talking to the administrator and others to assure them that despite some paperwork problems, Air Illinois was a good operation.

And in early November the NTSB hearing put the true story on the record.

Suddenly the GADO men were on the phone to FAA headquarters, asking for authority to shut the airline down. Their superior was amazed. "Do you guys know what you're saying? You gave them a clean bill of health. Now, because of NTSB evidence alone, you want to shut them down? Don't you have any idea what that says about your inspection capabilities?"

Instead, Washington sent a team of experts for a quick white-glove inspection and began dogging Air Illinois' every step, putting inspectors on every flight—as President Roger Street protested to the media what he considered "harassment."

On December 7, the BAC-1-11s were grounded temporarily.

And on December 14, Roger Street and others in the airline leadership were confronted with the long list of problems involving, primarily, record-keeping and were informed that the airline's operating certificate could either be returned voluntarily or be revoked.

Street chose, for public-relations reasons, to ground the airline and hand over the certificate. It would be the start of an off-again, on-again attempt to return to operation, which would end in failure the next spring. Air Illinois would continue to exist only as a charter carrier with the BAC-1-11s.

As is too often the case, the suspension of service came too late—especially for those whose lives ended in a cornfield near Pinckneyville.

Chapter 25

THE REAL COST

We have a problem. Although the Congress promised that airline safety would not be compromised by deregulation, it *has* been compromised. The "safety buffer" diminishes daily under the relentless force of economic reality and cost accountability.

What's worse, the government of the United States—our government—refuses to admit the obvious.

The aircrews who fly the airplanes, the mechanics who maintain them, the harried FAA air-carrier inspectors overburdened with impossible work loads, the air traffic controllers expected to perform perfectly 100 percent of the time, and the air-safety investigators trying to make sense of a dangerously unbalanced system all know the truth. Get them aside and they'll tell you in frightening detail what actually is happening in the airline industry. The maintenance folk know their maintenance stock has been reduced, their training shortened, their anxiety over the future heightened, and their morale decreased. The pilots can tell you horror stories of near-misses, management pressures, marginally qualified pilots sitting next to them, increasing fatigue on the job, and an overall decreasing margin for error. The FAA men can tell you of ulcers and unpaid hours by the hundreds, of carriers who lie but can't be caught, of superiors who (when violations are discovered) worry more about the financial health of the carrier than the lives of their passengers. (They can tell you as well of broken homes and divorces traceable to a work load fully three times what it should be.) And the air-safety investigators (as well as some of the board members of the NTSB) can tell you of immense frustration at trying to get the FAA in Washington to implement urgent recommendations for safety, to study weak areas, and to address the pressures deregulation has cranked into the equation.

But ask them when they're off duty and off the record, because, officially, there's nothing wrong!

The government denials of trouble are *not* the result of some conspiracy. These are all good, honest folk—the backbone of our government. But they have misled themselves by letting political support for deregulation color their understanding of the fatality and accident statistics—and that is a significant fallacy.

The accident rates tell a tale of yesterday. In no way in a constantly changing airline system do they provide irrefutable evidence that tomorrow won't turn out to be a bloodbath. They do not tell the reality of constantly reduced safety margins. They do not reflect the near-disasters that somehow were averted—but which are symptomatic, nonetheless, of a system in trouble.

Remember that the accident and death rates before the Grand Canyon crash were quite respectable up to May 29, 1956—yet 128 people died on the canyon floor the very next day.

Recall that more than two years had passed with virtually no Part 121 passenger deaths (a fact that elated the proponents of deregulation) until the moment in January 1982 when Air Florida's Flight 90 smashed into the ice-filled surface of the Potomac river—a darling of deregulation laid bare and bleeding at the feet of its congressional creator.

Here we are in the mid-eighties, and once again the "excellent" fatality rate for major Part 121 carriers has been trotted out by the Department of Transportation to buttress the argument that the airline industry is not experiencing the very decline in air safety that members of the industry can see so clearly.

But, of course, the argument is academic. There is nothing excellent about having any accidents or fatalities, however few in number they may be. And there is virtually no way that any American can tolerate that point of view as long as there is the possibility that it could be your body (or those of your family) that someone has to extricate from the wreckage of a destroyed airliner strewn over a residential area, in a frozen river, or through a forest of shattered trees.

Indeed we are blessed with the safest air-transportation system on the planet—a system of our own making. It is far safer to fly than to drive, statistically. But there is one profound difference: In a car you can turn off the ignition and get out. In an airliner, you are trapped. Ask Joseph Stiley what it feels like to be sealed in an airplane covered with snow, moving inexorably toward trouble as if in a slow-motion nightmare.

As long as the passenger is at the mercy of the airline, there is no such thing as an excellent accident rate, and there is no excuse for pretending that deregulation, the greatest disruption of the air-transportation system of the United States in forty years, has done nothing to unbalance its ability to fly passengers safely.

As it was, there were already more than enough problems with airline safety before 1978.

The industry, the NTSB, and the FAA were all far behind even the U.S. military in understanding the vital, inextricable role of human factors in the safety equation. Routine human-performance research into every accident was just beginning at the NTSB, but for the most part, those involved with the airline industry simply did not see significance in the fact that approximately 85 percent of the accidents resulted from human failure.

The established Part 121 airlines on the whole were not convinced that their people needed such things as assertiveness-training courses, or consideration of fatigue in aircrew scheduling. The predominant use of threat and punishment remained their prime method of motivating employees to comply with the rules. Pilots and maintenance people were ordered not to make mistakes. How they achieved that goal was to a large degree left up to them.

The commuter industry seemed perpetually in trouble—too many of the small commuter carriers always inching their way along a financial ledge with each new airplane or each new route, and always in danger of falling off into insolvency. (Even those commuters that had grown huge route systems faced a constant financial battle to stay in the black in the late seventies—and not all were successful.) With low salary structures, constant turnover, marginal training, often questionable management, and lengthy crew-duty times in a hostile low-altitude environment using (all too often) inadequate or antiquated aircraft, the commuter industry was in most cases totally unprepared to take the major role in the air-transportation system that the deregulation bill envisioned. As entrepreneurs they would expand and move into the opportunities anyway, but an inordinate number of accidents and passenger deaths would be the result.

Throughout the industry the idea of judgment training for aircrews was a foreign concept, as was the implementation of line-oriented flight training (LOFT), a NASA-sponsored program of realistic flight-simulator training.[1] Most major carriers were not terribly interested, and the commuters simply did not have a choice, since there were no simulators built to duplicate the majority of the aircraft used in the commuter market.

The NTSB up to 1978 simply had not learned to research the human causes of accidents and use the resulting discoveries to prevent repetitions. Political infighting and bureaucratic games, outright interdepartmental jealousies and lack of board support made things worse. In 1975 when Congress changed the NTSB's mandate and assigned it to investigate highway, rail, ship, and pipeline accidents as well as aviation mishaps, the result was a severe drain of resources from the Bureau of Aviation Safety and a demoralization of the staff. Partially as a result, the accident reports during the mid-seventies left much to be desired. As a consequence, when the NTSB recommended changes to the FAA, those recommendations too often fell on deaf ears.

The FAA, for its part (as was demonstrated so graphically with Downeast), was unable to uncover officially the sort of subjective managerial problems that breed accidents in airlines. If it couldn't write a rule to cover it, it couldn't control it, and it had no idea how to deal with inconsistencies in the rules and requirements—the interpretations of the FARs—enforced differently in the different regions. It was a situation of unnecessary confusion of which savvy operators could take advantage—and did. Politics, bureaucratic squabbling, and a penchant for promoting the industry severely retarded the FAA's ability to deal with the problems that existed, let alone the work loads that would be created with a free-market airline industry. And the ferment of labor unrest born of FAA abdication of management responsibilities in the nation's Air Route Traffic Control centers and airports was well on its way to an explosion (which would occur in the 1981 professional suicide of ten thousand controllers).

Pilots were still flying perfectly operable airliners into the ground and running low on fuel, maintenance men were still leaving wrenches in engines, and pilots were still playing blind man's buff with the "see and avoid" concept even under radar control.

The airline system of 1978 was not decrepit—just immature. It was not unsafe—it was simply not *as* safe as it could be.

1. Northwest Airlines was the first to implement a LOFT program, though United has taken it further and incorporated it as a regular training method.

What the industry *did* have was stability among the larger carriers—a stability that had nurtured their safety record since 1937 when the death of Senator Bronson Cutting in an airline accident galvanized the Congress and led to the enactment of the Civil Aeronautics Act of 1938. Since the industry was under the protective wing of the Civil Aeronautics Board from that date forward, no matter how inefficient an airline's management team might be, it had the assurance that its market share would be relatively secure. No other airline could muscle in on its bread-and-butter routes without an extensive hearing battle before the CAB, and thus its profits, if not predictable, were at least safe from profit-skimming competitive raids. This protection remained intact until the early seventies, when state-controlled upstart airlines, such as PSA in California and Southwest in Texas, first surfaced, later becoming models for Professor Alfred Kahn's airline-deregulation theories.

That stability meant fat union contracts, overstaffed and overpaid executive and management ranks (with vice-presidents as far as the eye could see), financed by strictly controlled ticket prices that were much higher per seat-mile than those of the cut-rate intrastate carriers.

But that stability also meant constant sources of finance for new aircraft, buildings, and equipment. It meant that financial institutions—even insurance companies—could trust the airline industry with their money because the routes themselves had great value, and thus there was always equity. Even if management failed completely, the CAB would arrange a hasty "shotgun wedding" with a healthy carrier and merge the essentially bankrupt carrier with minimal financial loss. That in turn meant that the airline personnel of the major carriers (Western, Eastern, Continental, Braniff) and the regional carriers (Frontier, Ozark, Piedmont) knew that while there could be a furlough in bad times, their airline would "always" be there.

Most important, that stability kept safety from being cost accountable! The ability to pass rising costs on to the consumer through CAB price regulation (which mandated that *every* airline flying a certain route had to charge the same price), meant that the carriers could build huge, modern, and effective training centers, invest in expensive flight simulators and maintenance equipment, keep large rotating stocks of spare parts on hand far in excess of FAA minimum requirements, and could say to the public with conviction, as Pat Patterson of United used to do: "There is no price too great for safety!"

Ironically, it would be clear to everyone but Patterson's United (when the deregulation issue was being hotly debated by the airline industry in Capitol Hill hearings) that denuding the industry of that stability would in fact make safety cost accountable.[2]

The executives of the U.S. airline system knew that if they could not pass their costs through to the ticket prices—if they were forced to compete in a free-market environment—every expenditure would have to be reexamined. Many of them were also quite aware of how much money their airlines poured into maintenance, training, and other safety functions—all of which would have to be justified if the protection of the regulated past was removed.

So many of the vital safety functions of an airline are of subjective value, and by definition very hard to justify (especially to a finance man who is new to the airline business and knows little of the reality of airline operations). When, for instance, is an aircrew sufficiently trained? If that point is reached when the crew meets FAA minimums, then nonmandated courses (such as assertiveness training, LOFT, professional standards) are a waste of money.

To Congress, however, such fears were merely the predictable cries of pain of executives who would have to learn how to run an efficient organization if they were to give the consumer more value for his dollar. They were the anguished yelps of union leaders who knew that deregulation would become the instrument by which the industry would finally wiggle free of sometimes oppressive and unresponsive union control. The senators and congressmen could not understand that the consumer was already getting immense and incalculable value from his dollar by flying with airlines that historically had spent a large portion of that dollar on air safety. The system was not perfect (as the many human-failure accidents of the seventies attested), but it was stable and working constantly toward improvement. That constituted value received.

The passenger was not told by Congress or the proponents of deregulation the ultimate truth about the enticing free-market pro-

---

2. United was the only major carrier to embrace and support deregulation. Of course there were ulterior motives. United had always felt itself discriminated against by the Civil Aeronautics Board because it seldom, if ever, received new routes. The CAB felt that United, the nation's largest airline, needed no additional expansion. United, on the other hand, was eager to break out of the traditional markets and expand to all fifty states plus international destinations—a feat it has since accomplished. United also had enough money at that point to assure itself internally that it would never break the tradition of funding safety functions as an end in itself. But recent history shows that cost accountability has nibbled away at United as well in areas such as mandated fuel reserves for United flights and other cost-conscious measures born of intense competition and marketing mistakes in the late seventies.

posal: If prices are cut, costs must be cut, and something more than executive salaries and union contracts will have to give. The cost of safety would be one of those affected items.

But Congress had convinced itself that just by ordering the FAA to prevent a decline in safety, there would in fact be no decline in safety. It was a belief based on a profound misunderstanding of the fact that the FAA could enforce only the legal minimums—and the mainstream of the airline system built so laboriously (and balanced so delicately) over nearly thirty years had *never* rested on the legal minimums.

This, then, is the central oversight of Congress in 1978. It did not realize that the airline industry standards existing at that time were far above the minimums required by the Federal Air Regulations, and that deregulation would, by economic imperative, pressure the industry to lower its standards toward the legal minimums. Since only the legal minimums could be enforced by the FAA, Congress's faith in the FAA's ability to prevent a decline was misplaced and essentially worthless. Thus the Deregulation Act itself mandated a lowering of the safety standards for U.S. commercial aviation across the board.

This is why there has been immense frustration among the dedicated people in the FAA's rank and file. They watch the system deteriorating around them, but they are powerless to do anything but enforce minimum standards.

The Federal Aviation Administration is staffed—in Washington, in the different regional headquarters, and in the field—by many of the finest, most hardworking and dedicated public servants in the U.S. civil service. There are exceptions, as in any human organization, but for the most part the people who are the FAA are superlative.

They are, however, employees of a creation of Congress, which is also a creature of the executive branch, responsible to the Department of Transportation (which is highly political) and through it to the White House. The FAA is also responsible to Congress and finds itself (in the form of its senior officials) called to testify quite often before various oversight committees and subcommittees on subjects that often put it in the unenviable position of defending generalized administration policy on highly technical subjects.

And as if such delicate relationships and responsibilities weren't enough, the FAA still has its original, contradictory mandate from the Federal Aviation Act of 1958 to "promote" aviation as well as to regulate it.

The FAA had become expert at the fine art of bureaucracy long before deregulation, learning to walk a tightrope on thousands of issues—including safety. Over the years, what the agency regarded as beyond its control, it tended to ignore (such as researching and addressing human-factor/human-performance problems in all phases of aviation). What was politically troublesome or costly to the airline industry, it tended to put aside or study to infinity (such as the need for distance-remaining markers on airport runways, backward-facing seats in airliners, and proper restraints for infants).

The desire to compromise and avoid major battles with industry quite often got in the way of its responsibility to enforce, as in the infamous gentlemen's agreement between McDonnell Douglas and the FAA in 1972 (the FAA is certainly not unique among federal agencies in that regard). Neither is the FAA immune to the usual rivalries that develop in any organization, spawning political games and infighting among the different regions and the different chiefs. All such traits, after all, are merely human.

But it was this flawed, human organization of hardworking, well-meaning people to which the Congress threw the impossible task of maintaining safety. After twenty years, which had seen solid accomplishment in developing the U.S. aviation system, the FAA still had a long way to go. As late as 1978 the FAA still couldn't adequately control the safety levels of the regulated airline system. Without the proper legal tools and the understanding of the subjective problems inherent in a human system, there was no way it would be able to handle a deregulated industry.

To meet the challenges thrown at it by deregulation, the FAA would have required a massive infusion of funds, significant alteration of its surveillance and monitoring authority, and a significant alteration of its philosophical approach to airline safety. Congress, however, gave it only the additional work load—and it was mind-boggling.

Whenever an airline gets ready to go into business, it must first be certified by the FAA. That sounds like a simple thing, but in fact, whether it's a 135 carrier or a 121 carrier, the number of FAA man-hours that have to be devoted to getting the fledgling airline through the preparation of manuals, crews, maintenance procedures and facilities, economic and reporting procedures, and literally thousands of other precertification details, is massive.

When deregulation threw open the floodgates, the gush of new applicants for certification immediately overwhelmed the air carrier inspectors who had to oversee such tasks. There was no way the air carrier inspectors could spend sometimes six months on certification efforts for a single carrier and handle a surveillance work load that sometimes included twenty to thirty Part 135 operators and one or more Part 121 airlines. It was humanly impossible to give more than paperwork treatment to many aspects of the job. The result: The airline system reverted from an industry under partial surveillance to an industry running on the honor system. As the Downeast and Air Illinois accidents were to illustrate so well, not all carriers were worthy of such trust.[3]

3. The reality that FAA overload caused inspection activities to get short shrift may be painful for the agency to admit, but its own people at many levels pull no punches (privately) in confirming just that. By early 1984 the situation had become sufficiently critical that Secretary of Transportation Elizabeth Dole (acting partially in reaction to the congressional pressure generated by Congressman Norman Mineta's Subcommittee on Aviation of the Committee on Public Works and Transportation, ordered a National Air Transportation Inspection (dubbed "NATI" for short), a two-stage effort involving a massive white-glove inspection of virtually every certificated 135 and 121 airline in the nation. It was an excellent move in the right direction for the first time in years. The inspection did identify a moderate list of carriers, 135 and 121, which would be more closely watched during the second phase, but even so, problem carriers slipped through the net. Some airlines that were creations of deregulation simply received more "help" in getting through the inspection than others. Others simply hid their problems from the FAA and got by. The NATI effort was based on several assumptions, among them (as stated in the NATI handbook): "The FAA's current inspection and surveillance practices are valid and will reveal, if any exist, air carrier system deficiencies." Unfortunately, that assumption would be proved wrong within the same year. The current inspection and surveillance practices may be valid for the majority of the carriers, large and small, which are honest and aboveboard, but for the darker few, those "practices" are totally ineffective.

During the NATI inspection, most certification activity came to a virtual standstill, and afterward the FAA ordered its people not to let certification activity shut down their surveillance responsibilities. All of this was evidence of the impossibility of covering all the required duties with the current FAA inspection work force.

Chapter 26

OUT OF CONTROL

"There's no way we can do it all. In fact, most days, there's no way we can do half of it. I've got Part One-thirty-five operators I haven't visited in nearly a year. Before the NATI inspection all the certification work was snowing us under. Now I'm not doing so much certification, but the flight checks and overall load is still impossible."

The disgusted senior air carrier inspector sat sideways in his chair, almost draped over one arm—exhausted. His tie was askew, his hair not quite combed, and his suit had a slept-in look. A small pyramid of cigarette butts filled the ashtray on his desk.

It was after 5:00 P.M. on a Friday, but he would be back the next day, Saturday, giving check rides in a flight simulator. The veteran FAA man was counting the days to retirement now, less than a year distant, but he couldn't slow down. There was simply too much to do.

Too many air carrier inspectors feel much the same way these days—as if they've been carrying the weight of the world on their shoulders, which is not too far from the truth. Indeed, every airline passenger in the U.S. airline system depends on their dedication and expertise to some extent. But there are too few of them, and

344

too many airlines to monitor, to provide any realistic assurance that the safety level they are charged with upholding is not being compromised.

In such an atmosphere, it would seem that the least important task of the ACI force would be the certification of still more new airlines, but up until the spring of 1984, a large percentage of their time was being spent on exactly that—certifying new carriers. That left too little time to properly monitor the existing ones, or for that matter to do any part of their job as thoroughly as it should be done. The result has been a disturbing lack of quality in all areas, including the process of deciding whether a new-entry or expanding carrier can be entrusted with the lives of the public.

Far too many air carriers, large and small, 135 and 121, have been allowed to lurch into the air with paying passengers on board who don't know (and have no way to discover) that they are trusting an airline which was never sufficiently "fit" to begin with. Passengers trust the FAA to discover such things, to find out when the owners are down to their last dollar, to spot tendencies to take the cheap and questionable shortcut, and to find out in time when an airline that appeared trustworthy during certification never really intended to follow the rules in the first place.

Of course, in some cases new airlines are not fully aware of the rules. Too many carriers have been granted certificates despite a total lack of commercial-aviation experience on the part of their owners and top executives. That, in itself, is not disqualifying, but when coupled with operational managers of limited experience, given titles but no authority (such as Jim Merryman, chief pilot for Downeast), the propensity for dangerous corner cutting is increased.

And in too many instances, the FAA is under intense political pressure to get the certification completed because a particular community needs the air service and is pulling every political string it can grab to "get the FAA off [the airline's] back so they can get into business."

But such problems are only the tip of the iceberg—a small part of the legacy of deregulation in terms of its impact on the FAA. Among the aberrant pressures the public never sees is the unbearable work load that has been created by the need to check and recheck the pilots who after 1978 began moving from carrier to carrier, airplane to airplane, position to position with frightening speed, generating the need for new FAA check rides with almost every move. There has been no possible way for the FAA's air carrier inspectors to keep up with all the demands for their time—

but if the FAA men aren't available to give certain types of check rides, the pilots who need them cannot remain legal and current.

The normal, logical, response to such a crisis of supply and demand would be to increase the supply—hire more inspectors. But the FAA has attacked the opposite end of the equation—and in the process spawned a human-factors problem by creating the Designated Pilot Examiner Program. That program holds the seeds of substantial, dangerous abuse of power, because it appoints company pilots to substitute for FAA inspectors in giving check rides and granting type ratings or other licenses to fellow company pilots. In other words, in response to the pressures of deregulation, the FAA has begun permitting certain airlines to monitor themselves.[1]

1. The philosophical origin of the designated pilot examiner concept is the DER, the designated engineering representative involved in the certification of aircraft designs in the factory as well as routine approval of manufacturing work. There is a long and mostly honorable history involving the DER system, and in fact, it would be functionally impossible for the FAA to do such work with its own people. Among the DERs there is a sense of honor and responsibility stretching back several decades. For the most part (although no more perfect than any other human system) it works.

The Designated Pilot Examiner Program is an entirely different matter. Traditionally, the FAA could certify various pilots in an airline to give the company's routine check rides, but when it came to granting initial approval of a pilot to operate as a captain in a type of airliner he had never before flown (a type rating), an FAA inspector was required. With deregulation and the incredible work load, the FAA manpower and the airline's demand for type ratings and other FAA-only checks were not matching. So the FAA began the Designated Pilot Examiner Program, qualifying carefully selected pilots from participating airlines to give type-rating checks and other checks formerly reserved to the FAA alone, and by so doing, save the FAA's time.

The program was based on the assumption that the FAA-anointed check airman would not cheat. He would not let his buddies slide by with an easy type rating, or be too harsh on someone he didn't like, or worse, bend to company pressure to pass or fail a particular pilot on a basis other than demonstrated merit. That assumption is accurate in most cases, but it inevitably involves human nature. Therefore, the predictable will occur—and has.

In 1978 a chartered DC-3 attempted takeoff with a gust lock installed on the elevator surface and crashed, killing twenty-two. FAA Civil Aeromedical Institute expert Jim Simpson was drawn into the NTSB investigation and was startled to discover that the captain—who had a waiver from FAA medical requirements because he was too short to reach the aircraft's rudder pedals—had received the waiver from a designated flight examiner. In fact, it turned out to be the same examiner who had approved the captain of another ill-fated flight that had crashed, killing the Lynyrd Skynyrd music group. But what was most startling was the hostile reaction of Simpson's own superiors in the FAA when he attempted to press for information on the examiner, the program, and the actions of the GADO office involved. In short, he was ordered to drop it immediately.

Not all the airlines wanted to have anything to do with the program, but with the increasing unavailability of FAA inspectors, many have had no choice. So it is growing uncontrolled, expanding throughout the airline industry, and granting even airlines that are prone to get around the rules (and have yet to be caught) the right to field their own designated pilot examiners and check themselves—without the inconvenience of having a harried FAA man looking over their shoulders.

And increasingly, harried is exactly what the nation's ACIs are.[2]

Ah, but the official view is that deregulation is working, so therefore there can be no overloaded FAA work force or accelerated retirements as a result of deregulation. What's an FAA administrator to do when the overall boss (the president) wants cutbacks, and his people are sinking beneath the waves of increasing work loads?

In former FAA Administrator J. Lynn Helms's case, the solution seemed fairly simple: Deny the problem exists while you change the philosophy of the FAA to allow the airlines vast freedom from FAA oversight. Admit internally, by cutting staff, that the industry is on the honor system, and at the same time tell the public (and Congress) that the nation's airlines—*all* of them—can be trusted to fly safely without a constant FAA presence.

Helms, who left office at the beginning of 1984, went so far as to engage in a bureaucratic charade with Congressman Norman Mineta (a Democrat from California), claiming that the FAA had enough air carrier inspectors to do the job adequately, and that in fact it could reduce the number of inspectors by more than one-hundred because a new computer system called ASAS (Aviation Safety Analysis System) would enable the average air carrier inspector to decrease his work load by more than 20 percent.[3]

2. The FAA responds that its inspector force is not overloaded for the current level of surveillance and enforcement activity expected of it. But that strikes at the heart of the problem. The current level of expected surveillance is not anywhere near adequate. If the inspector force were required to do the minimum necessary to really discover what is happening among the more than 400 certificated air carriers in the country, far, far more than 532 inspectors (the number of air carrier inspectors assigned during the NATI inspection) will be required.
3. Helms's testimony before the House Subcommittee on Aviation of the Committee on Public Works and Transportation and the subcommittee's questions and concerns were aired in a hearing on November 10, 1983, over Reductions in the Federal Aviation Administration's Airline Inspector Work Force and the Implications for Airline Safety. Copies are available from the U.S. Government Printing Office, and make revealing reading—especially in light of the fact that the capabilities of the ASAS system were totally misrepresented.

As Helms announced those conclusions, raised eyebrows marked the faces of FAA inspectors from Miami to Washington State. There were fewer than six-hundred inspectors to cover the entire range of Part 121 airlines to begin with, and the number of airlines had nearly doubled since deregulation. The ASAS system, as the FAA people knew well, was a computer system, only then in the *planning* stages, which, if it worked at all, was solely for the purpose of increasing an inspector's ability to retrieve information from a central computer data bank in Oklahoma City. Instead of sending a letter asking for a pilot's record, an inspector could (theoretically) pull it up on a computer screen. It would reduce paperwork slightly, but ASAS would never replace inspectors, their personal visits to the airlines they inspect, or their en-route flight checks of crew members. It was yet another attempt to offer a mechanistic solution to a human problem.

Helms had never been told that ASAS would replace inspectors. What he actually had been told by his people was that ASAS would *improve the efficiency* of the average inspector by 20 percent—*not* permit a 20 percent reduction in inspectors (which was Helms's conclusion). In addition, ASAS was not scheduled to be on line and useful until late 1984.

Nevertheless, Lynn Helms used the system as justification for quietly reducing the inspector force by nearly 20 percent—until Congressman Mineta and his subcommittee staff caught him red-handed.[4]

Lynn Helms had also pushed for a seemingly technical change in the FAA rules known as Rule By Objective. There were certain elements of the industry—especially those knowledgeable about human factors and human-performance considerations in a deregulated system—who were aghast at the RBO proposal.

What it meant was that airlines would be able to write their own procedural rules in many areas, such as maintenance and training, as long as they achieved their end "objective," which would be the

---

4. In fact, the program has been an embarrassing disaster in terms of its promised benefits. As of early 1985 it had become apparent that the software package purchased to run the program could not do the job. The FAA people who had headed the project were reassigned, new people were brought in, and the determination made that in order to get it to work, the FAA would have to purchase new computers. The scandalous waste of money was matched only by the realization that Helms's statements had amounted to little more than wishful thinking.

As of late 1984, ASAS was becoming a very bad joke around the FAA, and inspectors were still having to send letters to Oklahoma City for pilot records.

only thing mandated by the FAA. In other words, as long as its pilots could pass the check rides administered by its own designated pilot examiners, a particular airline could make its own decisions about exactly what courses the pilots needed and how much classroom instruction or simulator instruction would be required.

RBO, therefore, could become a license to cheat, steal, and take every economically attractive shortcut the airline's management could morally stomach.

At its most ludicrous extreme, the RBO principle could be reduced to the objective of having no crashes. As long as a carrier met that objective, the FAA would be uninterested in how it achieved it, and how it operated![5]

Helms had said on numerous occasions that the FAA was not supposed to be "the cop on the beat," but that is precisely what the passengers flying on U.S. airlines *expect* it to be! If not the FAA, then who?

And that, of course, brings up the subject of federal spending, which simply comes down to a question of what the public wants in the way of airline safety. How safe is "safe"?

In the end, only the public can decide.

If the American electorate's demand for a safer and more stable system becomes sufficiently loud, aviation spending will have to be evaluated on its own merits, and not be lumped in with the remainder of the federal budget. There is a thundering herd of sacred cows stampeding through Washington at budget time, all vying for federal dollars, but the governmental functions that directly affect your chances of reaching your destination alive when you fly must not be lost in the dust and confusion of that herd.

And the long list of problems plaguing the FAA will not be solved without budget increases. For instance, the ACI force desperately needs to be doubled (or better), and that cannot be done without more funding. In addition, the course at the FAA Academy in Oklahoma City used to train new inspectors needs vast improvement since it is currently little more than a "charm" school on how to conduct relations with the inspected air carrier. That too takes more money—and time.[6]

---

5. Though the Rule By Objective proposal was later withdrawn (and in fact was more of a trial balloon), it represented the ultimate lack of understanding of the FAA's important deterrent role in the system.

6. Even the most qualified of graduates—former airline pilots with direct experience in the system—need nearly two years of on-the-job training and experience after they leave the school. Unfortunately, many of their highly experienced teachers—the veteran ACIs—may be retired before the majority can get such training.

Money, however, is not the only answer. Many of the agency's problems arise because it is far too slow to change its methods—and substantial change in methodology often can be achieved without substantial expense. The increasing FAA work load has led to hasty solutions and a weak patchwork of disorganized remedies. For instance, there are major philosophical problems entailed in upgrading experienced general aviation inspectors to air carrier inspector positions—but that process has been accelerated to cover the increasing retirements of airline ACIs.[7]

In some regions, the functions of GADOs and ACDOs have been combined into a confused mishmash called a Flight Standards District Office (or FSDO, pronounced "fizz-doh"). Miami, Florida, the location of some of the worst inspection nightmares in aviation, is one office that has been converted to a FSDO.

While there are some excellent "supplemental airlines" (another name for large charter airlines, operating under Part 121) flying the latest equipment, there is also a collection of supplemental airlines

7. Despite the fact that the GADO (General Aviation District Office) men may have been "watching" commuter carriers for years, they do not necessarily embody what is known as the "121 philosophy" of strict compliance and standardized procedures. The fact that a middle-sized commuter airline has been assigned to a GADO for years does not mean that the GADO has ever really known what was going on in the heart and soul of that carrier (such as Air Illinois). If the carrier then expands, purchases a larger airplane, and begins operating under Part 121 as well as Part 135, the GADO men may be completely out of their league in terms of the sophistication of the flight equipment and the methodology of surveillance. Yet all across the nation, GADO inspectors have for years been assigned to surveil and monitor the compliance of such small airlines even after they become large airlines straddling the line between Parts 135 and 121 (such as Air Illinois). Now they are being converted to full Part 121 inspectors without adequate schooling.

In one recent case, a GADO inspector of many years' experience (who had been upgraded at his request to an ACI in charge of a small Part 121 operation), was called on the carpet by his chief after he issued four type ratings to pilots whose performance he had not observed. The pilots had flown a multi-hour training flight, but the inspector had stayed in the cockpit only a few minutes and had watched little of the rest of the check rides. Instead of being apologetic, the man was outraged. "I've been in this business long enough to know what I need to see in a pilot, and once I see it, that's all I need. Those boys were qualified!" The inspector could not understand that the philosophy of airline operations under Part 121 requires that there be no significant deviations from the rules. The health, welfare, and lives of the public are at stake, and there is no justification for seat-of-the-pants evaluations or good-ole-boy attitudes. Thus reprimanded, the new ACI returned to his job—and did the same thing once again. He was promptly returned to a GADO office—where he resumed supervising Part 135 carriers!

and cargo airlines operating antiquated aircraft with crews who come from all points on the spectrum of qualification. Many of these justify close and constant surveillance to say the least—and many of them are based in the Miami area. The combination of this fact with the confusion that typifies FSDOs has resulted in chaos. The number of inspectors assigned to surveillance is always being reduced by other projects and concerns, and the division of authority and responsibility is in a constant state of flux.[8]

Many of the nation's supplemental carriers are top-quality organizations, which stay far above the legal minimums (such as Flying Tiger and Transamerica), but many others are not. Nevertheless, the deregulatory work load has robbed all of them, the good, the bad, and the noncompliant alike, of FAA supervision. When such carriers use older airliners, there is a greatly accelerated need for FAA attention to the maintenance and airworthiness of such planes.

The Lockheed Electra is a prime example. In the space of seven months in 1984–1985, three Electras (two all-cargo and one passenger) crashed in the United States for apparently different reasons, and a fourth made a partial gear-up landing at Dobbins Air Force Base, Georgia, with hazardous material aboard.

The Electra has always been an excellent aircraft (despite its state-of-the-art wing-engine-mount problems in the sixties, which were solved after two fatal crashes). The military version, the P-3 (the type of aircraft that flew to Rockland, Maine, from Brunswick to help locate Downeast's missing Otter), is still in production and is highly reliable.

But the Electra first came into service in the late fifties. The Electras flying cargo and charter passengers today are elderly airplanes. It took the best and the brightest of the major airlines'

8. The Miami FSDO and its predecessor Air Carrier District Office have always been rather tough assignments. The collection of old airliners in the Miami area is too often put back into service by the worst of the cargo carriers and Part 121 supplemental (charter) airlines. In fact, that aspect of the non-scheduled industry in the Florida area is known in the airline business alternately as "Corrosion Corner" or "Cockroach Corner." The horror stories involving some of these operations and aging airplanes, like the DC-6, DC-7, old DC-8s and aging Boeing 707s, are legion in Miami aviation circles. In past years, the job of trying to keep track of—let alone inspect and oversee—the many different operations that swing in and out of business is very difficult. Sometimes it has approached the dangerous and the bizarre—as in the case of the cargo-airline owner in Miami who was in the habit of pulling out a loaded .38-caliber revolver and placing it on his desk whenever his FAA air carrier inspector showed up to discuss operational inadequacies. The man never threatened the inspector verbally, but the presence of the gun was enough.

maintenance departments and training departments to operate the Electra safely in the fifties and sixties—and despite the fact that those air carriers no longer use them, the airplane has not become any less complicated. In fact, it has become *more* complicated, because it has become older and more subject to metal fatigue, corrosion, parts-procurement problems, and maintenance-support difficulties.

Even though there was apparently no common cause among the three Electra disasters of 1984 (one of which killed seventy people in Reno, Nevada), it is a fact that the FAA has not had time to focus enough attention on this one aircraft and the supplementals that operate it.[9]

Throughout all the problems of time and manpower plaguing the FAA runs the thread of political concerns and special interests imposed on decent people. These men and women have careers to protect, and they cannot protect them by being whistle-blowers or adamant firebrands who constantly breach the unwritten code that

9. Improvement of aircrew training for the Electra has been one area where the FAA in Washington has been dragging its feet. When pilots hired to fly such older aircraft have many thousands of hours in them, that experience can make up for training deficiencies. But when (as is often the case) new, younger, and more inexperienced pilots are hired for such a complex machine, extensive ground and flight training is required. Yet only a few of the carriers that operate the Electra have their own ground training, and none of them have a flight simulator. In fact, there is only one known, operable Electra simulator in the world, and that is located in Seattle, Washington, at an independent airline-training school called Simulator Training Incorporated. The Electra aircrews throughout the nation desperately need continuous training and flight checks in that simulator (which does not have motion or a visual system, but which was officially a simulator in the eyes of the FAA when it was owned by Eastern Airlines in Miami), but the FAA has delayed for over three years in granting approval for the simulator to be used for proficiency checks. Without that approval, aircrews sent at great cost to train in Seattle would still have to reaccomplish their flight checks in the actual airplane when they got home. While some excellent and dedicated airlines such as Reeve Aleutian Airways (Alaska) and Gulf Air Transport (Louisiana) spend the money anyway in the interest of crew training and safety, most of the other carriers are not in a financial position to do so. The problem is that there are many emergency and abnormal procedures that cannot be safely practiced in the actual aircraft in flight but can be practiced safely in a simulator. Therefore, a simple FAA approval, which could have materially enhanced crew training in the Electra, has been ignored as the FAA studies the issue to death! It is still another example of how badly overwhelmed the FAA has become in the tidal wave of deregulation: It has an ever lessening ability to respond to the safety needs of the industry in a timely manner.

an FAA employee does not air problems outside the "family" of the FAA.

The punches that are pulled in the name of "promoting" aviation, and the enforcement actions that are watered down (or never begun) out of fear of making waves and creating trouble for one's boss, are simple reality in a bureaucracy. But such concerns have traditionally governed far too many day-to-day decisions related to safety. It is human nature, but it is certainly detrimental to the overall goal of air safety.[10]

The tragedies of Downeast, Air Florida, and Air Illinois among many more should demonstrate how difficult it is for the FAA to know what really goes on at the heart of an airline—and its relative inability to do anything about certain problems, even if they are discovered. But if the point had not been driven home by the accidents in Maine and Illinois, the fall of 1984 would provide still one more graphic reiteration.

In 1983, Carlos Giammetti, a pilot for Provincetown-Boston Airlines, also known as PBA, was fired. Shortly afterward, he showed up at a nearby office of the FAA with a long litany of information about allegedly falsified check rides, training records, and rule breaking of major proportions. The FAA men told him that if he would sign a sworn statement, they would investigate.

He refused.

So did they.

PBA continued to operate as the nation's largest commuter airline (and one of the oldest) with separate route systems in New England and in the Florida market. Principally, it was a Part 135 carrier, but it had taken immediate advantage of deregulation and had become a Part 121 carrier as well, flying Japanese-built Nihon

10. One recently retired air carrier inspector with more than twenty years of experience said flatly that in that period of time, he had only one boss who did not interfere with any of his field findings. "You wouldn't believe how many times one of my [chiefs] would come back to me and say, 'If we put out this [comment on the airline's violation] and somebody finds it on the record, it could cause the carrier trouble. Can't you rewrite this [so as to be innocuous]?'" Now, it must be noted that the normal and desirable cooperative relationship between the FAA and the carrier may make such a conversation quite proper on occasion. The inspector should be helpful to—and somewhat protective of—"his" airline. But the FAA inspectors are ultimately responsible to the passenger and his or her safety—not the welfare of the airline. If such conversations have become the rule and not the exception throughout the FAA, something is definitely in need of repair.

YS-11 twin-turboprop airliners along with a fleet of aging DC-3s and Martin 404s on its 121 routes, and a mixed bag of smaller piston and turboprop aircraft on its 135 runs.

On November 29, 1983, the chairman and chief executive officer of PBA, John Van Arsdale, Jr. (one of two sons of the founder, John Van Arsdale, Sr.), set out from Hyannis, Massachusetts, with passengers bound for Naples, Florida, in a company Martin 404 (a twin-engine, piston-powered aircraft similar to the old Convair 340).

Van Arsdale held an airline-transport-pilot rating and was qualified to be captain of the flight, but what was to happen within the next few hours would violate at least the intent of Part 121 philosophy. The story of that flight (according to the FAA) provides a startling glimpse of the attitude of PBA's leader toward compliance with the rules:

In short, the Martin 404 lost its hydraulic fluid forty minutes out of Hyannis, an emergency situation that affected the brakes, nosewheel steering, landing gear, flaps, and part of the flight controls.

Yet Chairman Van Arsdale pressed on, overflying his intended fuel stop in Winston-Salem, North Carolina, to land at Jacksonville, Florida, where another PBA aircraft would be waiting. All the way down the eastern seaboard of the United States, with every mile, Van Arsdale was flagrantly violating the rules by not landing what was legally a crippled aircraft.

But that was just the beginning.

In Jacksonville, Van Arsdale transferred the passengers and their baggage to a turboprop YS-11, and got ready to fly it to Naples, on Florida's southwestern coast (PBA's southern headquarters). But Van Arsdale was not type-rated in the YS-11. In fact, Van Arsdale had not received a proficiency check for the YS-11, nor did he have any qualified operational experience in the aircraft.

In fact, John Van Arsdale had never been trained to fly the YS-11, yet he was going to fly it to Naples one way or another— with passengers! After all, it was his airline.

His copilot and employee, Mike Connolly, told his boss he would not be a part of such an operation. According to the FAA, Van Arsdale threatened to fire Connolly and then fly the aircraft by himself to Naples. With that, Connolly gave in—even though he was only qualified as a copilot on the YS-11, and not as a captain!

They departed with Van Arsdale in the copilot's seat and the copilot in the captain's seat, and flew to Naples—where the chairman (according to the FAA) had the name of John R. Wales entered in the official logbook of the aircraft as the pilot in command. John Wales was the assistant chief pilot of PBA, but he had been nowhere near the YS-11 as it flew from Jacksonville to Naples. The entry was totally false—but the truth would stay hidden from the FAA for more than ten months.

In September 1984, PBA suffered a fatal crash, and the FAA (which had supposedly inspected PBA thoroughly during the NATI inspection in the spring) began to pay attention. Both engines of a PBA Cessna 402 light twin had stopped as it lifted off from Naples Airport. The plane crashed into the bay, and one passenger died several days later of injuries. The NTSB investigation quickly found the cause: Jet fuel, instead of gasoline, had been pumped into the airplane by fuelers!

The crash was the last straw to Carlos Giammetti, the former PBA pilot who had originally gone to the FAA and been ignored. He decided to try again. This time he agreed to give the FAA a sworn statement—despite his worry that his aviation career might somehow end up sabotaged by the powerful Van Arsdale family.

And this time, at long last, the FAA investigated.

Months later, on November 10, 1984, the FAA suddenly issued an emergency order of revocation and ripped up PBA's certificates to operate as an airline. What its investigation had found went beyond the reports of the fired pilot and resulted in thirteen pages narrating serious violations in maintenance, training, flight following, maintenance monitoring, aircraft inspections, unauthorized personnel performing maintenance, use of untrained and unqualified maintenance inspectors, failure to sign off airworthiness releases, and violations in weight and balance, manifests, and equipment.[11]

11. The FAA also let it be known that a PBA with John Van Arsdale, Jr., in the management would not be a PBA recertified. Van Arsdale resigned, protesting that the FAA's action was "sudden and arbitrary," exhorting his employees and fellow Floridians to shower FAA headquarters with protests, and claiming that "[the complaint] basically involves five people who did something they weren't supposed to [do]. . . . For this [the FAA] shut down the whole airline!"

Van Arsdale's brother, Peter (believed by the FAA to be far more interested in compliance), took over. A team of nine air carrier inspectors was sent to work through Thanksgiving, 1984, to try to get PBA requalified. The task was enormous, since literally every step of a normal certification had to be done from scratch, and all the pilots, cabin attendants, and maintenance people rechecked. PBA returned limited Part 135 operations to the air about December

The flight sequence of Chairman Van Arsdale, which the FAA confirmed was flagrantly illegal, was chronicled in detail (and a separate letter sent to him revoking his pilot's license).

But the most startling element revealed by the FAA's charges was the allegation that the vice president of operations, the chief pilot, his assistant, and three other management pilots had been issuing each other check-ride approvals when no check rides had been flown! The FAA document reported: "Each proficiency check report described . . . contains a fraudulent or intentionally false statement that the pilot successfully completed the appropriate proficiency check when in fact the check was not given."

According to the FAA, the management pilots of PBA had been saving time and money by flying check rides with a sharp pencil in what the FAA characterized simply as "fraudulent or intentionally false statements." Since each of those management pilots was a company check pilot, each and every one of the check rides they had given to PBA line pilots became instantly invalid.

There was still more. The predictability of abuse in the Designated Pilot Examiner Program, a worry dismissed by the FAA as a "remote" possibility, was confirmed in at least one instance.

PBA Vice-President of Operations John Zate, who resigned almost immediately (the man Van Arsdale would later point to as having been the instigator of policies that he, Van Arsdale, claimed to know nothing about), had become a designated pilot examiner. On November 1, 1984, long after the FAA had completed its investigation of PBA but ten days before the ax was to fall, John Zate issued a type rating in the DC-3 to PBA pilot Connie Scharr, certifying that a flight check in the DC-3 had been satisfactorily completed.

There was only one problem. There had been no flight check, and there had been no flight! The type rating, the "check flight," and the certification were fraudulent.

There are FAA-certified designated pilot examiners all over the country who would never have done such a thing. But human nature being what it is, the fact that abuse had occurred once simply indicated that under the economic pressures governing the industry, it would occur again—and the FAA would probably know nothing about it until and unless someone stepped forward with enough courage to blow the whistle.

1, 1984. Within six weeks one of its twin-engine Brazilian-made Bandeirantes would crash on takeoff in Florida when the horizontal stabilizer on the tail fell off. All aboard would die in that crash.

And that highlights the central problem.

The violations chronicled in that thirteen-page FAA notification had developed over at least the previous two years. The pilot who had originally come forward had been ignored. Even when he came back to sign a sworn statement (after people had died in the Cessna 402 crash), it took the FAA four months to complete an investigation.

The FAA took great credit for having waded in like a stern parent and taken away the misbehaving progeny's right to operate as an airline (until it could hurriedly recertify itself). Certainly the shutdown prevented passengers from being exposed in the future to the conditions the FAA considered dangerous, but there was a major, forgotten fact: PBA had been flying passengers all during the period of those violations.[12]

The revocation notice contained a final justification, which said: "The Administrator has determined that safety in air commerce and the public interest require the revocation of [PBA's] air carrier operating certificate. The Administrator further finds that an emergency requiring immediate action exists in respect to safety in air commerce and, accordingly, this Order shall be effective immediately."

But if PBA was sufficiently unsafe on November 10, 1984, to require an immediate shutdown, it was unsafe in the months—and perhaps years—before the cited violations. Therefore passengers had been flying on PBA aircraft for a long time, mistakenly trusting that since the FAA allowed them to operate, they must be safe.

That, however (according to the FAA's belated discovery), was not true. Husbands and wives, children, and entire families had been exposed to the very same compromised safety standards that

12. The same was true for Miami-based supplemental Rich International, which had been shut down by the FAA for one month in April 1984, until it brought itself back up to legal minimums. A similar incident occurred in Honolulu later in the year to South Pacific Island Airways (following a near-violation of Soviet airspace by an off-course chartered jetliner). The FAA had inspected it during the NATI sweep and found problems, but until the off-course incident, it was allowed to keep flying passengers. Only after it became obvious to the press and the public that major problems existed did the embarrassed FAA act. This is a pattern that has been repeated numerous times. Though the agency may be lauded for having the courage finally to pull the plug on a violator, the fact that the airline has been allowed to operate unfettered for lengthy periods while such problems are festering undiscovered or unaddressed by the FAA is scandalous. In fact, no one will ever know how many such problems exist until the system and the philosophy of FAA inspection and surveillance are changed drastically.

prompted the FAA to rip up PBA's certificates in November. They had trusted the FAA, and that trust had been compromised.

Where, in fact, *was* the FAA? Where was the surveillance and the monitoring that should have discovered these violations? Why did it take two years, a fatal crash, and a sworn statement, rather than just a tip, to spark an investigation?

The reason is the structure of the FAA's surveillance and enforcement capabilities. It never was adequate to ferret out a problem operator in the first place, and it has been badly flawed by the immense pressure inherent in a deregulated airline industry.

And that is the point. There are no "black hats" in the FAA trying to ignore safety problems. They are mostly good and dedicated people, overloaded, burdened with a system gone nearly berserk in terms of expanding airlines and changing demands, and saddled with a methodology and philosophy of surveillance and enforcement that is incapable of finding the PBAs before they require shutdown (and before they end up killing passengers).

Downeast proved it, Air Florida re-proved it, Air Illinois validated it, and PBA should be a call to arms.

The system *must* be changed!

Chapter 27

THE MONKEY WRENCH
IN THE WORKS

Deregulation has taken the imperfect system of 1978 and subjected it to immense pressures. Some of the sacrifices laid on the altar of the god of economics include the fledlging efforts to instill human-factor and human-performance awareness, training programs, and preventive measures in every corner of the industry. These improvements had not yet been fully under way by 1978. By 1980 they were all but dead in the water for one simple reason: Such efforts cost money.

The siphoning of experienced pilots from Part 135 commuter carriers to Part 121 carriers, which began immediately after the passage of deregulation, has forced commuter-airline managers to replace their experienced people by hiring cadres of raw and inexperienced general-aviation pilots. The complaint of Operations Vice-President Ray Schrammel of Air Illinois is all too typical: "It's very difficult for me to tell some young lad that here's a Twin Otter, you can fly around in that for the next couple of years as a first officer and then we'll make you a captain on it. When TWA's out there saying, hey, come over here and I'll give you this nice DC-9. . . . Whenever they start a hiring program, boy, you're in trouble."

The more inexperienced pilots (whose commuter-airline job is their first exposure to the airline industry) may well be the best and the brightest, but they still need guidance, extensive training, and tempering. They may be eager and willing to work for low wages under difficult conditions, flying at lower altitudes usually in small propeller-driven airplanes through all kinds of weather (turbojets can fly over the majority of bad weather), and putting up with crushing work schedules because they plan to join a major airline later on, but they cannot keep their passengers as safe as pilots who have logged many thousands of hours under the strict professionalism and controls of an established Part 121 carrier.

Even if the newer pilots in the commuters were given the same caliber of training as their more experienced counterparts in the major carriers, they would still not have the maturity, balance, and airline experience—and this problem can transcend age. Recall forty-year-old Captain Dwight French of Downeast, who was certainly a mature pilot, but who needed the stability of an airline that tolerated no deviations from the rules, and which provided superlative training. He received neither—and he died in a crash.

The massive disruption and turnover in the ranks of commuter pilots since 1978 has exposed the raw replacement crews to still another problem. When the commuters—especially the small commuters—were able to fly only the small piston-powered twins such as the Piper Navajo and the Cessna 402, the technology was not beyond their previous prop experience. But with complex turboprop mini-airliners such as the Beech 99 and the de Havilland Twin Otter, the commuters began operating aircraft that were more complicated than many of the piston airliners operated by the major carriers in the late fifties!

Now the situation has become more worrisome. In order to fill their order books with American customers, foreign aircraft manufacturers have been busily designing and building small airliners specifically for the commuter (now called regional airline) market. Aircraft like the Brazilian Embraer Bandeirante (called Bandits for short), the de Havilland Dash 7 (a four-engine turboprop), the Swedish Saab 340 (a turboprop), the Nord, the Short Skyvan 330, and many others are flooding the U.S. market—many of them sold at nearly cost and offered with financing rates and terms that no U.S. company can match.

But there are safety problems as well with such foreign craft. The complicated flight characteristics and systems can far exceed the training and maintenance capabilities of the commuter/regional carrier (recall Downeast and the Twin Otter, and Air Illinois with its Hawker 748).

The former commuter pilots gleefully going to work for major Part 121 airlines may find superlative training, a top-quality atmosphere of compliance with the rules, and a built-in maturing process that assures that before that pilot takes command as a captain, he or she will be well proven and well tempered.

And then again, they may not.

There are Part 121 carriers out there who will take fairly inexperienced pilots and rush them through the legal minimum requirements, never grasping the concept that perhaps their public license to operate as an airline mandates a higher level of care.

Of course, the majority of U.S. carriers operating under Part 121 as well as Part 135 *are* highly responsible. They tend to be excellent companies whose managers tolerate no rule bending and whose flight departments give their people the proper training, support, and supervision, and whose maintenance departments are aboveboard and above the minimums. Nothing sinister or cynical lurks in such corporate hearts. These carriers—the backbone of the system—would not knowingly reduce their level of safety—but that's just the problem. What such companies would never knowingly do, they may inadvertently do under the severe economic pressures of a free-market environment.

The cost-accountability pressure affecting virtually all of them creates a constant flow of decisions on every facet of the operation. Even the major carriers—the giants—must decide continuously what is really needed, and what is not: Perhaps the flight crews need only eighteen hours of flight-simulator time for a particular upgrade course. Perhaps the maintenance department doesn't really need to replace a particular part every four-hundred hours of service if the FAA maximum "pull" time is six-hundred hours. Perhaps the two-week ground school for the maintenance shop can be shortened slightly without any direct reduction of quality. Maybe the pilots can shorten their layover time or lengthen their duty days, since these are far below the FAA maximums.

In many cases the cost pressures lead airlines to ask for a relaxation of the FAA standards, or the granting of an exemption.[1]

1. The FAA has long had a rule regarding the maximum distance that a twin-engine airliner can fly offshore and away from a suitable landing field while over water. That rule was always sixty minutes (based on flying on one engine to the nearest emergency field) and it was derived from calculations of emergency capabilities of piston airliners decades ago. The European carriers have promulgated a ninety-minute rule (based on ninety minutes of flying with both engines operating), which allows them to operate the Airbus, the Boeing 767,

On and on the decisions are made—none of them alone creating an unsafe situation or compromising overall safety. However, over a long enough period of time, this trend, in the name of economy, becomes a process of erosion, leaching away at the sturdy foundation of airline safety built up so laboriously over such a long time. Bit by bit the standards decline in even the most safety conscious of companies.

The airlines that are most affected will deny it, of course, be-

---

and the Boeing 757 over the North Atlantic, even though such airliners have only two engines. U.S. airlines that fly these airplanes have put great pressure on the FAA to change the rule so as not to be at a competitive disadvantage. Several changes were required in the airplane by the FAA before they gave in (changes such as the addition of a hydraulic-powered emergency electrical generator and greater fire-extinguishing capability in the baggage compartments), but indeed, in the end, the pressure seems to have worked: The rule is in the process of being liberalized to permit a two-hour offshore distance from the nearest suitable emergency airport (based on the flight time with one engine operating). TWA has become the first U.S. carrier to begin North Atlantic service with the Boeing 767, and is eager for the new rule. The problem is that the definition of what is, and is not, safe in twin-engine overwater operations was made primarily by economic forces. Among North American operators of the 767, only Air Canada has conducted extensive simulator tests to prove its feasibility. Despite the honest misgivings of at least a few in the ranks of the FAA, the pressure from the airlines and from Boeing has been too great to resist. Time will tell whether the decision was right.

Similarly, two other major airlines began petitioning the FAA in 1983 to let them operate the same two aircraft with one of the passenger exit doors blocked off and inoperative. The 757 and 767 each have eight passenger entry/exit doors, four on each side. Normally, passengers board through only one, and due to the location of airport jetways, that is usually the front left door just behind the cockpit. But the aircraft were certified based on the amount of time it would take to successfully evacuate all the passengers on a full airplane using all eight doors (though they must demonstrate an emergency evacuation with a percentage of the existing doors blocked off). In normal operation, the airlines using the 757 and 767 have been having personnel problems: The flight attendants had been forgetting to disarm the automatically inflated escape slides on each door before opening the handle. The result was a rash of inflated slides, which take hours to replace or repackage. When that happens, the airline has to delay or cancel the flight. Numerous FAA men were dead set against granting an exemption for an inoperative escape slide/door combination, knowing that in an emergency evacuation under dangerous conditions (such as smoke or fire), passengers will head either for the door they came in or for the last one they recall seeing. The fact that such a door has been labeled "inoperative" is of no significance in a dire emergency. Nevertheless, because of high-level pressure in Washington, the FAA may end up granting the exemption anyway. If so, economic pressure, not safety considerations, will have been the motive force.

cause most of the leaders of such companies cannot see the overall effect (even if they could, it would be dangerous to sales to admit it). Reductions in the margin of safety are always made in the name of efficiency—never in the name of expediency. Yet the process is one of defining what is "safe," and how much "safety" is required, if the legal minimum is not enough. The decision, however, is not the carriers' to make. It is the passenger's decision—the public's decision—your decision.

Yet deregulation has, by promoting this insidious process in even the best of airlines, large and small, given the decision to managers and executives who have to keep their companies economically viable. Since each company will assume it has a right to exist, it will correspondingly assume that whatever cuts must be made to survive are justified. Seldom do the officials of the airline make the decision—write the definition—of what is "safe." They let the economics of their situation determine it.

And, of course, only the FAA (and the FARs it is required to enforce) stand in the way of that definition's sinking to clearly unsafe levels.

The massive, rapid expansion of many carriers—new and established—has caused a wide variety of pressures and abnormal situations. Air Florida and Air Illinois were merely two examples.

A carrier in a very great hurry to expand needs aircrew members as fast as possible. Seldom can such an airline afford to hire and train such people far in advance—the fast-paced changes in a free-market environment would doom it to be outflanked by the competition.

Therefore, pilots and cabin attendants are hired with great urgency at the last minute, and trained at the same pace—often in an environment of rushing and corner cutting designed to get them qualified and on the line as fast as possible by meeting the legal standards and not a bit more. This is not always the case, but far too often there is no time or money for additional training. There is no time or money for extensive indoctrination. All the "fine points" will be given to such people later on, "when things have settled down." Besides, they will have plenty of opportunity for "on-the-job training" (as Downeast's owner, Robert Stenger, used to remind his pilots when they asked for more flight training).[2]

2. The recent attempt of one purely deregulatory carrier to fly its inaugural flight on a new route in a Boeing 747 (the largest commercial airplane in the world) with three pilots who had done all their qualification and training in

363

If the pilots who are processed in such a manner are recycled airmen from bankrupt carriers who have logged many hours and developed substantial Part 121 experience, there is a reduced chance of trouble. But in many cases such rapid expansion pulls in at least a few (if not all) raw recruits; and sometimes undesirable individuals not necessarily fit to be airline pilots slip through the net—including individuals who have been fired from other carriers with good cause.

Hurried, minimal training for pilots, flight attendants, mechanics, and all the others who handle flights, provides the absolute antithesis of the type of preparation the major airlines provided before 1978.

But what can the FAA do? If the training meets the legal minimums, it has no authority under current regulations to compel the carrier to slow down because the *quality* of its product has been lessened. Recall the inability of the FAA (as enunciated by FAA chief John Roach testifying in the Downeast hearing) even to compel an airline management to "stop harassing its pilots."

With little to restrain the high-speed training except the minimal compliance rules, such training can end up at times becoming a variation of the old joke: Yesterday I couldn't even spell "airline professional," and now I are one!

---

what are called phase-three flight simulators (and had never flown the airplane itself), is a prime example of such lack of understanding. That crew and that airline were legal—but they were so frighteningly close to the absolute bottom in terms of qualification and knowledge that they could not find an FAA check pilot to fly with them. Because it was a first-line flight for all three (the captain, copilot, and flight engineer), they had to have what is known as a line check—a check ride administered by the FAA (or a designated examiner). Thus, either the FAA went along (since the airline had no qualified designated examiner at that time), or the flight could not be dispatched. But nowhere on the East Coast could the airline find an FAA man willing to be a part of such an operation. Finally, after much wrangling, the airline gave in to the request of its FAA principal operations inspector and flew a four-hour local training flight in the airplane so the FAA man could assure himself the three pilots were capable of handling the actual airplane. The phase-three-simulator program was never designed to be used in such a fashion. It was meant to be of assistance to stable airlines, which could use the extremely advanced, visual-simulation-equipped, six-axis flight simulators to fully qualify a pilot who would, in the vast majority of cases, be flying his first flight in the airplane on a passenger-carrying trip—but with at least one other pilot who was already experienced in the actual airplane. Even if all the crew members were to take their first flight in the aircraft in command of a revenue flight, they would be pilots of many years' experience and maturity who had simply upgraded (or cross-trained) to the new type of airliner. In the new-entry carriers, the experience level of such pilots—as with the 747 incident described here—can be minimal.

It is not just the new-entry carriers that are subject to reaping serious problems from intense expansion. No less an airline than fifty-year-old Braniff International—running toward the horizon with every new deregulated route it could grab in late 1978 and 1979—put an unbelievable strain on its training system and its operational expertise. Pilots increased their hours to an exhausting ninety hours (flight hours—not duty hours) per month for four months while no fewer than three-hundred pilots were hired and put through high-speed training. Flight attendants were hired with great urgency—some were literally hired over the phone—and their classes were trained in half the normal time with half the normal substance. In addition, the flight-crew members who were already employed found themselves moving to captain (in some cases directly from flight engineer), and to copilot in record time. It was the quality of the personnel already there and the quality of the pilots hired that kept Braniff from having a major accident during that period—but there were many opportunities for disaster.[3]

Rapid expansion also breeds the type of terrible record keeping former FAA ACI Al Koleno found at Air Florida: the confused procedures, the incomplete manuals, the confused or marginal maintenance, the clandestinely authorized shortcuts done to "help

3. On one occasion a copilot had to take control of a Braniff 747 from his captain (with the captain's grateful acquiescence) as they were maneuvering toward the runway in Hong Kong, threatening several buildings in the process (the Hong Kong approach is rather demanding and involves a low flight approach over the city). The captain had been rushed through upgrade training as fast as possible after more than ten years of comfortable flying in the much smaller Boeing 727. Suddenly he was an international heavy-jet captain, and it was clear he had needed more training time in reaching that goal.

In another instance, a younger Braniff captain penned a two-page open letter to his compatriots and left it on the bulletin board in the DFW airport crew room. The captain was normally a copilot who had been trained as a reserve captain. Reserve captains and reserve copilots could be upgraded with a single telephone call to fly a trip in their reserve position, when needed by the airline's crew schedulers. The letter, in essence, spoke of a late night flight in terrible winter weather at the end of an exhausting ninety-flight-hour month in which he was in command as a reserve captain who had not flown in the position for months. He was legal, but in his opinion, marginal, for what he had been asked to handle that night. In addition, he said, his copilot was a reserve as well who normally flew as flight engineer. And to top it off, the flight engineer had been hired three weeks before and was on his very first operational line flight with the company. It was, the captain wrote, "a quintessential case of the blind leading the blind. How we kept from having a major accident and killing ourselves from a stupid mistake I will never know! This kind of thing is ridiculous!"

the company get through the crunch," and every other permuta-
tion of human ingenuity in the face of pressure to get the mission
accomplished while damning the torpedoes. It may be an admirable American trait, but it has no place in commercial-airline operations, whether that attitude exists in the cockpit, or on the ground
with a vice-president of operations who, without thinking, barks at
a subordinate: "I don't give a damn *how* you do it, just do it!"

Under deregulation, the mantle of Part 121 operations, and the
concurrent fiduciary obligation to the public, has fallen on many
shoulders that are unfit, untrained, or unwilling to wear it properly. Sometimes a Part 135 commuter carrier makes the transition,
taking advantage of its "birthright" to deregulatory expansion
granted by Congress, and is unable to rise to the challenge. It may
be legal in all it's doing, but that does not guarantee an acceptable
level of safety.

Such carriers can be run by a fine group of executives who do
not know that there is a weak-link manager in the chain of command between the executive suite and the maintenance hangar.
Those same executives may be thunderstruck to learn later of massive illicit violations propagated right under their noses by well-
meaning, but misguided, loyalists out to get the job done for the
company. Too often they have been too busy expanding to pay
attention to what was going on in the ranks and on the operational
front lines.

The FAA's contention that the carriers bear the primary respon-
sibilty for staying safe, as well as legal, comes from the days of
regulated stability. It is a habit pattern unrevised by current realities. What happens, though (as in the case of Air Illinois), when
the company fails to uphold that responsibility? Then, in the view
of the vast majority of airline passengers, it is up to the FAA. The
ball is in its court, because aside from hairsplitting debates over
whose responsibility it should be, ultimately the passenger depends
on the FAA to keep the system (and his particular flight) safe.

However, just as with Air Illinois, the FAA may have no idea
what is really happening. The airline, whatever its size, may be out
of control on the operational level.

And make no mistake about it: An air-taxi outfit with a Part 135
certificate and an airplane that flies a regular schedule over a regular route is an airline—especially since deregulation has made
Part 135 carriers an integral part of the airline system.

It is far too common in this day and time to find yourself standing in a major airport (having just alighted from a Boeing 727 or a
Douglas DC-9 belonging to a major carrier), facing a tiny aircraft

flown by an airline with which you are completely unfamiliar. The reservationist, who sold you the connecting ticket that included this, never warned you that the last leg of your trip was on an Air Generic Piper Navajo (or another type of small aircraft).

Now, it may be that Air Generic is one of the majority of commuter carriers whose dedication to safety is second to none. It may be the type of commuter that will spend its last dollar to send its people to the best contract training courses (at other airlines or training operations), and that pressures its pilots to follow the rules religiously. Unfortunately, however, such a fine little airline is still a commuter/regional if it operates under a Part 135 certificate, and the rules and equipment realities for Part 135 carriers are simply not as stringent as those for Part 121—but they should be! The FAA has done much to upgrade them in the past five years, but to point to the historical record (in the same manner as the DOT does for 121 carriers), the accident statistics show that traveling on a Part 135 commuter/regional has been, in the recent past, two to three times as dangerous as flying on a Part 121 airline. Though the Part 135 industry has done much to improve since Downeast, the problems still exist and the crashes still occur. Since there is a constant turnover in the smaller Part 135 carriers, there is no reason to believe the situation will change appreciably in the near future.[4]

Yet these are the airlines that Congress anointed as the backbone of air service to the nation's smaller communities. Replacement of a Part 121 carrier—especially a well-established major carrier—by a Part 135 carrier that may have just come into being, is usually, on its face, a direct contradiction to the promise Congress made never to permit a decline in air safety at the hands of deregulation. Even the government's own statistics show it has allowed the safety level for such communities to drop by the very nature of the system.

4. The author does not want to leave the impression that Part 135 carriers (whether called commuters or regionals) are, by definition, not capable of being as safe as Part 121 carriers. There are many Part 135 airlines in the nation whose standards, and adherence to the principle of safety-before-profit, are beyond reproach. (In some cases, they exceed the levels of certain Part 121 carriers.) Their efforts have cost them untold millions of dollars over the expenditures of less-scrupulous counterparts, and that faithful carriage of the public trust should not be diminished by the acts and omissions of lesser 135 carriers. Nor should the excellent efforts of such smaller airlines be discounted simply because it is more difficult to put an airline with a smaller revenue base on a par in safety compliance, maintenance, and training with the best major 121 airlines. The fact that it is difficult does not mean that it cannot—and is not—being accomplished.

The ticket agent of the major carrier who sells you the connecting ticket that includes a leg on a commuter/regional airline never warns you about such things. (The commuter airline may even carry the name of that major carrier and operate as a feeder extension, but it is still a Part 135 carrier.) They never warn you because they simply assumed you knew. After all, it was your Congress that passed the bill that made all this possible!

Even if Air Generic turns out to be a new Part 121 carrier, flying under a Part 121 certificate, that is no longer a blanket assurance that it can provide transportation with the same level of safety that you could have expected in 1978.

Now, that does *not* mean by any stretch of the imagination that airlines that have come into being since deregulation are necessarily less safe than (for instance) United or Delta. In fact, all of them have met the technical minimums in every phase of their operation in the eyes of the FAA. The majority of those new (or greatly expanded) airlines operating under a Part 121 certificate are excellent operations, which have expended staggering sums of money to put together a thoroughly professional and safe airline. They may not be equal to the average Part 121 carrier of 1978 because of economic pressures, but as a group, they are decidedly not below the legal minimum definition of "safe" (though there are a few operators among them who do not have the dedication to safety that you have the right to expect).

The problem with the new-entry (or greatly expanded) Part 121 carriers, such as our mythical Air Generic, is their built-in *potential* for safety problems under deregulation. In some cases, such carriers hire only experienced pilots coming from other defunct Part 121 airlines (such as the highly experienced Braniff and furloughed Flying Tiger pilots who went to work as the backbone of Pacific East's Los Angeles to Hawaii operation in 1982). In too many cases, however, the pilot ranks of such carriers are sprinkled with (if not filled with) men and women who have a very low experience level (such as Captain Larry Wheaton of Air Florida)—many of whom are alumni of the commuter industry. If the airline has limited training capabilities (legal but limited) and the pressures of time and finances in an expansionary push are having an effect, that dearth of experience is neither the safest method of operating, nor is it what you, the passenger, think you're getting when you climb aboard with that $99 ticket.

Those men and women up front (indeed all throughout the operation) may be highly motivated, but it is also highly likely that they have had little or no exposure to crew-concept training, to assertiveness training, or to a formal course on their fiduciary and professional responsibilities as airline pilots and cabin attendants. If the major carriers, for the most part, do not have the time and money for such training, you can be assured the newer carriers will not.

The operation may be run by decent people of the highest integrity, but they may be abysmally ignorant of the nuts and bolts realities of running an airline. If such corporate leaders hire high-quality airline people who do have such knowledge, and install them as the managers of their operational side with sufficient authority to take the steps and spend the money necessary for the highest level of safety (without constant nitpicking from the executive office over every expenditure), their operation may be a model of excellence (and numerous examples out there are just that). But if the money managers attempt to run the airline operations themselves through sycophants who simply occupy the position to satisfy the FAA requirements (and who take most of their orders directly from the top), you have a potentially dangerous situation. Their ignorance of what elements are vitally important for safety will usually result in a substandard operation based entirely on economic efficiency. The definition of what is safe is answered in dollars and cents instead of experience and sense. If airline managers are not dedicated to safe operations first and foremost, the end result may not be an unsafe operation per se, but it will be decidedly *less* safe than the public has a right to expect.

In fact, new-entry carriers often do not have maintenance departments as such, but hire other maintenance facilities (usually from other airlines) on a contract basis. Sometimes literally all of the operation is done by other companies. Under deregulation, an airline that leases almost all services and equipment (and sometimes even aircrews) can get into operation within a short period of time. The training their people receive and the maintenance their leased airplanes are given may be vastly superior to what they could do themselves, but their coordination and in-house dedication to safety are open to question.

Such an operation is usually designed to make money. There's nothing wrong with that. But in the opportunity for such enterprising people to get into the airline business and make (or lose) millions, there is nothing to motivate the new-entry companies to spend tens of millions of dollars in nonrequired safety programs, as the major carriers have done for so long in a regulated environment.

The fact that such an operation cannot be counted on to contribute to the overall efforts toward airline safety brings into question the reason for its existence in the first place, and whether it is really in the national interest.

Recall, if you will, that several well-established carriers have gone bankrupt fighting the deregulated wars because they could not become efficient fast enough to stay up with new-entry carriers, which have been busily filling the sky with half-empty seats. According to the theory behind deregulation, if an airline can't learn to be efficient in a free-market free-for-all, it will not survive. Its death is an evolutionary thing, and should not be mourned.

Braniff (and to a lesser extent Continental) could not learn to play the game in their prebankruptcy incarnations, and bankrupted themselves. (Deregulation only gave them the opportunity to get in trouble—but without that opportunity their futures undoubtedly would have been brighter.) But should they have been dismissed so blithely?

The truth is that the contributions to air safety of a prebankruptcy Braniff or a prebankruptcy Continental have been voluminous over the years. Both ran well-established and practiced facilities that trained not only their people but the aircrews of many other carriers as well, under contract. Both had spent decades of time and uncalculated amounts of money sending their technical people and executives to industry meetings and safety seminars, participating and contributing to the give and take of safety information, procedures, ideas, and awareness that had served to build air safety into a common goal of all air carriers. No matter how bitter the competition between carriers, they could always (before deregulation) converse on matters contributing to air safety—and they did.

What of the mid-eighties?

Quite simply, that cooperation, if not dead, has been severely strained. Who has the time or money to spend on such meetings when the bottom line is red, the unions are being asked for concessions, the creditors are howling at the door for their money, new sources of credit and revenue have dried up, and yet another new-entry carrier is gearing up to skim off your best markets? Even new-entry carriers that have gone belly up have been replaced by more new-entry carriers, in what appears to be a potentially never-ending cycle (though venture capital for new-entry airlines has begun to dry up with the increasing number of failures).

There *was* value in the expertise of an established air carrier, and there was value in the industry's previous willingness to communicate and participate in studies that, to any nonairline finance man, would seem cost ineffective. But in the face of the national stampede to deregulate everything that moves, those values got trampled. The baby, it seems, sailed out the window with the bath water!

And there is a still broader issue, also inadequately considered by Congress in 1978, that affects the manufacturers of airliners in the United States and their tens of thousands of suppliers. Here we are in a continuing foreign-trade-deficit crisis in the United States ($123.3 billion deficit in 1984), and the influx of foreign-built airliners, including the excellent Airbus A-300 and A-310 and a host of aircraft aimed at the commuter market from Europe, Sweden, Brazil, Japan, and elsewhere, is threatening the sales of Boeing and Douglas jetliners. Lockheed has left the business of building commercial aircraft, and Douglas is building only the DC-9 derivative line, while Boeing has gone through some anxious moments trying to sell its 757 and 767 worldwide.

Part of the problem is battling the cut-rate prices of foreign jetliners. The other problem, however, is the inescapable fact that many of the new-entry carriers and newly expanded carriers are recyclers of old (sometimes tired) airplanes. When a cost-conscious new-entry carrier can lease an old 727 (or purchase one for $5 million) and use it to carry 145 passengers, it doesn't make much sense to buy a new Douglas Dash-80 for over $20 million to carry 165 passengers. Even the higher fuel bills for the 727 will be insignificant against the debt service that would be required for a new aircraft. (The recent drop in jet-fuel prices has made this logic more potent.) If our airline manufacturers are to have their horizons reduced, what happens to their extensive body of expertise and their previous world leadership in quality as well as sales? Boeing and Douglas, after all, taught the world to fly jetliners. But if the competition abroad (and at home) is too stiff, and our own market is too busy recycling old aircraft, buying diminutive versions of new aircraft, and acquiring foreign planes, some of that U.S. aerospace know-how will be imperiled.

And on the subject of experience, there is a body of thought that the expertise built up over the decades among the nation's airline employees—mechanics, pilots, schedulers, ground handlers, operations managers, and a hundred other positions—is worth something more than passing reference. Even if all those skills are easily replaced by newly trained individuals (and there is virtually no evidence for such a theory—in fact, quite the contrary in many positions), was the goal of lower ticket prices enough to justify the massive damage that has been visited on the employees of the airline industry?

The "theory" includes the claim that employees who lose their careers (if not their airlines) will be reabsorbed by the new carriers that take their places. Labor could not stomach that provision, so Congress in the deregulation bill tried to mollify labor and gain its support by putting in a latticework of benefits for those airline employees (with four years seniority in 1978) who might lose their jobs. Those provisions, however, were a sham.[5]

In many cases, only the presence of such dispossessed, experienced airline people floating around the airline job market has enabled new-entry carriers to get in the air with an acceptable safety cushion above the minimums. But from a human point of view, it should be obvious that mere reemployment in the industry does not even begin to compensate for the anguish (and the financial losses), the dislocations of families, the loss of seniority, and the dilemma of those men and women in their fifties who often are left unemployable.

There has been a wealth of experience and ability lost forever to commercial flying, and that experience has been replaced by recruits who must go through the same processes (and make many of the same mistakes—some of them fatal) to re-create the stability of the system we once had.

For those who do get a replacement airline job (often at a fraction of their former salaries), the future may hold more of the same. More than a few of those dumped on the job market by an airline bankruptcy have become two-time losers when their new employers have also gone belly up. How can we take national pride in such human tragedy as we clutch our cut-rate tickets (and pay extra for checking our bags)?

It's not just a matter of empathy, it's primarily a matter of safety. Certainly the passengers in Air Florida's Palm 90 would have done well to trade their lower fare for a more experienced

5. The "benefits," which were to accrue only to those lucky enough to have been hired before 1975, were subjected to the worst, most scandalous case of bureaucratic foot dragging in recent history. The administration did not want to spend the money, and it managed to avoid the issue for six years, until a federal court ruled the provision unconstitutional.

captain, as would the passengers of Air Illinois Flight 710 (who would have been better off if their captain had been given time off without pay—rather than encouragement—for his shenanigans in the past).

The airline industry built a safety record by achieving a delicate balance among wildly disparate interests. It was not perfect, it did not always work, but what we have now is far more unbalanced and unstable and unpredictable than what came before. How that amounts to an improvement in the name of cut-rate fares is difficult to see.

This is not to say that deregulation is bad or unworkable. This is to say that the Congress of the United States failed to think out all aspects of the problem in its hell-for-leather rush to create a free-market environment (and break labor's grip on the industry). Some of the problems become nothing more than sad or embarrassing footnotes. But when it comes to violating a solemn promise to the public by enacting a provision that would, on its face, guarantee a reduction in air safety, Congress committed a serious error—and that error must be corrected.

In nearly every municipality in the nation, the "traffic-light principle" seems to operate with disturbing regularity. To get a recalcitrant city to install a traffic light at a dangerous intersection, at least one person has to die there.

It took the deaths of 128 in the Grand Canyon crash of 1956 and more than 100 other deaths in subsequent midair collisions to get Congress to act at that time. But in 1982, not even the galvanizing deaths of 71 people in the Potomac River could motivate a reexamination of the airline system and of deregulation.

It's easy to forget if you were not involved—if you lost no friends or loved ones—if you don't have to suffer from the results of debilitating injuries for the rest of your life.

On January 13, 1983, one year to the day after Palm 90 failed to achieve sustained flight, some unknown soul in the Washington, D.C., area apparently decided that his fellow Washingtonians should not forget the significance of that tragic anniversary.

Dr. Janice Stoklosa, one of the human-factors investigators of the NTSB who had worked on the Air Florida accident, had rolled onto the Fourteenth Street bridge on her way into the city before she realized the significance of the hand-lettered posters attached in sequence to each of the light standards all the way across. As the lettering on the signs became clear—there at the very point of Palm 90's impact with the roadway—it sent cold chills down her back. On each sign were the now-familiar words from the voice recorder.

"We're stalling!"

"Larry, . . ."

". . . we're going down . . ."

"Larry!?"

"I know it!"

It was philosopher George Santayana who said: "Those who cannot remember the past are condemned to repeat it." Yet we never seem to learn—even from our own recent history.

How many bodies will it take this time?

Chapter 28

THE ULTIMATE VICTIM

It is entirely up to you!

That is, in a nutshell, the agony and the ecstasy of a free society. If the system needs changing, only an informed and complaining electorate will get it changed.

The level of safety in commercial-airline operations you have come to depend on no longer exists. It has diminished significantly.

The questions each citizen needs to ask and answer are quite simple: Is the current decreasing safety level in the airline industry acceptable to you? Are the massive changes and disruptions caused by deregulation worth the only apparent benefit: the cut-rate air fare? How much safety should you demand when you trust your life (or that of your family) to an airline?

There are those in President Ronald Reagan's administration who have seen the problems with greater clarity than they care to admit publicly, and who are trying to change the outcome.

The Secretary of Transportation, Elizabeth Hanford Dole, was the one who ordered the NATI inspection, and the one who has kept close tabs not only on congressional worries about the system but on the worries of the industry. She has done much to attempt to alter the course that was set, in part under her administration, by former FAA Administrator Lynn Helms (who spent too much time hacking away at budgets and personnel with a meat-ax, when he should have been listening to his subordinates).

In early 1984 as Helms left the FAA, Secretary Dole forwarded an excellent recommendation to the White House that retired navy admiral (and veteran pilot) Donald Engen, then a member of the National Transportation Safety Board, be nominated to the position. The president agreed, as did the Senate, and Admiral Engen became FAA administrator—one of the first to bring with him an in-depth understanding of the inseparable and vital role of human-factors and human-performance programs to the safety of transportation in general, and aviation in particular.

But Admiral Engen cannot do it all himself. Nor can Secretary Dole, or Congressman Norman Mineta. Engen alone is merely the policy leader of a massive organization of (on the average) highly intelligent men and women well versed in the fine art of living and surviving in a bureaucracy. For the administrator—as for the head of any government bureaucracy—to order is not necessarily to accomplish.[1]

But with all the problems in the system, what should be done? The reality is that there are no quick and easy answers that will

1. Some time after former Treasury Secretary William Simon left the administration of Gerald Ford, he recalled that during his pregovernment years as a senior corporate executive who traveled abroad quite a bit, he had been constantly incensed by the boorish and unfriendly attitude of customs agents to their own American citizens—something he observed every time he returned home. Simon determined during those years that if he ever joined the government in a position of power over the Customs Service, he would clean house with a vengeance. Sure enough, he said, as secretary of the treasury he got his chance (since the treasury secretary is also in command of the U.S.Customs Service). But much to his surprise, nothing he tried seemed to work. Though he was totally in command, he could not effect a change. His orders and memos were constantly deflected, niggled at, or buried in bureaucratic maneuvering, and he began to feel as if he were punching a marshmallow. Finally realizing that he was spending an inordinate amount of his time as secretary on an impossible quest, he gave up. The U.S. Customs Service remains pretty much the same today, with pockets of brilliance and pockets of withering arrogance-without-portfolio. The lesson is applicable to much of government. To be the head of an organization does not necessarily mean one can significantly change the organization in a short period of time—if ever. Bureaucracies tend to develop a life and methodology of their own.

return the system to even its 1978 level. Commercial aviation has become far too scrambled. Therefore, any simplistic call for complete across-the-board re-regulation of the industry and re-creation of the Civil Aeronautics Board would accomplish little—if anything—for air safety. At best it would freeze the current deteriorating state of affairs for the next five years; at worst it would be a suggestion soundly defeated as contrary to our noble national policy of minimizing government interference in the affairs of citizens and business alike.

And this touches one of the central misconceptions of free-market competition as applied to the airline system. The nation has been too preoccupied with purist theories taken to extremes. One side wants complete and total deregulation and minimal government interference in airline-safety decisions. To them, cutbacks in FAA air carrier inspectors and adoption of Rule By Objective proposals are the proper order of things. Let the airlines determine their own course. If they kill too many passengers, then the force of economics will discipline them, and mandate its own level of safety compliance.

That approach is dandy for a theorist, but if you happen to have the misfortune of being one of the passengers whose death is necessary to help that airline seek its proper economically dictated level of safety, the knowledge that you have helped free-market forces reign supreme will be of little solace to you in the final seconds before impact. Nor are such considerations likely to comfort your spouse or your children.

At the other extreme are those folks who want a total return to those thrilling days of yesteryear when an airline's leaders had the certain guarantee of being, in effect, wards of the state. No matter how badly they screwed things up, they could always get the government to help out by passing on costs through higher fares and by bailing them out of competitive problems. Of course, hand in hand with such a system went the reality that to start a new airline (or even get a new route) required an exhaustive and ruinously expensive multiyear trudge through the labyrinth of byzantine bureaucracy at its worst. Even safety considerations can suffer when a business-as-usual laziness infects an industry to this extent.

The art of compromise governs relationships in Congress, business, law, and throughout our society. Why is it that there has been no attempt at compromise where airline regulation versus deregulation is concerned? To utter a word against deregulation in industry or in government in the mid-eighties is to adopt a singularly heretical stance. Yet a middle ground is exactly what is needed before the deterioration of air safety can be addressed.

Air safety and free-market competition cannot coexist in a void. Given decontrol of both economics and safety, there will be little of either. There must be controls on safety, and those controls are not by nature a violation of free-market philosophies.

Basically Congress said that it would deregulate economic factors in the airline business, but would *not* deregulate safety. Yet by failing miserably to address the question of what deregulation would do to the system and by failing to beef up the FAA with an understanding that the FAA's surveillance and enforcement would have to increase in almost direct proportion to the degree of economic freedom, Congress virtually guaranteed that safety would be compromised, and that the FAA would lose control.

To repair that problem, it is not necessary to toss the new deregulatory baby and its murky, turbulent bath water out the window in imitation of the last mistake. Deregulation in at least some form can exist side by side with ever-increasing flight safety *provided* the tools are made available by Congress for the FAA (and the industry) to counter the predictable human reactions to the all-too-obvious economic pressures.

Suggestions on how to repair the problem come from throughout the industry, as well as from those folks in the FAA, NTSB, DOT, and others who have no official voice but have some excellent ideas, which at least need impartial examination.

For one thing, there probably should be no differentiation between the requirements for Part 135 commuter airlines and Part 121 airlines as long as the Part 135 carriers are to remain an integral part of the airline system. Just as a chain is no stronger than its weakest link, an airline system is no safer than its weakest element.

That, of course, threatens economic hardship on commuter airlines, which would have to spend huge sums to come up to 121 standards. But as long as such carriers feed to and from major 121 airlines and even bear their names (Pan Am Commuter, Piedmont Commuter, Delta Commuter) their requirements—including the requirements for their flight equipment—should be no lower than the minimum acceptable for 121. Congress has created this disparity by assigning Part 121 duties to Part 135 carriers. If there is an economic penalty to be paid, perhaps Congress has the duty to fund that penalty with public money!

Outside of the fact that the rules that govern Part 135 operators are inadequate, only a small percentage of the airlines themselves present a serious safety problem. But no longer is the commuter/regional industry the only part of the airline world that needs the FAA's attention. Too many of the Part 121 carriers—old and new—are becoming serious safety problems as well.[2] With the demise of the Civil Aeronautics Board (which used to make the determination of fitness), there is far too little concern by the government over whether a carrier is financially strong enough to stay safe. (In fact, there is no longer an adequate determination made of a carrier's fitness *before* granting a certificate.) The FAA has taken steps to accelerate its inspections of airlines suspected to be financially troubled, but there are no formal guidelines for determining what constitutes "financially troubled," nor is there much precedent for shutting down such a carrier just on the basis of *potential* financially mandated corner cutting. But as harsh as it may be, the fiduciary obligation of the carrier to the public must govern. If there is not enough money for full and complete maintenance and training, the carrier is, per se, more dangerous than the system should allow.

But what of the carrier that has learned to hide its cost cutting and corner cutting from the FAA, and has saved enough as a result to remain solvent on its face? How about the carrier, large or small, with a dedication to getting around the rules, or one that has an institutionalized contempt for the rules? What about a major airline that simply does not understand the rules because it has business-oriented leaders or economists running flight operations? At present there is no formal method of discovering which carriers might fit such categories.

It seems clear that the FAA's system of surveillance simply cannot find a Downeast, an Air Illinois, or a PBA in time. There are numerous ideas for improving the situation; some of them involve a radical departure from the methods of the past, and some of them are long-ignored common sense.

---

2. At this moment, the propensity for one new-entry carrier in particular to have a major fatal accident increases daily because of its attitude toward FAA enforcement. When the FAA attempts to come down on it for anything, whether a violation of the rules or merely a potential problem, the leadership of that carrier burns up the phone lines to its political angels in Washington in a too-often successful attempt to put great pressure on the FAA to "back off"! This has been going on for all the years of its existence, and shows no signs of abating.

For instance, it appears vital that this nation establish rapidly a no-fault reporting system for pilots, flight attendants, controllers, managers, ticket-counter personnel, and maintenance men. In other words, *anyone* in the airline business who encounters or uncovers a safety problem must have a way to bring that problem to the immediate attention of those who can (and will) do something substantive about it, without fear of retaliation. NASA has maintained a noble attempt at this with its Aviation Safety Reporting System (ASRS), but because of its inability to grant full immunity to categorize the reports according to aircraft, airline, control facility, or other specific identification, it provides mainly a background research tool—merely a precursor of what is needed today.

These things cannot be left to happenstance. The NTSB, for its part, should still for operational reasons have to filter its recommendations through the sieve of the FAA, but there must be created some way other than advance publicity with which the NTSB could force the FAA to take action when the issue is deadly yet the agency has refused to budge. Administrator Don Engen, coming as he has from the NTSB, has a unique ability to see both sides of this issue. But it would be up to Congress to make such a change in the laws.

As for the FAA, no amount of internal alteration of attitude, focus, or policy can increase the money that fuels that agency. That is the job of Congress, and under the Reagan administration, securing more money for any government agency (except the necessary funds for defense) will take as much an act of God as an act of Congress—yet the FAA is understaffed and underfunded, as is the NTSB. Not only is the FAA incapable of fielding enough air carrier inspectors under present funding, it is unable to properly support many of the other activities that should be a national priority.

Chief among these is massive research and development of urgent, substantive programs to incorporate a bedrock understanding of human-factors and human-performance considerations in every aspect of the FAA's work. It is a human system and the FAA must learn to deal with it, surveil it, discipline it, and encourage it in human terms. Programs to incorporate the results of nearly fifty years of research into human fatigue should be implemented (not studied to death). The fledgling programs on teaching, encouraging, and enhancing pilot judgment, which are now being developed deep within the FAA in Washington, should be lofted into the sunlight for all areas of aviation: private, commercial, and air carrier.

(NASA-Ames in California is a gold mine of information in this area, as are the appropriate branches of the armed services, but that too takes money and energy to set up.)

The FAA needs to standardize the policies of its nine regions and ninety other offices around the nation. (It probably should scrap the FSDO concept, and keep the functions of GADOs and ACDOs widely separated.)

It also needs a far better and more in-depth course of instruction for air-carrier-inspection personnel, whether at the FAA Academy in Oklahoma City or elsewhere.

It needs to develop and foster systemwide courses on professional responsibility and fiduciary trust, and for that matter it needs to become a teacher in many subjects.

The position of the FAA under a deregulated system often approaches that of teacher, but the system is not adequately prepared for the role. The "teaching" that the air carrier inspectors end up having to do to get a carrier certified or corrected on some point or other is a constant function. But this teaching needs to be institutionalized. In an industry in chaos, there needs to be some agency to hold the hand of the new-entry carrier, whether regional/commuter or large 121 operator, and help it bring all its operational people up to speed on far more than just the rules. This could be a highly effective bridge to FAA understanding of the internal workings of each carrier, and could further the trust and working relationship between FAA and the carriers.

Programs to exchange active airline pilots with FAA ACI personnel (a program tried once and dropped) could be refined and reactivated. An FAA reserve pool of talent from industry could be rotated in and out of government service much as the Associate Reserve Program of the U.S. Air Force Reserve operates with airline-pilot members.

And there are needed legal changes that only Congress can make. For one thing, it will probably be impossible to adequately assure the flying public that a Downeast-style situation is not governing the operations of an airline until a no-fault reporting system is in place. But more than that will be needed to catch a Downeast. The FAA must be able to legally shut down an airline for maintaining any style of management that tends to compromise safety. Certainly the thought of this much power in the hands of the FAA would strike fear into the heart of the airline industry, but perhaps that is exactly what is needed—the price that must be paid—for assuring the people that the few deregulatory carriers who would abuse their public trust by abusing their people could be dealt with properly. Otherwise, we have exactly what the FAA's John Roach testified to in the Downeast hearings: the total inability of the FAA to do anything about an airline that is (for instance) pressuring its pilots to break rules.

Perhaps the proper and effective method of getting an early warning of abuse within an airline is to interview the people leaving the employment of an air carrier, whether fired or otherwise. If all FAA-licensed personnel were required to undergo a debriefing under oath (which included a standard spate of carefully selected questions regarding specific abuses of which the man or woman had personal knowledge) or be banned from reemployment with a certificated air carrier, perhaps a Downeast or a PBA could be identified years before its problems had the chance to grow to lethal proportions. There would need to be very tight controls on such a procedure, but if it was industrywide, involved personal immunity, and forced the FAA to at least investigate what was turned up, the system might be the better for it. Certainly, carrying the public trust would seem to require a certain amount of "sunshine" on the activities of licensed airline employees as well as their managements.

And perhaps even tougher measures are needed to deal with the clever operator who knows how to clean up his act only when the FAA is around. Perhaps the FAA needs a clandestine-operations division to infiltrate carriers about which complaints constantly surface, but which never seem to be caught in the act. It may offend our values and smack of domestic spying, but if unlimited deregulation is to be retained and the public is still to be protected, there may be no other choice. At the moment the FAA is deaf and dumb with respect to operators who are masters of deceit.

And what of the airline industry itself? Must the FAA do all the work, and make all the changes?

Certainly the industry needs to rekindle the spirit of clubby cooperation that used to characterize it, but since that is difficult to imagine under the free-market system of today or tomorrow, it will need the government's help.

In this situation, suggestions have been made that NASA be given a more active role as teacher and researcher, doing what it has done so well with United and the Line Oriented Flight Training Program, among others. However the responsibilities might be developed and divided, it is obvious that the less affluent carriers (and the unconvinced carriers) are not going to stampede to NASA's door to sign up. The view of too many MBAs, running airlines without the benefit of operational airline experience, is that aircrews, maintenance people, dispatchers, and other technically qualified and FAA-licensed personnel are not professionals but merely easily replaced workers whose skills do not have to be honed above the legal minimums. To one who holds such a misguided philosophy, the idea of advanced LOFT training, assertiveness training, physiological training, is wasteful. In his airline, it will not be done unless it is required. Therefore, the only practical way (in our country's present deregulated free-for-all) to raise the level of training is to mandate it by Federal Air Regulation as *the* minimum acceptable. This area needs immediate study.

And, of course, with frightening figures on so-called near-misses making the news, and abject confusion ruling all too often in the nation's air-traffic-control system, a major effort needs to be made once and for all to employ modern human-factors and humanistic management techniques. The self-destructive spasm of more than ten thousand controllers who denuded the system of their services and their careers in 1981 must never be repeated—but unless the FAA understands at long last the need for a complete change in management philosophy in the centers and radar control rooms, the seeds of another labor catharsis (seeds that have already been planted) will someday bring forth a bitter harvest.

For one thing, the air traffic controllers must not be expected to operate at 100 percent efficiency 100 percent of the time. They are human, and they make mistakes. Controllers will forever mess up call signs and issue incorrect clearances, "lose" traffic through memory lapses or from fatigue, get angry and issue instructions that are totally improper, or simply get mixed up and run two airliners within a hairbreadth of each other in midair. These things have happened, and will happen again. Only by anticipating such mistakes can they be kept from becoming lethal. If the system is not designed to safely absorb hundreds of those mistakes per day, another tragedy specifically related to air-traffic-control error will result. This is human factors and human performance, and the need for thorough implementation of the lessons learned from such research is of the highest urgency.

There is another "great truth" in all these suggestions and concerns: There is no way that involved organizations, governmental or otherwise, can divorce themselves sufficiently from their own special interests and points of view so as to dispassionately consider how to repair a system in bad need of immediate help. This, unfortunately, includes the Congress, since bipartisan support for deregulation has muffled many a complaint on either side of the aisle during the past seven years.

But there are many, many brilliant minds out there in the aviation industry, the airline industry, the academic community, the scientific and medical community, and many more who could be of great service in doing a sweeping and thorough *non*partisan job of examining the current state of the industry and recommending what should be done to reform and repair it.

That sort of task could be initiated only by the executive branch, and it would take little more than a stroke of the president's pen. The appointment of a presidential blue-ribbon commission or task force on a twelve-month assignment to examine completely the entire airline system and to make urgent recommendations could render a great service to the airline industry and the nation. Such a commission would have to be made up of people who have no vested interests in the outcome, but it would have the unique ability to prepare a careful and nonpartisan study, which the industry would have greater reason to trust.

Deregulation and its profoundly disturbing effect in reducing the safety buffer (the margin for error that protects the airline passenger from the results of predictable human failure) must be put under the microscope immediately. After all, the sweeping change brought about by deregulation originated in the academic community with a shining theoretical ideal—the principle of free-market operation. Is that principle so weak that it cannot stand reexamination on its merits? It would seem that even deregulation would have everything to gain from a major effort to discover how to have our national cake and consume it too—how to have a free-market airline industry and airline safety at the same time. To achieve that, adjustment is definitely—and urgently—needed.

Unfortunately, presidential blue-ribbon commissions are not created by a very busy administration (engrossed in a drive to cut costs and government ranks) unless that administration is completely convinced that the American public will stand for nothing less. Similarly, Congress will not be moved to press for that or any other action without an avalanche of mail demanding action.

Therefore, the responsibility rests where it has always rested in this free society.

It rests with you.

The letters written by the airline passengers of the nation, their families, friends, and any interested residents of the ground over which airliners fly will determine whether the deaths of airline accident victims are to be totally in vain. If the airline business can learn and profit from its fatal mistakes by increasing that margin of safety for everyone, then there is some spark of consolation in the hope that future tragedies will be averted.

But it takes letters and telegrams and phone calls to congressmen and senators and presidents to get action. Not to reverse deregulation, but to make it safe.

And in the final analysis, how can that be a partisan issue?

There are those who persist in saying that the system is sufficiently safe, needs no reexamination, and is functioning exactly as it should. A few deaths every now and then, they contend, is the acceptable price for having the convenience of the system.

If you buy that justification, try explaining it to young Jimmy Merryman.

Ignore the problem—and maybe tomorrow someone will have to explain it to your family—or to you.

ACKNOWLEDGMENTS

Two years of fascinating (and sometimes disturbing) research have gone into the preparation of this book, and I have been left with an overwhelming collection of evidence—written and verbal—that airline safety is indeed in serious trouble. But the majority of that proof would have been beyond reach without the courageous cooperation and assistance of hundreds of people from within government and industry, some of whom have literally risked their careers in an attempt to get this story before the public. This work would never be complete, then, without acknowledgment of such substantial contributions (though many must remain nameless); nor would it be finished without some information regarding my methodologies in researching and writing the story, a brief chapter-by-chapter discussion of specific source material, and a listing of the more detailed findings of the NTSB in the Air Florida disaster.

Though I had quickly become aware of the Downeast accident in Maine, it took several NTSB staffers to point out the significance of the ground-breaking human-factors/human-performance research behind that investigation. That led me to Dr. Alan Diehl (now with the FAA) in Washington, who, with the approval of the FAA's public-information section, was of immense assistance in understanding what happened at Rockland.

In Maine, I was afforded the unselfish cooperation of such people as

Jim Merryman's brother and sister-in-law, John and Sharon Merryman of Brunswick; retired FAA Inspector Robert Turner; Maine Aeronautical Commission Chief Inspector Phil Simpson in Wiscasset; Innkeeper John Clapp and his wife, Stephanie Eaton Clapp, in Rockland; former Downeast mechanic Al Soleri; Reverend and Mrs. George Stadler and Johanna Stadler; survivor John McCafferty; Rocky Stenger, Robert Stenger's son; numerous other Downeast employees past and present; and many, many others in the Rockland and Augusta areas who helped fill in details and explain the personalities. Numerous telephone conversations nationwide with former Downeast pilots such as Al Brandano and Kurt Langseth were of great assistance as well.

The history of human-factors research and understanding in large measure began with the work of Professor Ross McFarland, and in Cambridge, Massachusetts, Emily McFarland, his widow, was most helpful in providing factual control and documentation, and introductions to other researchers and material in the field. The work of such people as Professor Stan Mohler, Dr. Graybiel, Dr. John Lauber of NASA, Gerrald Bruggink (prolific safety author and originator of human-factors research at the NTSB), the researchers at Wright State University, and many others have pioneered this vital but all-too-neglected subject.

Understanding the confusing human-performance disaster that was the crash of Air Florida Flight 90 involved many hours with many different individuals, including survivors. Senior aviation attorney Don Madole, of Speiser, Kraus, and Madole in Washington, D.C. (one of the world's preeminent plaintiff's-attorney law firms in aviation-accident litigation), provided invaluable help with hours of his time and the wholehearted assistance of his firm.

Charles O. "Chuck" Miller, owner of Systems Safety in McLean, Virginia, gave me several afternoons patiently explaining the long history of NTSB reluctance to embrace the subjective field of human factors. It was Chuck Miller, in fact, who provided one of the final pieces in the emerging puzzle: the recognition that ultimately airline management controls all aspects of airline safety.

Harold Marthinsen, longtime accident investigator for the Air Line Pilots Association, and Bill Edmunds, Jr., human-factors specialist for ALPA, have been unfailingly helpful in their readiness to answer questions and provide documents. Con Hitchcock of the Ralph Nader organization's Aviation Consumer Action Project; attorney Mike Pangia (of Smiley, Olson, Gilman & Pangia in Washington, D.C.), former FAA General Counsel; and many others outside government have given unselfishly of their time.

Roger Olian and Lenny Skutnik, both true heroes of whom we can all be proud, were generous with their time and the hospitality of their homes. Many of those who were there when Air Florida went down talked to me by phone, many others communicated through NTSB interviews and transcripts.

Bert Hamilton, who lived through the nightmare of Palm 90, spent hours in a Gaithersburg, Maryland, restaurant (over what seemed like gallons of coffee) reliving the painful memory for my tape recorder. And Joseph Stiley, who has become a respected friend, devoted not just hours but days to fine-tuning the details of the account you have read in Chapters 14 through 17.

It is hard for any of us to fully appreciate the wrenching nature of going so far back into such incredibly painful memories, but we were in agreement that in order to understand the deadly seriousness of this entire safety issue, we must put ourselves in the position of the helpless passengers—put ourselves in the cabin of Flight 90, for instance, with Stiley and Hamilton and all the others—and see through their eyes how utterly at the mercy of the system the passenger is. Our blind trust of the system as passengers means that it must never be allowed to accept anything less than perfection. This is why Joe Stiley went through it all again.

It is sobering to realize that survivors such as Joe Stiley will never be completely free of the scars, the back pain, and other physical and psychological manifestations resulting from survival of an airline crash. We hear that people have survived a crash, and we tend to assume that they will fully recover. Few, in fact, do. For those who live, the following years may be another nightmare of pain and operations, broken bones and casts, therapy and all too often despair, coupled with the recurrent question: "Why me, Lord? Why did I survive, while the others died?"

And for the friends and relatives of those who did not make it, the psychological pain is all too real, and all too lasting.

Captain James Marquis of Air Florida (now Midway Express) could have been defensive and safe, yet he gave me an entire afternoon in his Miami home for a lengthy and highly informative interview, and also provided me with voluminous documentation of the NTSB performance studies. Others connected with Air Florida were cooperative in various degrees, though some are still adamant in their unsupported insistence that Larry Wheaton and Roger Pettit committed virtually no errors in the operation of Flight 90. Clear across the nation in Oregon, former FAA ACI and Air Florida training chief Al Koleno provided interview time and copies of Air Florida's flight manuals, all of which I appreciate and all of which helped.

In Illinois, I owe a debt of gratitude to all the people who helped my probe into the crash of Air Illinois Flight 710. This includes the kind assistance of Officer Paul "Buddy" Day of the Pinckneyville Police Department (and the folks at the Perry County Sheriff's Office) who dropped everything on a busy afternoon to help me find the rural crash site. Jimmy Hill, who ended up with a broken turboprop engine in his backyard the night of the crash, spent hours showing me the cornfield and hillside where Air Illinois struck. In Carbondale, Ray Schrammel, a tough former marine major and operations vice-president for Air Illinois, took time out for an interview, as did Art Holtz (chief pilot for what remains of the airline). Numerous others responded with much appreciated detail, either in person or by phone, including former Air Illinois pilot Jeanine Urban (now a pilot for Federal Express) who devoted most of one evening to answering questions from her home in Memphis.

And in Naples, Florida, one FAA man (who shall remain unnamed) broke ranks and talked at length one rainy, tropical night before Thanksgiving about the hell-for-leather rush to recertify Provincetown-Boston Airlines (PBA) after its certificate revocation.

The men and women of the Federal Aviation Administration have provided the bedrock foundation for accuracy in this work. FAA people from literally all over the nation have contributed everything from short, informal interviews to long, revealing, and continuing flows of information (written and verbal) on the problems they see eating away at their agency. In fact, many of my most important sources came from FAA headquarters in Washington, and must, for obvious reasons, remain nameless. Their assistance in each case, however, has not been an act of mutiny but rather an honorable attempt to shed light on a rapidly deteriorating situation in the hope that the system can be improved. And indeed, as 1985 has gone by with Admiral Engen at the helm, progress has been made—though far too much remains unaddressed and at a crisis point. As so many have said to me, "I don't have a forum [to bring these problems to light], but you do. That's the only reason I'll help."

The assistance provided by the NTSB has been invaluable as well. NTSB investigators such as Ronald Schleede, until recently chief of the Human Performance Division, Gerrit Walhout, Steve Corrie, Dr. Janis Stoklosa, James Danaher, recently retired veteran investigator Frank Taylor, and several of the board members themselves have been consistently helpful, as have NTSB public-information staffers Brad Dunbar and Betty Scott. In addition to Dunbar and Scott's generosity in furnishing me with boxes of NTSB reports and material, Edith Brown of the public-inquiry section has helped me to secure microfiche copies of nearly ten thousand documents pertaining to the various accidents covered in this book, including thousands of pages of depositions and testimony.

Descriptions of locales throughout the book are, for the most part, from personal observation. There are some 54,000 miles of research travel behind this work, as well as close to two hundred hours of interviews, the majority of them taped (and some transcribed later for accuracy and clarity), most of them face-to-face, dynamic discussions of many hours duration involving great hospitality and sacrifice on the part of the interviewees in their homes or offices. Some, of necessity, have been held in airports and government hallways, and many others could only be accomplished by phone.

The situations, accidents, incidents, characters, and controversies presented in these pages are real without exception. Where cockpit scenes involve crew members who did not survive to tell their stories in person—or to defend their actions—I have used only the transcribed words from the cockpit voice recorder, and by a combination of careful research into cockpit layout, instrumentation, and checklist procedures, filled in the description of the actions that of necessity took place. (For instance, if the NTSB determined that the crew extended the flaps, someone's hand *had* to have manipulated the flap handle on the control pedestal.) Pilots current in each particular aircraft have helped keep the details of such scenes accurate, as has my own experience.

The account of Downeast Chief Pilot Jim Merryman included several references to his thoughts at various times before the fatal flight to Otter Point. Merryman had communicated such introspective thoughts and feelings to friends and family before May 30, 1979, and they have graciously assisted me in presenting them in context.

Attorney Ted Kubiniec, a victim of United Flight 718's crash in the Grand Canyon in 1956, was, as presented, a passenger seated on the left side of that airliner. Since no one on Flight 718 lived to tell the tale of those last moments, I limited the descriptions involving him to the sights and sounds and feelings that any of the passengers on the left side had to have experienced.

No one but the author and his editor have had control over any aspect of this work, but many men and women in government and industry have taken their time to read, comment, and suggest corrections to the manuscript, and their efforts have been indispensable in producing a book of factual and conceptual precision.

My fellow officer-pilots of the U.S. Air Force Associate Reserve (MAC) 446th Military Airlift Wing at McChord Air Force Base in Tacoma, Washington, have helped immensely in keeping me abreast of the front-line operational status of the industry in general, and of their carriers in particular, including American, American West, Alaska, Cascade, Eastern, Frontier, Horizon, People's, Republic, TWA, United, and Western.

In addition, the following individuals deserve special mention and heartfelt thanks for helping to read, edit, and suggest changes and corrections in the developing work, or for just providing support:

Bo and Anita Corby of Seattle
Richard Davenport, senior FAA Air Carrier Inspector, Seattle
Jim and Jeanne Nance of Aurora, Colorado
Maris Catalano of Tacoma, Washington
Lowell and Connie Allen of Garland, Texas
Patricia A. Davenport of Kent, Washington (a fellow writer)
Gary and Gerda Franc of Tacoma, Washington
Major General and Mrs. Edmund Lynch of Austin, Texas
Neil and Mary Carolyn Campbell of Lubbock, Texas
Reverend Gerald A. Priest of Marshall, Texas
Mr. and Mrs. Charles C. Chapman of Jacksonville, Texas
Mr. and Mrs. Ben Kanowsky of Dallas, Texas
Southwest Airlines Captain and Mrs. A. J. Magliolo of Dallas

I am grateful to Howard Cady, my editor at William Morrow, to his assistant, Dawn Drzal, and to Joan Amico, my copy editor, for all their help.

Finally, and most important, the uncounted hours devoted to this project by my wife, Bunny Nance—my editor of first resort—as well as the support of my family (daughters Dawn Michelle and Bridgitte Cathleen, and son Christopher Sean), have guided me every inch of the way. Bunny and I are a team—even if she does throw out some of my most cherished prose every now and then.

For those who may wish to follow my research, selected NTSB report citations and some additional, significant bibliographic references are listed below. NTSB Blue Cover reports can be purchased from the NTSB, 800 Independence Avenue S.E., Washington, D.C. 20003, or reviewed in person in the public-inquiry section. In most cases, the full record of the investigation and reports of the various NTSB groups can be found at NTSB headquarters on microfiche. Please keep in mind that the Blue Cover reports contain merely summaries and conclusions based on the much larger record.

Chapters 1–4

1. The NTSB report on the Augusta crash of Downeast Captain Dwight French on August 19, 1971, is File 3-0369, AAR-72-6, adopted December 29, 1971.

2. The NTSB Blue Cover report on the Downeast Twin Otter crash is NTSB-AAR-80-5, adopted May 12, 1980. The entire investigative file and hearing record may be viewed at NTSB headquarters in Washington, or copies obtained on microfiche.

3. The Air New England accident Blue Cover report is NTSB AAR-80-1.

4. An excellent newspaper series on Downeast, Air New England, PBA, and deregulation appeared in the Providence *Sunday Journal,* Providence, Rhode Island, beginning August 22, 1982: "Ticket to ride: The Commuter Airlines," by Doug Cumming.

Chapter 5

1. A comprehensive review of the work of Dr. McFarland is contained in an undergraduate thesis titled: "The Human Factor: The Harvard Fatigue Laboratory and the Transformation of Taylorism," by Bard Clifford Cosman, Harvard University, March 1983.

2. The critical issues of flight-crew fatigue have been well treated recently in a paper titled: "Of Iron Men and Wooden Ships: The Effect of Sleep Loss on Pilots," by William J. Price of San Jose State University, 1982.

Chapter 6

1. Three excellent sources on the Grand Canyon crash are: *The New York Times,* page 1, July 1 and 2, 1956; *Lawsuit,* by Stuart M. Speiser

(New York: Horizon Press, 1980), Chapter 3; the formal accident report of the Bureau of Accident Investigation of the CAB, which may be obtained through the NTSB.

2. Some of the hearings sparked by the Grand Canyon crash, which subsequently led to passage of the Federal Aviation Act of 1958, can be found in the record of the U.S. Senate Committee on Interstate and Foreign Commerce, Subcommittee on Aviation, sessions of June 5 and 16, 1958.

Chapter 7

1. The letters received by Al Diehl and reproduced in this chapter may be found—along with many more—in the full record of the Downeast accident of May 30, 1979.

2. The NTSB Special Study of Commuter Airline Safety 1970–1979 (including analysis of the Downeast and Air New England crashes) is NTSB-AAS-80-1, adopted July 22, 1980.

Chapters 8–9

1. The listing of pitch-up/roll-off incidents involving the Boeing 737, and the history of Boeing's written notices on the subject, may be seen along with all the technical findings in the complete NTSB record of the Air Florida crash of January 13, 1982.

Chapters 11–12

1. The complete record of the Cambridge hearings into the Downeast accident may be obtained from the NTSB.

Chapters 14–20

1. Understanding the myriad details of the Air Florida crash, and reading through the hundreds of pages of interviews and depositions, requires perusal of the entire NTSB investigation file of Docket Number 477, January 13, 1982.

Chapters 21–23

NTSB Blue Cover Report Citations
 1. Eastern 401, Everglades, FL, 12-29-72; NTSB-AAR-73-14
 2. Texas International, Mena, AK, 9-27-73; NTSB-AAR-74-4
 3. Pan Am, Pago Pago, American Samoa, 1-30-74; NTSB-AAR-74-15
 4. Eastern, Charlotte, NC, 9-11-74; NTSB-AAR-75-9
 5. TWA, Berryville, VA, 12-1-74; NTSB-AAR-75-16
 6. PSA, San Diego, CA, 9-25-78; NTSB-AAR-79-5
 7. Collision between a Pan Am Boeing 747 and a KLM 747 at Tenerife, Canary Islands, March 27, 1977. This is a report of the Spanish Government. The NTSB public-information section may be able to provide copies. See also the excellent report of the ALPA Engineering and Air Safety Division, available through ALPA, Washington, D.C.

8. United, Portland, OR, 12-28-78; NTSB-AAR-79-7

9. National, Escambia Bay, FL, 11-9-78; NTSB-AAR-78-13

10. Eastern, O-ring incident, Miami, FL, 5-5-83; NTSB-AAR-84-04

Chapter 24

1. The NTSB report into the Air Illinois accident of October 1983 is still pending and, when released, will be available from the NTSB.

2. For extensive local coverage of the details of Air Illinois, see the newspaper *The Southern Illinoisan,* Carbondale, Illinois, issues of October 12, 13, 14, 15, 16, and 17; November 28, 29, and 30; December 1, 2, 4, 5, 6, 8, and 15, 1983.

Chapter 26

1. Copies of the FAA's shutdown order and certificate revocation action against PBA are obtainable from the FAA Southern Region Headquarters Public Affairs Office in Atlanta, Georgia.

Additional Selected Human Factors/Human Performance Source Material

1. "An Evaluation of NASA's Role in Human Factors Research" (Washington, D.C.: National Academy Press, 1982).

2. "Formalizing a Human Performance Program," a paper by Al Diehl, Ph.D., Senior Air Safety Investigator, Human Factors Division, NTSB, June 1980.

3. *Research Needs for Human Factors,* National Research Council (Washington, D.C.: National Academy Press, 1983).

4. "Recent Trends in Human Factors Analysis in Aviation Accidents," by C. O. Miller, a paper presented to the Florida Bar, Nov. 5, 1982.

5. "Pilot Judgment: Current Developments in Evaluation and Training and Future Issues in Aviation Cases," by Michael Pangia, *Journal of Air Law and Commerce,* 1983, pp. 237–262.

6. "Information Transfer Problems in the Aviation System," NASA Technical Paper 1875, September 1981.

7. "A New Cause of Legal Action," by C. O. Miller, an article in *Hazard Prevention,* September–October 1984.

8. "Aviation Accident Investigation: Functional and Legal Perspectives," by C. O. Miller, *Journal of Air Law and Commerce,* 1981, pp. 237–293.

9. "Guidelines for Line Oriented Flight Training," NASA Conference Publication 2184, 1981.

10. "Resource Management of the Flight Deck," NASA Conference Publication 2120, Proceedings of a NASA/Industry Workshop held at San Francisco, June 26–28, 1979.

NTSB Conclusions (Findings) from the Air Florida Accident Report, NTSB-AAR-82-8 (referenced on page 282).

1. The aircraft was properly certificated, equipped, and maintained in accordance with existing regulations and approved procedures.

2. The flight crew was certificated and qualified for the scheduled domestic passenger flight in accordance with existing regulations.

3. The weather before and at the time of the accident was characterized by subfreezing temperature and almost steady moderate-to-heavy snowfall with obscured visibility.

4. The aircraft was de-iced by American Airlines personnel. The procedure used on the left side consisted of a single application of a heated ethylene glycol and water solution. No separate anti-icing overspray was applied. The right side was de-iced using hot water and an anti-icing overspray of a heated ethylene glycol and water was applied. The procedures were not consistent with American Airlines' own procedures for the existing ambient temperature and were thus deficient.

5. The replacement of the nozzle on the Trump de-icing vehicle with a nonstandard part resulted in the application of a less concentrated ethylene glycol solution than intended.

6. There is no information available in regard to the effectiveness of anti-icing procedures in protecting aircraft from icing which relates to time and environmental conditions.

7. Contrary to Air Florida procedures, neither engine inlet plugs nor pitot/static covers were installed during de-icing of Flight 90.

8. Neither the Air Florida maintenance representative who should have been responsible for proper accomplishment of the de-icing/anti-icing operation, nor the captain of Flight 90, who was responsible for assuring that the aircraft was free from snow or ice at dispatch, verified that the aircraft was free of snow or ice contamination before pushback and taxi.

9. Contrary to flight manual guidance, the flightcrew used reverse thrust in an attempt to move the aircraft from the ramp. This resulted in blowing snow which might have adhered to the aircraft.

10. The flight was delayed awaiting clearance about 49 minutes between completion of the de-icing/anti-icing operation and initiation of takeoff.

11. The flightcrew did not use engine anti-ice during ground operation or takeoff.

12. The engine inlet pressure probe (Pt2) on both engines became blocked with ice before initiation of takeoff.

13. The flightcrew was aware of the adherence of snow or ice to the wings while on the ground awaiting takeoff clearance.

14. The crew attempted to de-ice the aircraft by intentionally positioning the aircraft near the exhaust of the aircraft ahead in line. This was contrary to flight manual guidance and may have contributed to the adherence of ice on the wing leading edges and to blocking of the engine's Pt2 probes.

15. Flight 90 was cleared to taxi into position and hold and then cleared to take off without delay 29 seconds later.

16. The flightcrew set takeoff thrust by reference to EPR gauges to a target indicator of 2.04 EPR, but the EPR gauges were erroneous because of the ice-blocked Pt2 probes.

17. Engine thrust actually produced by each engine during takeoff was equivalent to an EPR of 1.70—about 3,750 pounds net thrust per engine less than that which would have been produced at the actual takeoff EPR of 2.04.

18. The first officer was aware of an anomaly in engine instrument readings or throttle position after thrust was set and during take-off roll.

19. Although the first officer expressed concern that something was "not right" to the captain four times during the takeoff, the captain took no action to reject the takeoff.

20. The aircraft accelerated at a lower-than-normal rate during takeoff requiring 45 seconds and nearly 5,400 feet of runway, 15 seconds and nearly 2,000 feet more than normal, to reach liftoff speed.

21. The aircraft's lower-than-normal acceleration rate during takeoff was caused by the lower-than-normal engine thrust settings.

22. Snow and/or ice contamination on the wing leading edges produced a noseup pitching moment as the aircraft was rotated for liftoff.

23. To counter the noseup pitching moment and prevent immediate loss of control, an abnormal forward force on the control column was required.

24. The aircraft initially achieved a climb, but failed to accelerate after liftoff.

25. The aircraft's stall warning stickshaker activated almost immediately after liftoff and continued until impact.

26. The aircraft encountered stall buffet and descended to impact at a high angle of attack.

27. The aircraft could not sustain flight because of the combined effects of airframe snow or ice contamination which degraded lift and increased drag and the lower than normal thrust set by reference to the erroneous EPR indications. Either condition alone should not have prevented continued flight.

28. Continuation of flight should have been possible immediately after stickshaker activation if appropriate pitch control had been used and maximum available thrust had been added. While the flightcrew did add appropriate pitch control, they did not add thrust in time to prevent impact.

29. The local controller erred in judgment and violated ATC procedures when he cleared Flight 90 to take off ahead of arriving Eastern Flight 1451 with less than the required separation.

30. Eastern 1451 touched down on runway 36 before Flight 90 lifted off; the separation closed to less than 4,000 feet, in violation of the two-mile [separation] requirement in the Air Traffic Control Handbook.

31. Runway distance reference markers would have provided the flightcrew invaluable assistance in evaluating the aircraft's acceleration rate and in making a go-no-go decision.

32. The Federal Aviation Administration's failure to implement adequate flow control and the inability to use gate-hold procedures at Washington National Airport resulted in extensive delays between completion of aircraft de-icing operations and issuance of takeoff clearances.

33. The average impact loads on the passengers were within human tolerance. However, the accident was not survivable because the complex dynamics of impact caused the destruction of the fuselage and cabin floor, which in turn caused loss of occupant restraint. The survival of four passengers and one flight attendant was attributed to the relative integrity of the seating area where the tail section separated.[1]

---

1. To Joe Stiley and Nikki Felch, the phrase "relative integrity" was a cruel joke. The majority of the extensive, partially debilitating, and hideously painful injuries to their legs and ankles occurred when the metal legs of their seats collapsed. The collapse came not on the second impact, estimated at nearly twenty times the force of gravity, but on the first impact with the bridge—a far milder jolt. In order to save weight, aircraft manufacturers use the lightest possible seat-support structures, stressed for only a few "gravities" forward force. The lethal combination of forward-facing seats on support legs that may collapse under the force of an otherwise survivable crash is something few passengers think about. The average airline passenger mistakenly *assumes* those seats are safe in high-impact loads, not realizing that economics dictates the degree of safety. The question is not whether economic considerations have a place in such design, but how far such considerations should be allowed to go in lightening seat structures. Aircraft couldn't fly with invulnerable seats perched on heavy latticeworks of steel, but neither should passengers be given flimsy structures to save weight and money for the airline so their tickets will cost less. Joe Stiley had to learn how to walk again after six months of constant pain in a brace, and Nikki Felch ended up with one leg shorter than the other, all due to economy of seat construction.

INDEX

Acker, C. Edward, 151, 152, 154, 269
Adams, Donna, 204, 222
Aeronautical engineers, 85
   human-factors/human-performance
     concern and, 84, 85–86, 134
Aerospace Medical Association, 78
Aerospace Medicine, 70*n*
Aérospatiale, Ltd. (European aircraft
   consortium), 134
Air-accident life-insurance policies, 90
Airbus A-300, 371
Airbus A-310, 371
Airbus Industrie (European aircraft
   consortium), 134
Air Canada Boeing 767 cockpit
   automation failure, 318–319
Air Carrier District Offices (ACDOs)
   of FAA, 350, 351*n*
Air carrier inspectors (ACIs) of FAA,
   150, 152
   active airline pilots as reserve pool
     for expansion of, 381
   Aviation Safety Analysis System
     and, 347, 348
   Carter administration curbs and,

     113, 119–120, 301*n*
   as FAA Academy teachers, 349*n*
   inspection procedures of, 111–112
   NATI inspections and, 343*n*, 347*n*
   need for in-depth instruction
     courses, 381
   121 carriers manuals and procedures
     and, 158
   pilot type-rating check rides and,
     157–158
   supplemented airline surveillance of,
     351–353
   work overload (inspector fatigue) of,
     189–190, 342–343, 344–353
Air Coordinating Committee, 100
Aircraft manufacturers
   ALPA and, 78–79
   FAA designated engineering
     representatives (DERs) and,
     138
   foreign competition and, 360, 371
   human-factors/human-performance
     concern and, 74, 76*n*, 78–79,
     80*n*, 81, 86–87, 131*n*
   McFarland and, 83–85

399

404

414